WANDERING
STARS

About Planets and Exo-Planets
An Introductory Notebook

WANDERING
STARS

About Planets and Exo-Planets
An Introductory Notebook

George H. A. Cole
University of Hull

Imperial College Press

Published by

Imperial College Press
57 Shelton Street
Covent Garden
London WC2H 9HE

Distributed by

World Scientific Publishing Co. Pte. Ltd.
5 Toh Tuck Link, Singapore 596224
USA office: 27 Warren Street, Suite 401-402, Hackensack, NJ 07601
UK office: 57 Shelton Street, Covent Garden, London WC2H 9HE

British Library Cataloguing-in-Publication Data
A catalogue record for this book is available from the British Library.

WANDERING STARS
About Planets and Exo-Planets: An Introductory Notebook

ISBN 1-86094-464-7
ISBN 1-86094-476-0 (pbk)

Printed by FuIsland Offset Printing (S) Pte Ltd, Singapore

Preface

One of the most exciting discoveries of the 20th century was that of companions of broadly planetary mass orbiting stars similar to our Sun. This was not a chance, one off, discovery. More than a hundred and fifty planetary companions are known around other solar-type stars at the present time and more are being discovered month by month. The discovery of these companions, called here exo-planets, brings to life the vision of a complete Universe in which the Earth is not a unique object but one of a class of cosmic bodies. It also raises the emotive thoughts of other living creatures elsewhere perhaps observing and describing the Universe as we do on Earth. Does this bring to reality the speculations of thinkers over many centuries some of whom have forfeited their lives for indulging in so free ranging a philosophy? This introductory notebook tries to answer such questions and aims to provide a base from which the reader can develop a personal study of these issues. Of obvious interest to astronomers, these are also of great importance to a much wider audience. This notebook consists of thirty linked introductory essays designed to explain in simple terms what has been discovered and the background to this modern work.

Although this is designed to be a generally non-mathematical account, some of the arguments are simplified and are made more precise by the limited use of elementary mathematics, including elementary calculus. We also stress the value of dimensional analysis which is simple and direct in its application. The "powers of ten" index is also used: thus 100 is written 10^2 and one million as 10^6, and so on. There is a short non-mathematical summary at the end of each essay so the reader wishing to avoid what mathematics there is, and there isn't a great deal, need not feel disadvantaged. The continual reference elsewhere for details of parts of arguments is rather off putting in practice and tends to break the flow of the argument. To avoid this, we have tried to make each essay self contained even though this will lead to some repetition.

The account is in six Parts. The first describes the importance of gravity for planetary systems. The second considers the Solar System in a little detail as the one planetary system we know better than any others. Magnetism is an important influence in the Solar System and is included next. The general properties of stars like our Sun, which form the heart of a planetary system and provide the energy to support much of it, are considered in Part IV. Next we review the general features of the exo-planet systems discovered so far. This is followed by comments on the possibilities for life in the Universe as a whole with possible relationships between inanimate and animate matter. Finally there is a Glossary intended to be developed by the reader as a personal dictionary for the subject. We also list some of the more important space highlights over recent years. Some problems are included together with solutions. The illustrations throughout the notebook have been assembled from a collection some of whose origins has been lost by the author. I hope those sources not mentioned will be kind and understand.

I hope you find the notebook interesting and useful. Its success will be due in no small measure to the members of the Imperial College Press, and especially Kim Tan and her colleagues, for their most excellent and understanding transformation from manuscript to finished book.

G.H.A. Cole

g.h.cole@hull.ac.uk

Hull, 2005

Contents

Part II.   General Features of the Solar System

Part III.   Magnetism within the Solar System

17. Intrinsic Magnetism of the Earth   257

18. The Earth's External Magnetism   273

19. The Magnetism of the Other Planets   284

Part IV.   Stars as a Continuing Source of Energy

20. Evolution of Stars   299

Part V. Exoplanets

Part VI.   Exo-Biology

Part I

OBSERVATIONS REVEAL GRAVITY

The early observations that led to the empirical Kepler's laws of the motion of a planet orbiting the Sun are reviewed. The laws of motion of Galileo and Newton, with Newton's universal law of gravitation, provide a theoretical basis for Kepler's laws. The extension of the arguments to include more than one planet is included.

OBSERVATIONS ON COAL GRAVITY

1

Early Observations

The Sun dominates the sky by day but the stars and the Moon dominate the sky by night. It is true that the Moon can sometimes be seen by day but only very faintly.[1] The Moon and stars in their various phases can be studied comfortably whereas the Sun, with its bright and fiery appearance, is very difficult to study directly. Consequently, early observations were concerned more with the Moon and stars than with the Sun. We give here a very short review of the early work in so far as it refers to planetary science.

1.1. Stars and Planets

On a clear moonless night a person with normal eyesight can expect to see between five and six thousand stars. They remain in a standard set of fixed patterns across the sky, and have been portrayed since antiquity in terms of groups called the *constellations*. The stars of any group are generally not really linked together but appear linked only as seen from the Earth. They are, in fact, at very different distances from us but seem related rather like trees in a forest. Very careful observations show that the stars do, in fact, move to a small extent. This is called the *proper motion*. These very tiny movements were first detected by Halley in 1718. He showed that the positions of the stars Aldebaran and Arcturus had changed since the positions were catalogued in antiquity. The movements of the stars are generally very small (often some $0.1''$ of arc or less) and have been detected widely only since the accuracy of astrometry has been improved by the introduction of modern astronomical instruments. In excess of 100,000 such cases are now known. The proper motions of the nearer stars have recently been measured with high accuracy in the Hipparchus space mission of the

---

[1]A bright star whose position in the sky is known can also be seen by day through a small telescope against the bright background of the sky.

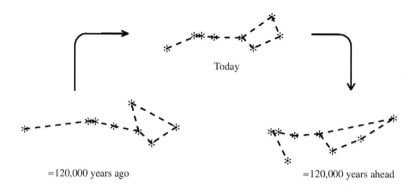

Fig. 1.1.   The changing face of Ursa Major over a period of about 240,000 years. The top arrangement is as seen today, bottom left as it would have appeared 120,000 years ago and bottom right as it will appear 120,000 years in the future. Its everyday use as a guide to the north pole is only temporary.

European Space Agency. It has shown that the stars are rushing through space often at speeds that are very high by our everyday standards.

The small proper motions of the stars obviously will cause the constellations to change their shapes over time, although the time scale may be very large. As one example, Fig. 1.1 portrays the changes of the constellation Ursæ Majoris (the Great Bear or Big Dipper) over the period of some 240,000 years. The shape well known today was not there in the past and will not be there in the future. It can be noticed, however, that one pair of stars in the constellation are actually related and keep their link through the changes. This is a *double star*. The use of the constellation as a guide to the northern pole is only temporary on a cosmic time scale, and the marking of the pole by a star is also a transient feature. There is no star marking the southern pole at the present time. The same temporary form applies to the other constellations as well. The familiar pattern of the heavens was not the same for our very earliest ancestors and will not be the same for our distant successors. The changes are, however, so slow that they remain fixed to all intents and purposes from one generation to the next.

There are five points of light in the sky, separate from the stars, that the ancient observers realised to be special because they do, in fact, move noticeably across the constellations and are clearly not part of them. These were called then the *wandering stars* or the *planets* as we know them today (the word planet comes from the Greek word for wanderer). Whereas the stars flicker in the sky, it is said that the planets tend not to and that they can be recognised in practice immediately this way. They move across the sky following very much the same path as the Sun, called the *ecliptic*.

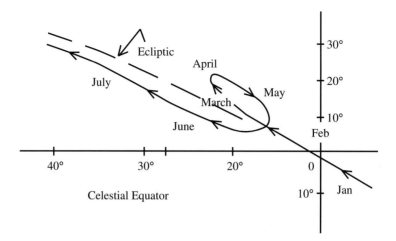

Fig. 1.2. The celestial path of Venus over a period from January to July. The motion is from right to left and moves from south to north of the celestial equator.

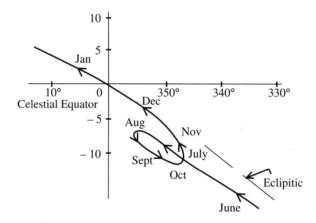

Fig. 1.3. The trajectory of Mars during the period June to January. The motion is from right to left. The ecliptic is also shown.

Whereas the Sun moves across the constellations in a simple path, that for the planets is more complicated. The path for Venus is shown in Fig. 1.2 and that for Mars in Fig. 1.3. There is an interesting feature. Although each planet moves from east to west in the sky, there is in each planetary path a part, lasting several months, where the motion is retrograde, that is moving the other way, from west to east. The retrograde motions of the two planets differ. Whereas that for Venus covers four months, that for Mars covers three. The different speeds across the sky allow the completion of a passage across the sky in different times for the two planets. The speed of

Venus across the sky allows it to complete one passage in 250 days but Mars takes 465 days. Jupiter and Saturn take much longer times, respectively 5 and 10 years. The movement of the planets through the constellations is not the same for them all and this suggests they are not linked together.

Another set of observations is significant here. This describes the relations between the planets and the Sun. Whereas the motions of Jupiter and Saturn appear to move independently of the Sun though remaining on the ecliptic, the motions of Mercury, Venus and Mars seem linked to the annual movement of the Sun. The planet Mercury always stays close to the Sun as it moves across the sky, rising barely half an hour before sunrise or setting less than a half an hour after sunset. This makes the planet very difficult to observe by eye.[2]

1.2. Interpretations of the Observations

The earliest star chart yet known, but which unfortunately hasn't survived, is that of Hipparchus (c 127 BC). The earliest catalogue that has survived is the Almagest (c 137 AD) of Ptolemy with 1028 entries, including the five wandering stars. He attempted an explanation of the heavens as he saw them and offered a model of the observed motions of the Sun, Moon and planets which survived more than 1000 years. He made four assertions. The first is that the Earth is at rest (this seemed very obvious and was hardly an assertion to him). The second was that the Earth is at the centre of the firmament, as indeed it appears to be. The third that the Moon orbits the Earth and beyond that Mercury, Venus, Mars, the Sun, Jupiter and Saturn (in that order). The orbits are all circles lying in a common plane about the Earth. Finally, the stars in their constellations lie on the surface of a sphere, centred about the Earth, beyond the Sun and planets. This is the celestial sphere which rotated due to a supernatural action. The first of Ptolemy's assumptions seems so eminently reasonable — surely, any motion of the Earth would be recognised immediately — people would be knocked over and every moveable thing would move of its own accord. The third assumption involved the culture of the times which regarded the circle as the perfect figure, following the wisdom of the classical Greeks. The fourth assumption was obvious because there was no concept at that time of depthto the realm of the stars — they do appear to lie on the inside of a spherical ceiling. This is yet another example of the fact that you cannot always trust your eyes.

[2]A number of eminent observers of the past admitted to never actually having seen it.

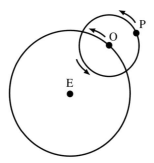

Fig. 1.4.   A planetary circular path, with centre E, augmented by the epicycle with centre O. The observed motion of the planet is point P on the epicycle.

The orbits of the Sun and Moon appear simple to an observer on the Earth but we have seen that the motions of the planets involve the retrograde motions over part of their paths.[3] These special features had to be accounted for. This was done by introducing a so-called epicycle structure for the orbit, with the planet following a circular path whose centre is itself constrained to follow a circular path centred on the Earth.

Such a scheme is shown in Fig. 1.4. The selection of the orbits and the epicylces were chosen purely to fit the observations as closely as possible. This is an empirical approach with no theoretical background. Two cycles alone may not be able to reproduce the observations with very high accuracy. Better accuracy could have been obtained by adding more epicycles but this generally was not done.

It is interesting to notice that had Ptolemy moved consistently along this approach he would have achieved a very close empirical fit to the motions he sought and very likely have discovered the analysis which is now called the Fourier expansion.[4] It represents an actual curve by a series of sine or cosine terms with harmonic frequencies and with a magnitude for each term designed to fit the initial curve as closely as possible. This could have guided the analyses to a true path for the planet and would have advanced this branch of mathematics by 1,700 years but it was not to be. Mathematics, like language, is dependent on the perceptions of the times.

The Ptolemy model might appear artificial but the arrangement of the Sun, Moon and planets was made according to the times they take to move through the constellations. It remained the accepted wisdom until the 15th century. It was then that the Polish cleric Nicholas Copernicus made the first

[3]The paths of Jupiter and Saturn show similar effects.
[4]It is interesting that Pythagorus (6th century BC) introduced the idea of celestial harmonics.

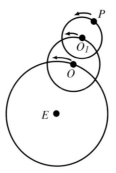

Fig. 1.5. Multiple epicycles to account more accurately for the motion of planet P.

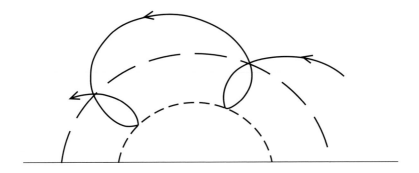

Fig. 1.6. The epicycle system for Venus.

moves to rearrange the model. The essential feature was to recognise that the
retrograde motions of the planets are apparent and not real and are due to
the motion of the Earth itself about the Sun. It follows at once that a moving
Earth cannot be at the centre of the System. The differing appearance of the
paths of Venus and Mars suggested that Venus should be placed *inside* the
Earth orbit and Mars *outside*. The epicycle structure become considerably
simplified if the Sun and Earth were interchanged, making the Sun the centre
of the System orbited by the planets. The Earth then became the third body
outwards orbiting the Sun. Copernicus got the order of the plants correctly
and was even able to predict the correct relative distance scale for the System
using geometrical arguments. It was not possible to determine the actual
distances between the bodies by simply using geometrical arguments, an an-
noying feature which is still true today. When gravity was later recognised as
the controlling force for the system of the planets it was realised that it is necessary to make only one accurate measurement of the distance between two

objects in the System to be able to find the rest. The opportunity to do this came four centuries later, in 1932, with the determination of the distance of the asteroid Eros. Today the distances are found directly by radar measurements — the first using this technique was, in fact, the Earth–Venus distance.

The Copernicus model still supposed the planets to move in circular orbits about the Sun in the same plane with the Moon orbiting the Earth. There were still some discrepancies between calculated data and those from observations. As a result, it was still necessary to include some form of epicycles to predict the motions of the planets with some accuracy. The distinction between the Ptolemy approach and that of Copernicus was in many ways philosophical and open to argument. The correctness of the Copernicus arguments could only be established finally by observations using a telescope. This had to wait until the early 17th century.[5]

1.3. Sun, Moon and Earth

The ancients knew that, wherever they were, very occasionally, the Moon moves "behind" the Earth, into the shadow. The brightness of the Moon fades and this is seen everywhere as an *eclipse of the Moon.* It is not an uncommon sight for the Moon to be dimmed from its normal brightness. Alternatively, the Moon sometimes moves in front of the Sun cutting out its light and converting day into night for a few minutes. This is an *eclipse of the Sun* and the circumstances vary from one eclipse to the next. The duration of the dark period is quite variable but is always quite short. The full eclipse is seen only locally: a partial eclipse, where only part of the Sun is covered, is seen over a wider area. The eclipse forms a wide path across a region of the Earth although the path of full obscurity is narrow. The solar eclipse arises from a fortuitous relation between the sizes of the Sun and Moon and their distance apart. This coincidence is unique to the Sun and Moon and is not repeated anywhere else in the Solar System.

The disc of the Moon fits very closely over the disc of the Sun but not quite. The result is the temporary blocking out of nearly all of the Sun's light. A small quantity does reach us from round the edges but the amount varies from one eclipse to the next. Two examples are given in Figs. 1.7a and 1.7b. The diameter of the Moon appears rather larger in Fig. 1.7a when it covers much of the solar disc. The result is the appearance of bright

---

[5]The crucial observation was to find that Venus can show a crescent form but Mars never does. This means the Venus orbit is inside the Earth orbit while that of Mars is outside. Such an observation could not be made without at least a low-powered telescope.

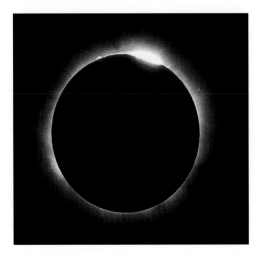

Fig. 1.7a.   The solar eclipse of 1999 showing the diameter of the Moon is closely the same as that of the Sun, as seen from Earth (the image was taken by Lviatour of the French expedition).

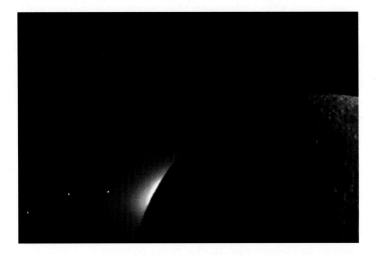

Fig. 1.7b.   A photograph taken during the Clementine Mission of the Moon (lit by reflected light from Earth) with the Sun behind showing just the corona. Also visible are the planets Mercury (nearest to the Sun), then Mars and finally Saturn. It is clear that all these bodies lie closely in a single plane, the ecliptic. (BMDO & NASA)

streamers and "beads" of light (Bayley's beads) around the circular edge. In early times eclipses were the only times when the outer atmosphere of the Sun was visible for study. This difficulty persisted until the advent of automatic space vehicles when it became possible to observe the Sun from space

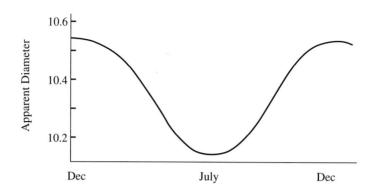

Fig. 1.8. Measurements of the apparent size of the Sun, made in London. It is seen that the maximum diameter occurs in the winter in the northern hemisphere.

without obscuration by the Earth's atmosphere. The Sun has been under constant investigation from space using the automatic Solar and Helioscopic Observatory (SOHO) of the European Space Agency. Measurements are made 24 hours of the day, 365 days of the year. Our knowledge of the Sun and its environment has grown enormously from these measurements.

The different appearances of eclipses suggest that the relative spacings of the Earth, Moon and Sun might change with time. In fact, measurements of the diameter of the Sun from Earth show that there is a periodical change with season during the year. It is found that the size increases and decreases about the mean periodically during the course of the year, every year. This can only mean that the distance of the Sun from the Earth decreases and then increases during the year from December to December. This is seen in Fig. 1.8 showing measurements made in London over the course of one year, from January to January. If the size of the Sun is fixed these measurements can only mean that the distance between the Earth and the Sun changes during the year. From the measurements it seems that the total change is about 3.8%, or 1.9% about a mean value.

Measurements of the apparent size of the Sun during the course of a year as observed from the Earth show clear differences between summer and winter. The diameter is less in the northern hemisphere during the summer months, and smaller in the winter. This implies that the Earth is nearer to the Sun in the winter and further away in the summer. It might have been expected to be the other way round — it is in the southern hemisphere.[6]

[6]Comparable measurements can be made for the Moon showing the size of that body varies throughout the lunar month. The diameter of the Moon varies by about 4.3% about the mean diameter during a lunar month corresponding to an elliptic orbit with eccentricity about 0.041. The phases of the Moon complicate the measurements of the diameter.

1.4.   The Shapes of the Orbits

We have seen that the circular path of the planets was accepted up to the 17th century because the classical Greek thinkers had said this is a perfect shape and the Universe must be perfect. It is clear from Fig. 1.8, however, that the path of the Earth relative to the Sun cannot be a perfect circle; otherwise the plot there would be a straight line and not a periodic curve. In the 17th century, Johannes Kepler realised that the shape could actually be found from observations. He devised a beautifully simple geometrical way of achieving this. The path turned out not to be a circle.

(a) The Earth and Mars

Kepler showed that a study of the location of the planet Mars over a Martian year allows the shape of the Earth's orbit to be found. A study of the location of the Earth then allows the shape of the orbit of Mars to be found. We follow Kepler and consider the Earth first.

The starting point is when the Sun, Earth and Mars lie on a single straight line, a situation said to be one of opposition (see Fig. 1.9, point S, E_0 and M_0). The time for one Earth orbit is 365 days, meaning the Earth covers $360/365 = 0.98° \approx 1°$ per day, on the average. The Earth moves substantially faster than does Mars. The time for one Mars orbit is 687 days, giving a mean speed of $360/687 \approx 0.53°$ per day. In the time it takes for one Mars orbit Earth will have completed $687/365 = 1.88$ orbits or $676.8°$. This is

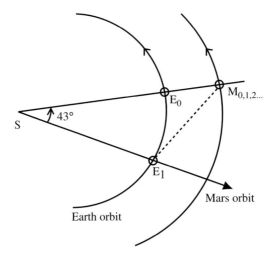

Fig. 1.9.   The geometry of the orbits of Earth and Mars.

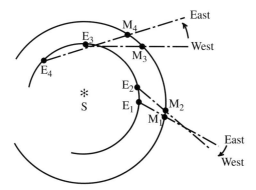

Fig. 1.10.   The motions of the Earth and Mars, accounting for the retrograde motions.

closely 43° less than two Earth orbits. At the end of one Mars orbit the Earth is at the position E_1 while Mars is again at M_0. $\angle SM_0E_1$ can be inferred because the position of the Sun is known and that of Mars can be measured. $\angle E_1SM_0$ is 43° so $\angle SM_0E_1$ follows immediately. The place where the line from M_0 with $\angle SM_0E_1$ crosses the line SE_1 gives the location of the Earth. The same method is applied again and again so building up the shape of the orbit of the Earth. While the shape follows from this procedure the distance scale remains arbitrary because the distance SE_1 is itself arbitrary. The shape of the Mars orbit follows in a comparable way, but using Earth as the datum. The starting point is again the opposition for Earth and Mars but it is now Earth that is held at a fixed location. The shape of the Mars orbit follows and again the distance scale is arbitrary.

(b) The Results

Kepler found that the orbit of the Earth is closely circular but with the Sun not at the centre of the circle. His calculations using the observations of Tycho Brahe gave the value $\frac{d}{R} \approx 1.8 \times 10^{-2}$. The situation is shown in Fig. 1.11. The Earth is represented by the point P on the circle with radius R and centre C. The Sun is located at point S with $SC = d \ll R$. Select the point Q along the line SC extended such that $CQ = SC = d$. The point Q is called the equant. A motion of P about Q with constant angular speed ν provides a rotation of P about the point S of almost exactly the observed form for the Earth about the Sun. This construction was first introduced by Ptolemy. The rate of sweeping out area by the line SP is very closely constant throughout the orbit. For motion about Q this is $\frac{1}{2}\nu R^2$. For the perihelion point A it is $\frac{1}{2}\nu(R-d)(R+d) = \frac{1}{2}\nu(R^2 - d^2)$. For the aphelion point B it is

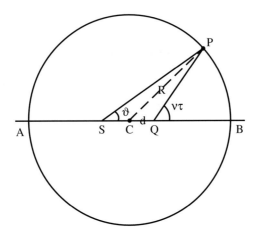

Fig. 1.11.   The equant for the Earth.

$\frac{1}{2}\nu(R+d)(R-d) = \frac{1}{2}\nu(R^2-d^2)$ which is the same. At each of the two points on the orbit perpendicular to C it is $\frac{1}{2}\nu(R^2 + d^2)$. Because $\frac{d}{R} \approx 3 \times 10^{-4}$ it follows that these four expression are the same to very good approximation. The motion of the point P about the gravitating centre S, and which mirrors closely that of the Earth about the Sun, sweeps out orbital area uniformly to all intents and purposes. The non-central location of the Sun is strange. Kepler applied the same equant approach to Tycho Brahe's data for Mars but found small though important discrepancies between calculations and observations. Clearly, the Mars orbit departs too much from the circular form and he abandoned the equant approach for that case. A more general approach is needed.

(c) Enter the Ellipse

The task of finding the curve that will satisfy the orbital measurements is formidable if it is tackled in an ad hoc way. This, in fact, is the way Kepler approached the problem. This is surprising with hindsight because Kepler was well aware of Greek geometry and the analyses gave several clues to guide him had he realised. The ellipse is, after all, a very special form of a circle. Cutting a cylinder perpendicular to the axis will provide a circular cut: cutting it at an angle will provide an ellipse. Alternatively, cutting through a right circular cone parallel to the base will provide a circular cut whereas a cut at an angle provides an ellipse. The elliptic shape is shown in Fig. 1.12.

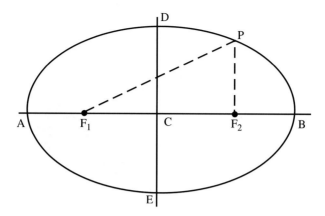

Fig. 1.12. An ellipse with major axis AB and minor axis DE. F_1 and F_2 are the focii. If AB = DE the shape is a circle.

The long axis AB marks the major axes with semi-major components ($\equiv a$) AC and BC. The perpendicular minor axis DE has semi-minor components ($\equiv b$) EC and DC. The points F_1 and F_2 are the focii from which the ellipse can be generated: the distance F_1PF_2 (where P is any point on the curve) is the same for any point on the ellipse. The ratio $(a - b)/a \equiv e$ is called the eccentricity. If $e = 0$ the ellipse degenerates to a circle: if $e = 1$ the ellipse becomes a straight line. It follows that $F_1C = F_2C = a(1 - e)$ while $AF_1 = BF_2 = ae$. It follows from the figure that $F_1PF_2 = 2a$.

1.5. Kepler's Laws of Planetary Motion

Kepler showed that the Copernican view of the Solar System is correct and that the motion of Earth and Mars about the Sun can be represented to high approximation by an elliptical path without the need to introduce epicycles. Over a period of rather more than two decades he studied with untiring patience the motions of the then known planets and devised three laws describing the motion of a planet about the Sun.[7]

Law 1. The path of a planet about the Sun is an ellipse lying in one plane with the Sun at one of the foci.

The orbit may be closely circular but the Sun is not located at the centre of the circle unless the orbit is actually circular when the two foci merge.

---

[7]The 1st and 2nd laws were published in 1609 and the 3rd was published in 1619.

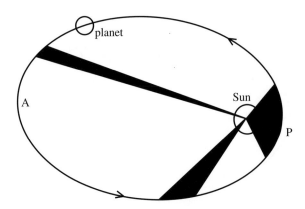

Fig. 1.13.   Kepler's Second Law. The line joining the planet to the Sun sweeps out equal areas in equal times. The perihelion point is P while the aphelion point is A.

Law 2. The motion of the planet in its orbit is not uniform. The motion is such that the line joining the planet to the Sun sweeps out equal areas in equal times.

This describes the fact that the planet moves fastest at perihelion, when it is closest to the Sun, and slowest at aphelion when it furthest away. This is shown in Fig. 1.13. The difference in motion between the region of perihelion and the region of aphelion is very clear. The distinction becomes less, of course, as the eccentricity of the ellipse decreases, the shape become more circular.

Law 3. The square of the period to execute one orbit is proportional to the cube of the semi-major axis of the orbit.

The first and second laws are qualitative but the third law is quantitative, making a specific result applicable to all cases. There is no need to resort to epicycles to obtain the symbolic form: if T is the period of the orbit while a is the semi-major axis, the law says that $T^2 = Aa^3$, where A is a constant, the same for all planets of the Solar System.

1.6.   Galileo's Law of Inertia: Newton's Laws of Motion

Independently of the empirical work on planetary orbits in Europe, Galileo, in Italy, made a fundamental discovery. He conducted experiments involving the rolling of a sphere under the force of gravity down planes of different slopes. He found the sphere accelerates under gravity down the slope but that the acceleration is less the smaller the angle. The local gravity is, of course, vertically downward but it is the component down the plane that is relevant to the motion. The component becomes zero when the angle of the slope becomes zero, that is when the plane is horizontal. In that case Galileo found that the acceleration of the sphere becomes zero as the force of gravity becomes zero. He recognized that the same would result from the action of any other force. This led him to generalize his results and enunciate the *Law of Inertia* according to which a mass will remain at rest or move with constant speed in the same straight line if it is not acted on by a force. The correctness of this law was known to ice skaters in the past but space missions have shown it to be universally true. This law represents a major advance in our understanding of nature and is still central to many arguments.[8]

A generation later, Newton developed Galileo's work. He accepted the law of inertia as the basis for the behaviour of a particle not acted on by a force. This is now called his 1st law of motion or perhaps more properly the Galileo-Newton law. With that as the standard he then postulated the effect of the action of a force. It is to accelerate a particle. What is the scale of the effect? Newton said that the force is balanced by a corresponding change of momentum with time, in the same direction as the force. If m is the mass of the particle and $\mathbf{v}$ its velocity (with magnitude and direction), its momentum is $m\mathbf{v}$. The change of momentum, $d(m\mathbf{v})$ during some vanishingly small time interval dt per unit time interval is $d(m\mathbf{v})/dt$. Newton defined the force, $\mathbf{F}$, acting on the particle as equal to this change of momentum per unit time, that is

$$\mathbf{F} = \frac{d(m\mathbf{v})}{dt} = m\mathbf{a},$$

where $\mathbf{a} = dv/dt$. This is known as Newton's second law of motion. The second form of the law follows because the mass m is a constant.

The world is not made up of point particles but of bodies with size. In order to move his argument to include such bodies he introduced a 3rd law. This can be stated as requiring a reaction to every action. The consequence is that every particle in a finite volume balances the action of every other particle to provide a stable, finite, body. It is possible to find a centre where all the mass can be supposed to reside, called the centre of mass and to apply the other laws of motion to this finite body as if it were a point. Other concepts appear because the finite body can spin as one example but these effects can be accounted for.

---

[8]Kepler could not take his laws further because he did not know about inertia.

The laws of Galileo and of Newton are descriptions of how the world is found to behave and are just as empirical as Kepler's laws of planetary motion. The test of their usefulness is whether they lead to predictions of events that agree with what is observed. Galileo's law of inertia has survived but Newton's world picture, with an absolute three-dimensional space and a separate absolute (universal) time has not for arguments of high accuracy. Nevertheless, for planetary problems the Newtonian world is fully adequate and provides appropriately accurate descriptions of the world.

1.7.   Newton's Law of Gravitation

One force at the Earth's surface known before Newton is gravitation and it is important to include this in the armory of mechanics. A major advance was made by Isaac Newton in his proposed law of universal gravitation. This is expressed by a mathematical formula that was offered for universal application. The law has proved accurate to a high degree and is central to planetary studies involving the motion of planetary bodies. It was already clear before Newton's day that the force must fall off with distance as the inverse square of the distance from a source. The gravitational intensity passing through an enclosing spherical source will remain the same as the surface expands because the source is unaltered. The area of the sphere of radius r is πr^2 so the intensity per unit area must decrease as $1/r^2$. The problem then is to specify a proportionality factor to make this dependence precise. The novel thing about the law is that the force is mutual between two masses, m_1 and m_2 and that the mass for gravitation is the mass found from inertia experiments. The equality of gravitational and inertial masses is an assumption for Newtonian theory that is supported entirely by the many experiments that have been conducted over the last three hundred years. This means that the proportionality factor will depend on $m_1 m_2 / r^2$. This has the dimensions $(\text{mass})^2/(\text{length})^2$ and not that of force. To correct the dimensions Newton introduced the universal constant of gravitation G with dimensions $(\text{length})^3/(\text{mass})(\text{time})^2$. The full law of gravitation then becomes

$$F = -\frac{Gm_1 m_2}{r^2}.$$

The force acts along the line joining the centres of mass of the particles. The negative sign is present to denote an attractive force. The three laws of motion together with the law of force provide the basis for the quantitative description of the gravitational interaction between bodies. The force accelerates each mass. For the mass m_1 and mass m_2 the accelerations are

$$a_1 = \frac{Gm_2}{r^2}, \quad a_2 = \frac{Gm_1}{r^2}.$$

The larger mass is accelerated less than the smaller mass.

1.8. A Passing Encounter without Capture

Kepler's laws as usually stated apply to a closed orbit where one body orbits another under gravity. The gravitational energy is greater than the kinetic energy of the motion so the kinetic energy of the orbiting body is not sufficient to allow it to break away. There is the alternative case when one body passes another of greater mass but with a kinetic energy that is always greater than the gravitational energy of their mutual interaction. What path will the passing body follow? It will certainly be attracted to the more massive body provided its distance of separation is not always too great.

The whole encounter will be described by the conservation of energy in which the sum of the kinetic and gravitational energies is constant throughout. Initially, the incoming body is so far away that the gravitational energy is zero so the total energy throughout the encounter is the initial kinetic energy of the particle. As the distance between the body decreases the gravitation energy becomes stronger and the kinetic energy of the incoming body decreases. It has its smallest value when the two bodies are closest together. The kinetic energy is always greater than the gravitational energy, even when the bodies are closest together so the incoming body is not captured. It moves away with increasing speed as the distance between them increases once again. The outward path is a mirror image of the inward path and the body finally reaches a large distance away with the same energy that it had when the encounter started. The mathematical details will not be given here. The path followed can be shown to be a hyperbola and is sketched in Fig. 1.14. The passing body is B and is coming from below.

The star, presumed stationary, is S. The initial direction of B is changed, to move towards S by gravity. After the closest approach B passes off in a direction different from the initial one. The energy at the beginning and the end of the encounter is the same. The body B is moving with the same

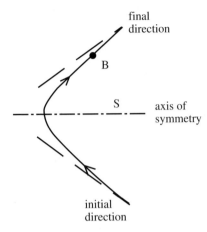

Fig. 1.14.   The hyperbolic path of a close gravitational encounter between the body with the indicated initial direction and a star S. The angle between the initial and final directions will be greater the stronger the encounter.

speed but in a different direction. The same analysis will apply to a small satellite passing a planetary body.[9]

Although Kepler's laws of planetary motion are usually stated for a closed, orbiting, body it is true that a passing encounter can also be a planetary motion and can be included as a corollary. Kepler was concerned with captured bodies in orbit, and not with passing encounters which were not observed.

*

Summary

1. Of the five to six thousand stars visible in the night sky five are special in that they move rapidly across the sky. These are the planets known to antiquity.

2. Ptolemy in the 1st century AD set a model Solar System in which the Sun, Moon and planets orbited the Earth as the centre of the Universe.

3. He described the motions of a planet in terms of a circular path, certainly one primary one and perhaps one or more smaller, secondary epicycles.

[9]This also applies to the encounter between two charges bodies of opposite electrical charge, usually a central proton with positive charge and a passing electron with negative charge.

4. Copernicus in the 15th century proposed the correct arrangement with the Sun at the centre, the Moon orbiting the Earth and the planets orbiting the Sun. He deduced correctly that Mercury and Venus orbit inside that of the Earth while the remaining planets orbit outside.

5. Kepler found the correct shapes of the orbits of Earth, Mars and Jupiter by observation and showed they are ellipses.

6. Kepler inferred the three laws controlling the motion of a planet in an elliptical orbit, his laws of planetary motion.

7. The motion of a body moving past another without capture is a hyperbola.

8. Other empirical discoveries at the time were the law of inertia by Galileo which led Isaac Newton to develop his three laws of particle motion.

9. The laws of motion can be applied to astronomical and planetary problems only if a simple law is known for the gravitational interaction between two bodies. Newton's empirical inverse square law provides such a simple expression.

10. The discoveries up to the 17th century, important for planetary science, have now been collected together.

2

A Planet and a Sun:
The Role of Gravity

[This is a rather technical chapter and involves some elementary calculus. It may be passed over in a first reading provided the Summary is understood.]

The empirical observations reported in I.1 must be related to the action of forces according to the laws of mechanics. In particular, Keplers' laws of planetary motion must be related to the controlling force of gravity. This is the topic now.

It will be assumed in the analyses that the Sun and a planet can each be represented by a point particle to a very good approximation. To see that this is indeed the case, consider the example of the Sun and the Earth. The two bodies are 1 AU apart. The Earth's radius is of the order 4×10^{-5} AU while the solar radius is of order 4×10^{-3} AU. These are very small quantities in comparison with the distance apart. To make a more homely comparison, if the Earth–Sun distance is represented by a 1 metre length, the radius of the Sun will be sphere of radius 4 mm while that of the Earth would be a sphere of radius 0.04 mm.

2.1.   Specification of an Elliptic Orbit

In order to investigate the special features of a planetary moving in an elliptic orbit it is necessary to have a mathematical expression for the ellipse so that it can be drawn on paper. This will put Kepler's first law in a quantitative form. The elliptical path is shown in Fig. 2.1, with the focus F_1 as the centre of the force. The semi-major axis is a and the semi-minor axis is b. The distance CF_1 is ae where e is the ellipticity of the path. The point A is the perihelion of the orbit and B is the aphelion point. The distance AF_1 is $a(1 - e)$. Suppose P is a point on the ellipse. Let $F_1P = r$ and $F_2P = r'$. The characteristic property of the ellipse is that

$$r + r' = 2a \quad \text{or} \quad r' = 2a - r\,. \tag{2.1}$$

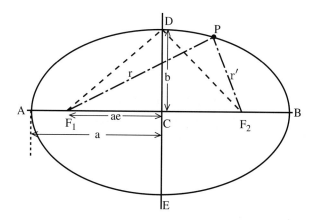

Fig. 2.1.   The ellipse with foci F_1 and F_2 with semi-major axis a and semi-minor axis b. The ellipticity is e. The definition of the curve makes $F_1D = DF_2 = a$. The $b^2 = a^2(1-e^2)$ and $AF_1 = a(1-e)$. If $e = 0$, $AF_1 = a$ so that F_1 and F_2 collapse into the centre C and the figure becomes a circle of radius a.

This can be regarded as one possible definition of the figure and is one way of actually drawing the shape. We now introduce the polar coordinates (r, θ) to specify the arbitrary point P on the orbit, with $F_1P = r$ and $\angle CF_1P = \theta$. The law of cosines allows us to relate r and r' according to

$$r'^2 = r^2 + (2ae)^2 - 4ear\cos\theta = 4a^2 - 4ar + r^2$$

where the second expression comes from Eq. (2.1). Rearrangement gives easily

$$r(1 - e\cos\theta) = a(1 - e^2)$$

that is

$$r = \frac{a(1 - e^2)}{1 - e\cos\theta}. \tag{2.2}$$

This formula can be expressed a little differently using the triangle F_1DC in Fig. 2.1 and introducing b. Then

$$b^2 = a^2 - (ea)^2 = a^2(1 - e^2)$$

so that (2.2) becomes alternatively

$$a(1 - e^2) = \frac{b^2}{a}$$

that is

$$\frac{1}{r} = \frac{a(1 - e\cos\theta)}{b^2}. \tag{2.3}$$

The expressions (2.2) and (2.3) are alternative forms for the equation of an ellipse. If a and e are specified the ellipse can be drawn in terms of r and θ. Kepler's 1st law, obtained from direct observation, asserts that this curve is the one followed by a planet moving in orbit around the Sun. A further property of this curve will be found in a moment.

2.2. Equal Areas in Equal Times

Keplers' 2nd law states that the line joining the point planet in an elliptic orbit to the Sun as the centre sweeps out equal areas within the orbit in equal times. What physical condition is this expressing? To see this consider a portion of the orbit shown in Fig. 2.2. The planet is initially at the point P distance r from the Sun at point C and making an angle θ with some arbitrary fixed line. During a time interval Δt it moves the distance Δr to the point P' increasing the angle by the amount $\Delta \theta$. The distance of P' from C is $r + \Delta r$. The planet has moved the distance Δr along the orbit. The small distance Δr is given $\Delta r = r\Delta\theta$. The small area swept out by PP' during the time Δt is the area

$$\Delta A = \text{CPP}' = \left(\frac{1}{2}\right) r\Delta r$$

$$= \left(\frac{1}{2}\right) r(r\Delta\theta) = \left(\frac{1}{2}\right) r^2\Delta\theta \, .$$

The rate of sweeping out area is the area swept out during a small time interval, that is $\Delta A/\Delta t$. Therefore $\Delta A/\Delta t = 1/2\, r^2\Delta\theta/\Delta t$. This becomes the instantaneous rate of sweeping out area in the limit that Δt approaches zero to refer to a specific time. This can be written finally

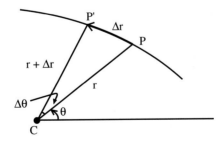

Fig. 2.2.   The area swept out during the small time interval $\Delta\theta$ by the line joining the planet P to the centre of the force, C, distance r. The planet moves to point P' during the time interval Δt, distance Δr.

$$\frac{dA}{dt} = \frac{1}{2}r^2\frac{d\theta}{dt} = C. \tag{2.4}$$

Kepler's second law states that C is constant.

This is simply restating the law but it is possible to go further. The angular momentum of the body of mass m in orbit is defined by $J = mvr$, where r is the distance joining the body to the centre of the force. The velocity is the distance travelled in a time interval along the direction of motion. In the terminology of Fig. 2.2 this is

$$J = mvr = mr\frac{\Delta r}{\Delta t} = mr\left(r\frac{\Delta\theta}{\Delta t}\right) = mr^2\frac{\Delta\theta}{\Delta t}. \tag{2.5}$$

As before, considering the angular momentum at a specific time we let Δt become indefinitely small. Then J takes the form

$$J = mr^2\frac{d\theta}{dt}. \tag{2.6}$$

This has the same form as the expression (2.4) for the sweeping out of areas so we can write $2mC = J$. Because m and C are constants, it follows that J is constant as well, which is to say that the angular momentum is constant. This is the new interpretation to be given to the 2nd law. Kepler's 2nd law of planetary motion, telling of the constant rate of sweeping out area, is nothing other than a statement that the planet moves in its orbit such that the angular momentum is constant. When the planet is in the neighbourhood of perihelion it is nearest to the Sun and so r has its smaller values. The speed of the planet in this region must have its higher values to meet the requirement of the angular momentum. At aphelion the distance from the Sun has the largest value so the motion of the planet must have its least value. In this way the variation of the speed of the planet in it orbit can be understood entirely. What is said for the Sun and a planet applies also to the orbit of any mass about a star.

2.3. Consequences of an Elliptical Path: The Inverse Square Force Law

The motion of the planet in the elliptic orbit is an accelerated motion and, according to the laws of mechanics, must involve the action of a force. The force must be between the Sun and the plant and so be a central force. The problem now is to isolate the force. The centre of force is a focus of an ellipse. The particular force involved is found by calculating the radial acceleration, a_r, of the mass in the orbit. In finding an expression for the acceleration it

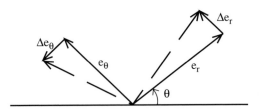

Fig. 2.3.   The behaviour of the changes of the unit vectors Δe_r in the direction of r and Δe_θ perpendicular to it. It is seen that Δe_r is in the direction of positive θ while Δe_θ is in the direction of e_r but in the opposite direction (negative e_r).

is necessary to realise that the particle is moving round the orbit so that the directions of the unit vectors specifying the motion are also changing. This is seen in Fig. 2.3. There are, then, two separate terms. One relates directly to the change of velocity in the direction of the force, that is towards the focus. The second relates to the change of the directions of the unit vectors and so gives a contribution in the direction of the centre but involving the angular motion. The expression for the acceleration a_r towards the focus is, then, given by the expression

$$a_r = \frac{d^2 r}{dt^2} - r\left(\frac{d\theta}{dt}\right)^2 . \tag{2.7}$$

The distance r changes round the orbit but we wish to have an expression which contains only the semi-major axis a and the orbital details through the constant angular momentum. The distance r can be eliminated by re-membering the expression obtained earlier for the elliptic orbit, (2.3). There are two terms and we must find expressions for each term.

(i) The First Term in (2.7)

To find an expression for $\frac{d^2 r}{dt^2}$, first differentiate (2.3) to obtain

$$-\frac{1}{r^2}\frac{dr}{dt} = \frac{ea\sin\theta}{b^2}\frac{d\theta}{dt}$$

or

$$-\frac{dr}{dt} = \frac{ea\sin\theta}{b^2} r^2 \frac{d\theta}{dt} .$$

This can be simplified by remembering (2.4) which is that $r^2\frac{d\theta}{dt} = 2C$. Consequently

$$\frac{dr}{dt} = -\frac{2Cea\sin\theta}{b^2} .$$

Differentiating once again to find the acceleration towards the centre gives

$$\frac{d^2r}{dt^2} = -\frac{2Cea}{b^2}\cos\theta\frac{d\theta}{dt}.$$

Use (2.4) once again to eliminate $\frac{d\theta}{dt}$. The result is

$$\frac{d^2r}{dt^2} = -\frac{4C^2ea\cos\theta}{b^2}\frac{1}{r^2}. \tag{2.8}$$

It also follows from (2.4) and (2.6) that $2C = (\frac{J}{m})^2$. This, then, is the first term in (2.7).

(ii) The Second Term in (2.7)

This is $r(\frac{d\theta}{dt})^2$. The equal areas law states that $r^2\frac{d\theta}{dt} = \frac{J}{m}$ where m is the mass of the orbiting particle. Then $r^4(\frac{d\theta}{dt})^2 = (\frac{J}{m})^2$ so that

$$r\left(\frac{d\theta}{dt}\right)^2 = \left(\frac{J}{m}\right)^2\frac{1}{r^3}. \tag{2.9}$$

(iii) Combining Together

The expressions (2.8) and (2.9) can be inserted into (2.7) to give the final expression for the acceleration in the direction of the centre of the force. This is

$$a_r = -\left(\frac{J}{m}\right)^2\frac{ea\cos\theta}{b^2}\frac{1}{r^2} - \left(\frac{J}{m}\right)^2\frac{1}{r^3}$$

$$= -4C^2\left[\frac{ea\cos\theta}{b^2} + \frac{1}{r}\right]\frac{1}{r^2}. \tag{2.10}$$

This might seem, at first sight, a very complicated expression for the acceleration but in fact it can be simplified greatly. It might seem that the force law has got a complicated distance law but this isn't so. Remember the equation for the ellipse (2.3). According to this

$$\frac{ea\cos\theta}{b^2} + \frac{1}{r} = \frac{a}{b^2}.$$

Inserting this expression into (2.10) reduces the form of the acceleration to the more compact expression

$$a_r = -4C^2\frac{a}{b^2}\frac{1}{r^2}. \tag{2.11}$$

This is an inverse square dependence of the acceleration on the distance from the Sun. The force F between the particle and the centre is the central force $F_r = ma_r = -4C^2 m \frac{a}{b^2} \frac{1}{r^2}$.

According to Eq. (2.11) the force has an inverse square dependence upon the distance and the negative sign shows that the force is attractive between the mass and the centre. This is, in fact, the gravitational interaction.

It can be concluded that the elliptic path of the planet around the Sun is the result of the inverse square law of force between the two bodies with the centre of force being at one of the foci. An inverse square law of force with the centre at the geometrical centre will lead to a different type of orbit.

2.4. The Semi-Major Axis and the Period of the Orbit: The 3$^{\rm rd}$ Law

The arguments so far contain even more information. As an example, the planet is in orbit under the Newtonian gravitation law and this gives an alternative expression for the acceleration resulting in the planet in orbit. For the mass M at the focus F$_1$ we can write the acceleration of the orbit as

$$a_r = -\frac{GM}{r^2}.$$

This must be the same as (2.11) so that

$$GM = 4C^2 \frac{a}{b^2}. \tag{2.12}$$

Apparently, the mass M is related directly to the figure of the ellipse through a and b.

The area A of the ellipse of semi-major axis a and semi-minor axis b is $A = \pi ab$. It is also known that C is the constant rate of sweeping out area. With the orbital period T, the time for one orbit is given by

$$T = \frac{\pi ab}{C} \quad \text{or} \quad C = \frac{\pi ab}{T}.$$

Substituting this expression into (2.11) gives

$$GM = 4\pi^2 a^2 b^2 \frac{a}{b^2} \frac{1}{T^2} = 4\pi^2 \frac{a^3}{T^2}.$$

This can be written in terms of the time period for the orbit as

$$T^2 = \frac{4\pi^2}{GM} a^3 \, .$$

This is a theoretical expression for Kepler's 3rd law of planetary motion. It means that

$$\frac{T^2}{a^3} = \frac{4\pi^2}{GM} = \text{constant} \, . \tag{2.13}$$

Measurements of the ratio $\frac{T^2}{a^3}$ allow the mass at the centre of the force to be calculated if the value of the universal constant of gravitation is known from observation. The data for several planets can be averaged to improve the accuracy of the calculation. This is a truly remarkable result. For the Solar System, observations of the orbits of the planets allow the product GM for the Sun to be found. Taking $G = 6.67 \times 10^{-11}$ m^3/kgs^2 the mass of the Sun itself follows. Surely, Kepler never suspected that his laws would lead to a method for weighing the Sun.

2.5. Two Immediate Consequences

Our understanding has deepened and two new features of the laws of planetary motion become clear.

(i) The analysis applies to elliptic orbits generally including circular ones where $a = b$. The eccentricity of the orbit does not enter the expression so that all orbits with a common semi-major axis will have the same time period for the orbit. Circular and elliptic orbits are all included.

(ii) Second, the Kepler statement of a mean orbit of the planet is made precise — it is indeed the semi-major axis of the orbit. This can be related to the maximum distance, r_{max} and the minimum distance r_{min} of the orbit from the centre of the force. Then

$$r_{\text{max}} = a(1 + e), \quad r_{\text{min}} = a(1 - e)$$

giving

$$a = \frac{r_{\text{max}} + r_{\text{min}}}{2} \, . \tag{2.14}$$

The semi-major axis is the arithmetic mean of the maximum and minimum distances. This is obvious for a circualr orbit where $r_{\text{max}} = r_{\text{min}}$ and the semi-major axis is the same as the radius of the orbit.

2.6.   The Energy in an Elliptic Orbit

We have seen that the orbital period depends only on the value of the semi-major axis. We can now show that this is also true of the energy in the orbit. Newton's law of gravitation does not involve dissipation. Provided the planet is orbiting in free space and not in a dissipative medium it will continue to orbit without loss of energy. It follows that the energy in the orbit is constant.

The planet moves along the orbit and so has no motion perpendicular to it. This means that its kinetic energy has the form K.E. $= \frac{1}{2}mv_\theta^2$. The potential energy of gravitation is P.E. $= -\frac{GMm}{r}$, remembering the negative sign for an attractive force. The total energy in the orbit is the algebraic sum of these terms (retaining the negative sign in the sum) so that

$$E = \frac{1}{2}mv_\theta^2 - \frac{GMm}{r} = \text{constant}. \tag{2.15}$$

Because the energy is constant in the orbit, meaning that it the same everywhere, we can evaluate the expression at the point that is easiest for us. We choose to evaluate this expression at the aphelion position, that is where $r = r_{\max} = a(1 + e)$, because at that point of the orbit the velocity is purely transverse to the force.

(i) The Kinetic Energy

The expression for this is

$$\text{K.E.} = \frac{1}{2}mv_\theta^2 = \frac{1}{2}mr^2\left(\frac{d\theta}{dt}\right)^2$$

$$= \frac{J^2}{2mr^2} = \frac{J^2}{2ma^2(1 + e)^2}. \tag{2.16}$$

since $r^2(\frac{d\theta}{dt}) = \frac{J}{m}$. It was seen previously that

$$C = \frac{\pi ab}{T} \quad \text{and} \quad J = 2mC$$

where the second form comes from the relation $b^2 = a^2(1+e)^2$. Consequently

$$J^2 = \frac{4\pi^2 m^2 a^2 b^2}{T^2} = \frac{4\pi^2 m^2 a^4(1 - e^2)}{T^2}.$$

where the second form comes from the relation $b^2 = a^2(1+e)^2$. This expression can be inserted into (2.16) but we must remember that

$$(1 - e^2) = (1 - e)(1 + e)$$

(multiply the brackets out if you are skeptical) so that finally

$$\text{K.E.} = \frac{GMm(1 - e)}{2a(1 + e)}. \tag{2.17}$$

(ii) The Potential Energy

This is very easy to write down. It is, in fact,

$$\text{P.E.} = -\frac{GMm}{r_{max}} = -\frac{GMm}{a(1 + e)}. \tag{2.18}$$

(iii) The Total Energy

The total energy is the sum $E = \text{K.E} + \text{P.E.}$ and from (2.17) and (2.18) we have

$$E = \frac{GMm}{2a}\frac{(1 - e - 2)}{(1 + e)} = -\frac{GMm}{2a}. \tag{2.19}$$

The general expression for the energy is then (2.15) in the form

$$E = \frac{1}{2}mv^2 - \frac{GMm}{r} = -\frac{GMm}{2a}. \tag{2.20}$$

It is seen from this expressions that the total energy depends only on the value of the semi-major axis, a. The negative sign is present showing that the attractive gravitational energy is greater than the kinetic energy of motion so the planet is trapped.

It is also the same as the energy of a mass m in a circular orbit of radius $2a$, equal to the major axis of the ellipse. The centre of the force is now the centre of the circular orbit.

2.7.   Specifying the Orbit from Observations

If a planet is observed to be orbiting the Sun it is necessary to calculate the orbit. Suppose a planet is observed to have a velocity v_i at a distance r_i from the centre of force. What are the details of the orbit? The same problem arises alternatively when a body is launched into space under gravity. The orbit is shown in Fig. 2.2.

The starting point is to compare $\frac{1}{2}mv_i^2$ with $\frac{GMm}{r_i}$ to make sure that the body is in a closed orbit. There will be a closed orbit only if $\frac{GMm}{r_i} > \frac{1}{2}mv_i^2$. To determine the orbit start with (2.20)

$$E = \frac{1}{2}mv^2 - \frac{GMm}{r} = -\frac{GMm}{2a} .$$

This involves the single unknown a and can be solved to yield the semi-major axis of the orbit.

More information comes from the angular momentum l which is a conserved quantity. Explicitly, the initial angular momentum is

$$l = r_i m v_i \sin \theta$$

and so is known from the initial conditions. The values at perihelion and aphelion can then be expressed in the forms

$$l = m v_p r_p = m v_a r_a ,$$

giving

$$\frac{1}{r_p} = \frac{m v_p}{l} , \quad \frac{1}{r_a} = \frac{m v_a}{l} .$$

Inserting (2.18) into (2.17) gives a quadratic equation for the speed v from which the values of r_p and r_a follow immediately. We know that

$$r_p = a(1 - e) \quad \text{and} \quad r_a = a(1 + e)$$

so, since a is known, the eccentricity e follows. The polar equation for the orbit can now be written in the form

$$r_i = \frac{a(1 - e^2)}{1 - e \cos \theta_i}$$

and the orbit is fully specified.

2.8.   The Different Possible Orbits

Although the procedure outlined in the last section is simple in principle, in practice there can be difficulties if the total energy is close to zero. Different orbits arise from different relative values of the kinetic and potential energies. These are as follows.

(i) P.E > K.E.

Gravitation now controls the motion and the body is held close to the centre of force, in a closed elliptical orbit. The motion is about one focus and the ellipticity is large if the speed of the body is small. As the speed increases, the eccentricity increases, the foci moving together until, at a critical speed, the foci coincide and the path is circular. For larger speeds the ellipse is more open and the perihelion and aphelion points are reversed.

(ii) K.E. = P.E.

The total energy is now zero and the body is neither constrained nor is free. The body will move in a parabolic path about the centre of force, the speed decreasing with the distance away. The body approaches rest at an infinite distance away after an infinite time.

(iii) K.E. > P.E.

In this case the gravitational energy has a limited role, the mass following an unbound, hyperbolic, orbit about the centre of the force.

It will be appreciated that these separate conditions are clearly defined away from the condition (ii), but that if the K.E. is closely similar to the P.E. it is not easy to distinguish between these three cases.

*

Summary

We have derived a number of very important results, fundamental to the description of two bodies in orbit under the mutual action of gravitation.

1. The 2nd law of planetary motion, with equal areas of orbit being swept out in equal times, was found to be an expression of the conservation of angular momentum throughout the orbit.

2. The 1st law, describing the planets orbital path as ellipse, was found to be an expression of the gravitational inverse square law of attraction between the planet and the focus of the ellipse where the Sun is located. A law of force between the planet and the geometrical centre of the ellipse does not follow an inverse power law.

3. Introducing the law of gravitation explicitly into the arguments of the elliptic path provides the 3rd Kepler law exactly. It also follows that the

size of the semi-major axis is the arithmetic mean of the maximum and minimum distances of the planet from the Sun.

4. The total energy of the planet in its orbit is conserved so the sum of the kinetic and gravitational energies is constant throughout the orbit. The energy depends only on the value of the semi-major axis and not on the ellipticity of the orbit.

5. The theory for the elliptic orbit can be extended. For the elliptic orbit the gravitational energy must be greater than the kinetic energy, so that gravity dominates. It is possible for the kinetic energy to be greater than the gravitational energy. In this case the kinetic energy dominates and the planet is not held to the Sun. The path of the planet now is a hyperbola. It is possible for the two energies to be exactly equal in which case again the planet is not captured but the orbit is a parabola.

3

Several Planets: The Centre of Mass

3.1.   More Than One Planet

The Solar System is composed of nine planets orbiting the Sun with orbits lying in a thin flat disc of radius nearly 40 AU (that is a distance a little less than 6×10^{12} m). Beyond that is a region of small particles which extend the System substantially. Kepler's laws refer to a single planet in orbit about the Sun under the action of gravity, although they can be extended quite generally to apply to any body orbiting another under gravity. Just as the force of gravity acts between the Sun and each planet separately, so a corresponding force of gravity acts between each pair of planets. Of course, the interactions between the planets must be weak for Kepler's laws to have been found empirically. Nevertheless, Kepler's arguments can apply only approximately because the planetary orbits must in principle be slightly open ellipses. This more general heliocentric theory cannot yet be approached analytically and its description must involve numerical analyses.[1] We explore this situation now.

The Sun is much more massive than any planet — even Jupiter, the most massive planet, has no more that one thousandth the solar mass. The gravitational interactions between the planets must, therefore, be correspondingly weak in comparison with that between the Sun and each planet. The extent of this weakness is clear from Table 3.1. Here the individual planets are listed in the first column and their acceleration in their orbits towards the Sun is listed in the second column. The perturbing planets are set out in the rows. The third and the succeeding columns contain the perturbation interactions with the other planets. The interactions are compared with those of the Sun

---

[1]Modern mathematics still cannot treat the interactions of more than two bodies simultaneously. This extended problem is part of the so called N-body problem. Limited aspects of the 3-body problem have been given an analytical description.

Table 3.1. The relative gravitational effects of the planets on each other compared with the influence of the Sun, all cases for opposition. The influenced planet is listed in column 1 while the influences of the other planets as a fraction of that of the Sun occupy the remaining columns.

The primary influence	Planet–Sun interation Accel. due to Sun (m/s^2)	Acceleration due to perturbing planet (as fraction of solar-planet acceleration)							
		Mercury	Venus	Earth	Mars	Jupiter	Saturn	Uranus	Neptune
Mercury	3.7×10^{-2}	–	3.3×10^{-6}	1.3×10^{-6}	4.2×10^{-8}	5.5×10^{-6}	5×10^{-7}	1.7×10^{-8}	1×10^{-9}
Venus	1.2×10^{-2}	6.7×10^{-6}	–	1.9×10^{-5}	1.1×10^{-7}	2.5×10^{-5}	1.9×10^{-6}	5.9×10^{-8}	2.8×10^{-8}
Earth	6.0×10^{-3}	3.3×10^{-7}	3.1×10^{-5}	–	1.3×10^{-6}	6.3×10^{-5}	4×10^{-6}	1.3×10^{-8}	5.5×10^{-8}
Mars	2.6×10^{-3}	2.5×10^{-7}	8.3×10^{-6}	2.3×10^{-5}	–	1.5×10^{-4}	9.5×10^{-6}	2.9×10^{-7}	7.1×10^{-7}
Jupiter	2.2×10^{-4}	1.7×10^{-6}	3.3×10^{-6}	5×10^{-6}	6.7×10^{-6}	–	4×10^{-4}	5.9×10^{-6}	2.2×10^{-6}
Saturn	6.0×10^{-5}	1.4×10^{-7}	2.5×10^{-6}	3.8×10^{-6}	4.8×10^{-7}	5×10^{-3}	–	4×10^{-5}	1.1×10^{-5}
Uranus	1.5×10^{-5}	1.3×10^{-7}	2.5×10^{-6}	3.9×10^{-6}	4.2×10^{-7}	2×10^{-3}	1.1×10^{-3}	–	1.4×10^{-4}
Neptune	6.7×10^{-6}	1.2×10^{-7}	2.5×10^{-6}	3.3×10^{-6}	3.3×10^{-7}	1.4×10^{-3}	6.5×10^{-4}	3.3×10^{-5}	–

for the condition of planetary opposition where the two planets are at their closest. This gives the maximum gravitational interaction between them. Thus we see that the effect of Jupiter on the Earth is 6.3×10^{-5} of that of the Sun on the Earth, which means the acceleration suffered by the Earth due to Jupiter is $6 \times 10^{-3} \times 6.3 \times 10^{-5} = 3.8 \times 10^{-7}$ m/s^2. On the other hand, the effect of Jupiter on Saturn is 5×10^{-3} times that of the Sun on Saturn, that is $6.0 \times 10^{-5} \times 5 \times 10^{-3} = 3 \times 10^{-7}$ m/s^2. It will be seen from the table that, in general terms, the interactions between the major planets are stronger than those between the terrestrial planets. All the interactions are small in comparison with that of the Sun which means that the planets move about the Sun as if they are independent to a very high approximation. It is this fact that allowed Kepler to discover his laws of planetary motion in the form that he did.

To gain a feel of the effects of the gravitational perturbations, consider the mutual interactions of Jupiter and Saturn. The orbital period of Jupiter is very nearly 12 years while that for Saturn is nearly 30 years, and the orbits are closely circular. During 1 year, Saturn moves through 1/30 of its orbit while Jupiter moves through 1/12 of its orbit. These changes are small and we can suppose these planets have remained still during this time. The directions of the accelerations can then be taken as constant, along the line joining the centres of the planets. It was seen from Table 3.1 that the acceleration of Saturn due to Jupiter is 3×10^{-7} m/s^2, and this can be supposed constant over an Earth year. The distance that Saturn will move in this time along the line with Jupiter can be calculated from the standard formula of mechanics for the distance, $s(\text{Sat})$, under constant acceleration g. For the time t

$$s(\text{Sat}) = \frac{1}{2}g(\text{Sat})t^2$$

and $g(\text{Sat}) = 3 \times 10^{-7}$ m/s^2 with $t = 3.15 \times 10^7$ s. Then, $s(\text{Sat}) = 3/2 \times 10^{-7} \times (3.15 \times 10^7)^2 = 1.49 \times 10^8$ m $= 1.49 \times 10^5$ km. This is a large distance in Earth terms, but is insignificant in comparison with the radius of the Saturn orbit, which is 1.4×10^{12} m. There is a corresponding response of Jupiter. The acceleration of Jupiter due to Saturn is $g(\text{Jup}) = 9 \times 10^{-4}$ m/s^2. The distance $s(\text{Jup})$ travelled in one year is $s(\text{Jup}) = \frac{1}{2} \times 9 \times 10^{-8} \times (3.15 \times 10^7)^2 = 4.5 \times 10^6$ m $= 4.5 \times 10^4$ km. For comparison, the radius of the corresponding orbit is 7.8×10^{11} m so the movement is about one-millionth % of the orbit. These are maximum displacements because the movements of the planets take them away from the condition of opposition, where their distance apart is a minimum. This means the interactions are at their greatest and from then onwards will decrease. Clearly the ellipse of Kepler is not strictly closed but sufficiently so for the orbit to remain stable.

A more accurate location of these planets follows from the calculation of the accelerations of the two planets with the corrected locations. The displacements will be smaller. This is the next approximation covering the second year. The new locations can be used to act as the base for the third year correction, and again the effects will be smaller. The procedure can be repeated to obtain any level of accuracy. There is no comprehensive theoretical structure to allow this calculation to be achieved directly and numerical work is limited by levels of accuracy involved. This may not always be the case but is likely to remain so for the foreseeable future. A comparable analysis can be developed to include the orbits of Uranus and Neptune thereby adding a further refinement. This is the sort of calculation made by LeVerrier and independently by Adams when they were attempting to understand the variations of the observed orbit of Uranus. The Titius-Bode law also gave information of a possible orbit radius for a new planet. The result of all this enormous effort was the prediction of the new planet Neptune.

3.2.　Jupiter, Mars and the Asteroids

There is a special relation between Jupiter and the asteroids orbiting inside. The important point is that the asteroids are a group of many small particles with a random motion as well as a mean solar orbit. Both Jupiter, as an exterior planet, and Mars, as an interior planet, attract each asteroid to disturb its simple orbit. As an example, we will consider the orbit of the asteroid Ceres.[2]

This is a spherical body of mass about 1.43×10^{21} kg in a closely circular orbit round the Sun with a mean orbital distance of 4.139×10^{11} m. It is surrounded by smaller bodies in the asteroid belt. The main influence on its motion is the Sun: it has an acceleration $a(S) = 7.7 \times 10^{-4}$ m/s^2 towards the Sun. The next influence is Jupiter at opposition: the acceleration towards Jupiter is $a(J) = 9.53 \times 10^{-7}$ m/s^2. Finally, it has an acceleration towards Mars at opposition of $a(M) = 1.72 \times 10^{-1}$ m/s^2. The influence of Jupiter is strong every 11.86 years. The force between Ceres and Jupiter can be supposed constant for about 1/2 year $= 1.59 \times 10^7$ s. During this time Ceres will be pulled out of its orbit by about 1.2×10^5 km, a very small distance in

---

[2]This was the first to be discovered by G. Piazzi at Palermo in 1801.

relation to its orbital size. Mars presents a different problem in that the time of opposition is rather shorter. The orbital period of Mars is 1.88 years and it might be supposed that the distance between the two bodies is constant for about 1/10 year — 3.17×106 s. The distance that Ceres is pulled away is quite small, about 3 or 4 kilometres and this can amount to about 50 km in a year. A similar story can be found for each asteroid in the belt although the actual distances involved will be less.

The major influence on the individual asteroid motions after the Sun is Jupiter but the immediate effects of the movement due to Jupiter arise from the neighbouring asteroids themselves. The motions cause the asteroids to collide with each other thus changing the trajectories and perhaps also shattering larger ones to produce smaller members. The collisions will also force smaller asteroids out of the belt, some moving towards the inner Solar System and others towards the outer regions. This change in the number of "free" asteroids can have important consequences in the bombardment of planetary surfaces by small bodies. Such effects could account for occasional large meteorites hitting the surfaces of the terrestrial planets.

3.3. The Centre of Mass: Two Masses

The effect of the interactions is to displace the planets from their independent elliptic orbits. Although the displacements may be small, we saw they cause the orbits not to be closed ellipses. Just as the planets influence each others' motion so too the planets influence the Sun as the apparent gravitational centre. Copernicus supposed the Sun actually to be the centre of the System but this can be so only to a certain approximation. The heliocentric theory requires some correction even though it may be very small. The centre is, in fact, the centre of mass of the System.[3]

Suppose there are a set of masses $m_1, m_2, m_3, \ldots$ with the total mass $M = \Sigma_i m_i$. The masses are at distances r_1, r_2, r_3, from the origin of co-ordinates. The distance R from the origin such that $MR = m_1 r_1 + m_2 r_2 + m_3 r_3 + = \Sigma_i m_i r_i$ is the centre of mass of the group of masses. An alternative specification of the system can be made with the centre of mass as the centre of co-ordinates. The distances of the masses are then expressed as $(R - r_i)$ and $\Sigma_i m r_i$ is replaced by $\Sigma_i m(R - r_i)$. Thus, the distance $r_1 - r_2$ for two masses about the origin can be written instead as $(r_1 - R) + (R - r_2) = (R - r_2) - (R - r_1)$, about the centre of mass.

---

[3]Copernicus would certainly have agreed had he known about the concept of the centre of mass.

The motions of two masses under gravity can be considered also in terms of the centre of mass. Because there are no external forces acting, the centre of mass remains stationary in a fixed co-ordinate frame, according to the law of inertia. Each mass causes an acceleration of the other, the larger mass inducing the larger acceleration in the smaller mass. These accelerations can be accounted for in the centre of mass co-ordinates. Each mass executes an elliptic path about the centre of mass, the semi-major axes being inversely as the mass, the greater mass having the smaller semi-major axis. The periods of the two orbits are the same as are the ellipticities. The two orbits are sketched in Fig. 3.1 for a planet in orbit about a central star located at the common focus of the two ellipses. The motions of the two masses are shown, each synchronised with the other.

The ratio of the amplitudes of the orbits can be found by considering the moments of the masses. Let the mass of the Sun (or a star in general) be M_S and of the planet m_p. The corresponding distances from the centre of mass are R_S and r_p. The geometry is shown in Fig. 3.2. For balance, $M_S R_S = m_p r_p$ so that $R_S/r_p = m_p/M_S$. Each mass orbits the centre of

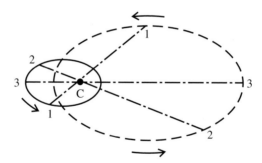

Fig. 3.1.   The paths of two masses orbiting under gravity with a common focus C acting as the centre of the force. The greater mass follows the solid orbit while the smaller mass follows the dashed orbit. The corresponding motions 1, 2, 3 are shown in each orbit.

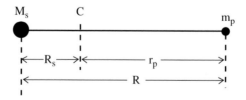

Fig. 3.2.   The large mass M_S and the smaller mass m_p in relation to the centre of mass of the pair, C.

Table 3.2. The gravitational effects of the planets Jupiter, Saturn and Earth on the motion of the Sun. R_S is the semi-major axis of the Sun's orbit about the centre of mass of the pair, τ_C is the time for one orbit of the Sun/planet, and v_S is the velocity of the Sun in its induced orbit.

Planet	R_S	τ_C	v_S
	(m)	(yrs)	(m/s)
Jupiter	7.16×10^8	11.86	12.5
Saturn	1.6×10^8	29.46	2.7
Earth	1.5×10^5	1.00	0.02

mass with the same orbital period τ. Supposing a circular orbit for each body,

$$\tau = \frac{2\pi R_S}{v_S} = \frac{2\pi r_p}{v_p}$$

where v_S and v_p are respectively the speeds of the larger body and the smaller body in their orbits. For the Solar System the effects of the planets on the Sun have been seen to be very small are small The major influence is Jupiter and $\frac{M_S}{M_J} \sim 10^3$ while for the Earth, $\frac{M_S}{M_E} \sim 10^6$. The effects of Jupiter, Saturn and Earth on the Sun are listed in Table 3.2. It is seen that the semi-major axis of the orbit induced in the Sun is 7.16×10^8 m. The radius of the Sun is 6.96×10^8 m, so that the solar orbit induced by Jupiter is 1.04 solar radii. This places the mass centre of the Solar System just outside the solar surface. Saturn has a similar effect but the induced solar orbit has a radius rather less than 1/4 the solar radius. The effect of the Earth is entirely insignificant. The combined effects of the planets orbiting together is the simple superposition of the separate effects. In fine detail, the Sun executes a rather complex motion about the mass centre of the System. Because the planetary orbits are closely circular the motion is the superposition of simple harmonic motions of different periods and amplitudes: the principal periods are 11.86 years and 29.46 years. If the orbits had greater ellipticities the superposition would have been of curves differing from sine curves.

This approach would not be possible for masses of comparable magnitude in gravitational interaction. The extension of the arguments presents more than technical difficulties. While the motion of one or two particles under the action of a force can be treated fully analytically, the extension to more particles has not yet been achieved.[4] The actual motion of two particles can be regarded as the relative motion of one particle with respect to the other, reducing this to all intents and purposes to a one body problem. This sort of trick is not possible for more particles than two. There the approach has been one of successive approximation involving numerical computation. The rounding-off errors of the computation involved restrict the time over which future orbits can be safely predicted. This may be as little as a few hundred thousand years. Numerical work is a dangerous approach to the detailed analysis of orbits.

3.4. Transfer Orbits

The orbits of two bodies orbiting a centre of force are distinguished by the energy of the orbits. If the energy of a mass in orbit can be varied it will be possible to transfer it from one orbit to the other. The energy in an orbit is inversely proportional to the magnitude of the semi-major axis, the larger the orbit the less the energy. In moving from one orbit to another the mass is transferred first to an orbit of intermediate size and then transferred to the final orbit. If the movement is to a smaller (interior) orbit, the energy of the mass must be increased at each transfer meaning the mass is *retarded* into the new orbit. If the movement is to a larger (exterior) orbit the mass must be increased at each stage. This can be seen in the following way. To make the discussion precise let us suppose a space vehicle of mass m is to be transferred from an Earth orbit to an orbit around Venus. For convenience we assume the orbits are circular as shown in Fig. 3.3. The essential statement is the energy equation for an orbit which was derived in I.2, Eq. (2.16):

$$\frac{1}{2}mv^2 - \frac{GM_{\mathrm{S}}m}{r} = -\frac{GM_{\mathrm{S}}m}{2a}. \tag{3.1}$$

with M_{S} being the mass of the Sun.

The radius of the Venus orbit is 0.72 that of the Earth orbit. The radius of the transfer orbit must, therefore, be 1.72 times that of the Earth. For the Earth, the distance round the orbit is $d_{\mathrm{E}} = 2\pi r_{\mathrm{E}}$ and $r_{\mathrm{E}} = 1.49 \times 10^{11}$ m. The time to complete the orbit is $T_{\mathrm{E}} = 1$ year $= 3.17 \times 10^7$ s. The mean

---

[4]If ever contact were made with "exo-people" who live in a world with several bodies of comparable mass, we might be told how to solve this problem. We might teach them how to deal with two-body problems!

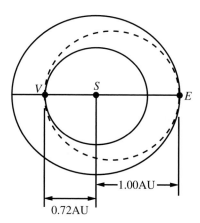

Fig. 3.3.   The orbits of Earth and Venus, assumed circular, with the circular transfer orbit linking them together. The mass to be transferred from Earth orbit is set into the transfer orbit at E and taken out at V by adjusting the orbital speeds at each point.

speed in the orbit $v_E = 2\pi \frac{1.49 \times 10^{11}}{3.17 \times 10^7} = 2.95 \times 10^4$ m/s. The mean speed for the transfer orbit at point E is

$$\frac{1}{2}mv_T^2 - \frac{GM_S m}{r_E} = -\frac{GM_S m}{1.72 r_E}.$$

This gives

$$v_T^2 = \frac{GM_S}{r_E}\left(2 - \frac{2}{1.72}\right) = 0.84\frac{GM_S}{r_E}.$$

But from (3.1)

$$mv_E^2 = \frac{GM_S m}{r_E}$$

giving

$$v_T^2 = 0.84 v_E^2.$$

Finally

$$v_T = 0.92 v_E = 2.71 \times 10^4 \text{ m/s}.$$

To enter the transfer orbit from the Earth orbit, the speed must be reduced by the amount $(29.6 - 27.1) = 2.5$ km/s. This is a very moderate retardation and would need little fuel.

The same argument will apply in moving from the transfer orbit to the Venus orbit at point V. The Sun is again the centre of the force. The major axis of the transfer orbit is of length $1.72r_E$ and the distance from the Sun at the point V is $0.72r_E$. The conservation of energy then is expressed in the form

$$\frac{1}{2}mv_V^2 - \frac{GM_Sm}{0.72r_E} = -\frac{GM_Sm}{1.72r_E}$$

so that

$$v_V^2 = \frac{GM_S}{r_E}\left(\frac{1}{0.36} - \frac{1}{0.86}\right) = -1.62\frac{GM_S}{r_E}$$

$$v_V = 1.27v_E = 3.75 \times 10^4 \text{ m/s}.$$

The orbital speed for Venus is 3.49×10^4 m/s so the probe will be moving slightly too fast — by 37.9–34.9 km/s = 2.6 km/s in fact. These speed changes are small: any attempt at a direct path between Earth and Venus would involve the expenditure of substantially greater energy.

The time taken to make the transfer can be found from Kepler's third law. The ratio of the major axis for the Earth about the Sun is 2 AU and 1.72 for the transfer orbit. The ratio of the time periods is, then, $\frac{T_T}{T_E} = (\frac{1.72}{2.00})^{3/2} = 0.80$. This gives the orbital time for the transfer orbit $0.8 \times 365 = 292$ days. The time to go from E to V is one half this time, that is 146 days ≈ 0.4 year. This is the time for the transfer.

The same arguments can be applied to other transfers. Those for the planets Mercury, Venus, Mars, Jupiter and Saturn are listed in Table 3.3.

The velocity increments to achieve the transfer orbits are given in the third column and the increments to move from the transfer orbit to the final planetary orbit are given in the fourth column. Notice that the velocity increments are negative for planets nearer the Sun than the Earth (the

Table 3.3. The velocity increments for the transfer of a mass from an orbit about the Earth to an orbit about other planets. The times for transfer are also given. All orbits are presumed to be circular. (+ denotes a velocity increase and − a velocity decrease.)

Planet	Orbital radius (AU)	Velocity increment to transfer orbit (km/s)	Velocity increment to final orbit (km/s)	Orbital period (years)	Time for transfer (years)
Mercury	0.39	−7.3	−9.8	0.24	0.29
Venus	0.72	−2.5	−2.6	0.61	0.4
Mars	1.52	+2.8	+3.6	1.9	0.71
Jupiter	5.20	+8.6	+5.9	11.9	2.73
Saturn	9.51	+10.2	+9.4	29.4	6.02

interior planets) but positive for those further away (the exterior planets). The times to achieve the transfers are given in the final column. All the orbits are presumed circular but in practice the elliptic orbits must be taken into account. Again, the timings must be chosen so that the move from the transfer orbit can take place with the second planet at the correct place.

The concept of transfer orbit is designed to involve the minimum of energy. Another method of conserving energy over a transfer between planets with intermediary steps, such as Earth to Saturn, is to use the gravitational pull of individual planets to change the direction of the space vehicle and also to change its speed. This is rather like a sling-shot approach. The path of the vehicle is hyperbolic at each encounter until the last one. It is, of course, necessary to wait for the window when all the requisite planets are aligned in the correct positions.

3.5. Tidal Forces

No body is indefinitely rigid so the effects of the gravitational interaction with a second body will be to distort the figure of each. An immediate example is the tides in the terrestrial oceans due to the presence of the Moon. A second example is the internal heating suffered by the satellite Io of Jupiter as it moves through the gravitational field of the parent planet. These effects arise because the bodies involved have a finite size.

(a) Extension Along the Line of Masses

Consider two masses, M_1 and M_2. Let their radii be respectively, R_1 and R_2, and their distance from centres R. The masses may be similar but it is supposed that $R \gg R_1, R_2$. We will find first the effect of the mass M_2 on the mass M_1 (see Fig. 3.4). The acceleration a_M induced in M_1, acting at the centre C, due to the gravitational attraction of M_2 is

$$a_M = -\frac{GM_2}{R^2}.$$

The acceleration at the surface facing $m = M_2$, point A, is

$$a_M(A) = -\frac{GM_2}{(R - R_1)^2} = -\frac{GM_2}{R^2(1 - \frac{R_1}{R})^2}$$

$$= -\frac{GM_2}{R^2}\left(1 - \frac{R_1}{R}\right)^{-2}.$$

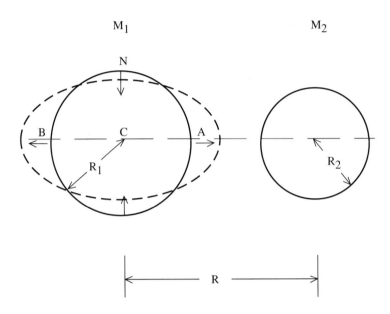

Fig. 3.4.   Showing the gravitational effect of one spherical mass (M_2, radius R_2) on another (M_1, radius R_1) distance $R(\gg R_1, R_2)$ away. The arrows show the effect in M_1 of stretching along the line of centres and compression perpendicular to it, causing a small distortion of the spherical shape. The mass M_1 has a corresponding effect on M_2.

But $R_1 < R$ so the bracket expression can be expanded using the binomial theorem as

$$\left(1 - \frac{R_1}{R}\right)^{-2} \approx 1 + 2\frac{R_1}{R}$$

so that

$$a_M(A) = a_M\left(1 + \frac{2R_1}{R}\right).$$

The difference of accelerations between the near point A and the centre of the body C is

$$a_M - a_M(A) = a_M - a_M\left(1 + \frac{2R_1}{R}\right) = 2\frac{GM_1}{R^3}. \qquad (3.1a)$$

A comparable story applies to the opposite face, B, except that there

$$a_M - a_M(B) = a_M - a_M\left(1 - \frac{2R_1}{R}\right) = -2\frac{GM_1}{R^3}. \qquad (3.1b)$$

This is an extraordinary result. First, the interaction is proportional to $1/R^3$ when it is based on gravitation which is an inverse square law of force.

Second, the interaction is in two parts: one is the movement of A *towards* M_2 and the other is a movement of the opposite side B *away* from M_2. The body M_1 is stretched along the line $M_1 M_2$.

The physical explanation is sensible. The gravitational force across the body becomes smaller from front to back because the distance is increasing. The acceleration induced in the body is correspondingly greater at the front than at the back. The acceleration is midway between the two at the centre. Viewed from the centre of mass the part nearer the other body is accelerated towards the attracting body while the part away from the body *appears* to be pushed away from it. The same effect is found in the other body.

(b) Compression at Right Angles

There are compression effects associated with the extensions. Referring again to Fig. 3.4, the acceleration at the point N towards the mass M_2 is

$$a_{M_2} = -\frac{GM_2}{(NM_2)^2} = -\frac{GM_2}{(R^2 + R_1^2)} \approx -\frac{GM_2}{R^2}\left(1 - \frac{R_1^2}{R^2}\right)$$

again using the binomial theorem. The component of acceleration in the $M_1 M_2$ direction is $a_{M_2} \cos \theta = a_{M_2} \frac{R}{\sqrt{R^2 + R_1^2}}$. Using the binomial expansion to the first term in the distance ratios the component of the acceleration becomes $\frac{GM_1}{R^2}$ which is the same as that of the centres of mass. There is, therefore, no acceleration of N relative to C in the direction $M_1 M_2$. The component of the acceleration in the direction NC is

$$a_{NC} = \frac{GM_2}{R^2 + R_1^2}\frac{R_1}{\sqrt{R^2 + R_1^2}} \approx \frac{GM_2}{R_1^3} .$$

This is one half the acceleration (3.1a,b). There is, then, a compression in the directions perpendicular to the line joining the centres of mass of the two bodies. An initially spherical mass will be distorted into an ellipsoidal form, although the distortion may be small.

Comparable effects would follow if a third mass were present and it can be appreciated that the distortions of the bodies involved can be very complicated in detail, although the details may be rather fine.

(c) Energy Dissipation

Two bodies can maintain a stable link together only if the force of gravity is balanced by another force. The centripetal force due to orbital motion is the one relevant here, the two bodies moving around each other about the common centre of mass. The orbits will generally be elliptical so the separation distance will vary periodically. Consequently, the tidal forces will vary so a static equilibrium shape cannot be established. The materials of the masses will not be indefinitely rigid and so will yield to the changing tidal forces. The changes of shape are associated with internal dissipation of energy which manifests itself as heat — the temperatures of the masses increase proportionately to the degree of distortion. The effect is often very small but in certain circumstances can be large, where the material density is sufficiently high and the perturbing forces sufficiently strong. One example is the satellite Io of Jupiter. Here the elliptical shape of the planet combines with a resonance effect of other satellites to maintain a substantial heating of the interior of the satellite leading to volcanic eruptions on the surface.

Heating is not the only possible effect. If the surface is covered with a liquid (for instance, the oceans of the Earth) dissipative effects can become important. The viscous forces can affect the rotation of the pair and, because, angular momentum is conserved, the relative separation of the pair can be affected. This has certainly occurred in the Earth–Moon system. Friction has reduced the rate of rotation of the Earth forcing the Moon to increase its orbit appropriately. For the Moon, its rotation has been reduced to showing always one face to the Earth.

3.6.   The Roche Limit

Tidal forces have an extreme effect for the near approach of a body of appropriate size which can be disrupted by the gravitational forces. The smallest orbit that such a body can follow with stability is called the Roche limit. This comes about as follows.

In Sec. 3.5a it was seen that the effect of tidal forces on a companion body is to stretch the material along the line joining the centres of masses of the two bodies, there being induced a force in the direction of the larger mass and away from it. This force is greater the larger the mass of the companion but more especially the nearer the companion is to the larger mass. There is a corresponding compression at right angles to the extension. These effect induce strains in the companion and in particular the stretching is opposed by the strength of the material. While the strength is greater than the stretching force the companion is stable but once the stretching force becomes greater the companion will be unstable and will break up. The stretching force increases with decreasing distance between the larger mass and the companion. There is, then, a minimum distance from the larger mass where the companion remains stable. This minimum distance is called the Roche limit, R_R.

Because it depends on the materials involved, there is no simple formula specifying the Roche limit in general cases. One case is for two liquid bodies, the primary body having the mean density ρ_P while the secondary body has a mean density ρ_S. If R_P is the radius of the primary body, theory gives the result $R_R = 2.46(\rho_P/\rho_S)^{1/3}R_P$. In general, $\rho_P \approx \rho_S$ so that $(\rho_P/\rho_S)^{1/3} \approx 1$. This gives the simple rule that the Roche limit in this case is about 2.5 times the radius of the primary body. This result does not, in fact, apply to a solid secondary body but is often quoted in that context.

Some physical insight can be gained by considering the energies involved in the secondary body, designated 2, orbiting the primary body, designated as 1. Two energies are of interest now. One is the energy of interaction between the bodies, E_I. This can be written approximately as $E_I \approx \frac{GM_1m_2}{R}$, where R is the distance between the line of centres of the bodies. The second energy is the gravitational self-energy of the secondary body, $E_S = \frac{Gm_2^2}{r_2}$. If the atomic mass of the secondary body is written Am_p where $m_p = 1.67 \times 10^{-27}$ kg is the mass of a nucleon, the secondary body will contain approximately $N_2 \approx \frac{m_2}{Am_p}$ particles. The self-energy per particle is $\varepsilon_2 = \frac{E_S}{N_2} = \frac{GAm_p}{r_2}$. The interaction energy per particle of the secondary body is $\varepsilon_I = \frac{GM_1Am_p}{R}$. The secondary body will remain stable provided the self-gravitational energy per particle remains greater than the interaction energy per particle, that is, so long as $\varepsilon_2 > \varepsilon_I$. The secondary body will be unstable if this is not so, that is, if $\varepsilon_2 < \varepsilon_I$. The intermediate distance between the bodies can be associated with the Roche limit.

Each planet has a spherical region around it where its gravitational influence is stronger than that of the central star. The planet then becomes the major influence on a satellite even though the central star may provide a substantial perturbation. The region where the planet is the primary gravitational force is called *the sphere of influence* of the planet.

*

Summary

1. Kepler's laws refer to a single planet of low mass orbiting a centre of gravitational force of high mass. Whereas the Sun is at least 1,000 times more massive than a planet, there are several planets. Kepler's laws must

apply to this situation, albeit approximately, because they were deduced empirically from observations of the planets.

2. The gravitational effects between the planets provide a very small perturbation on the Sun-planet motion for each planet.

3. Jupiter and Mars are major influences on the motions of the asteroids.

4. Many bodies together move relative to the mass centre of the set of bodies.

5. The centre of mass of the Solar System is at a point just outside the solar surface. This implies that the Sun orbits the centre of mass, together with the planets, although the solar orbit has a very small semi-major axis.

6. Moving from one planet to another by spacecraft can be achieved most economically (in terms of energy) using transfer orbits. This is an elliptical orbit linking one planet to another. The speed of the initial planet in its orbit is augmented to place the craft in the transfer orbit at the perihelion point. Once there it moves without power to the aphelion point. It moves into the orbit of the final planet by adjusting its speed.

7. The gravitational effect of one body on another is to raise tides in each. These act on the line joining the centres and both extend each body on the inward side *and* on the outward side.

4

The General Structure of a Planet

What is a planet? This is a difficult question to answer directly but is important to have it clear for the arguments to follow in the lectures to come. The range of materials that have been identified as composing the Solar System are likely to be available in other star/planet systems. Different combinations of elements will lead to special planetary characteristics. We review various possibilities now.

4.1. Several Energies

A planet is an independent body in equilibrium under its own gravity. This statement has some important consequences. More especially, if gravity is pulling the material inwards it is necessary for other forces to be present to oppose this unremitting inward pull of gravity. For equilibrium the forces must be in a stable mechanical balance, providing a fixed mean size for the body. The are a number of influences associated with a planet which we list now. It will be found that only two are crucial to the structure while others, that might seem important, are of secondary importance. Although a planet is a large body it will be found most convenient to view its behaviour from the point of view of the atoms of which it is made.

(a) Gravity

The planet is supposed to have a mass M and characteristic linear dimension R (later to be identified as the radius). Each atom within the volume attracts every other atom by gravity. The result is a self-energy for gravity, E_g, of value

$$E_g \approx \frac{GM^2}{R}$$

where as usual G $(= 6.67 \times 10^{-11}$ m^3/kgs^2) is the Newtonian constant of gravitation. The average mass of a constituent particle is Am_p where A is the atomic number and $m_p = 1.67 \times 10^{-27}$ kg is the mass of a nucleon. No distinction is made between proton and neutron masses, the two being regarded collectively as nucleons. The number of atoms in the mass M is $N = M/Am_p$. The average gravitational energy per atom, e_g, is therefore

$$e_g = \frac{E_g}{N} = \approx \frac{GM^2}{R}\frac{Am_p}{M} = \frac{GMAm_p}{R}$$

$$= 1.14 \times 10^{-37} A\frac{M}{R} \text{ (Joules, J)}.$$

To gain some idea of the practical magnitude of this energy, consider a silicate body of Earth size, with mass $M = 6 \times 10^{24}$ kg and $R = 6 \times 10^6$ giving $M/R \approx 10^{18}$ kg/m. The atomic mass A could be about 26 (for silicates). The energy $e_g \approx 3 \times 10^{-18}$ J. As another example, for a body of solar mass $M = 2 \times 10^{30}$ kg and radius $R = 6 \times 10^8$ m giving $e_g \approx A \times 10^{-16}$ J. The Sun is composed very largely of hydrogen so we take $A = 1$, giving a value of the energy some hundred times larger than for the Earth. At the other extreme a small body with mass 10^{16} kg (perhaps an asteroid) and characteristic length 50 km has the gravitational energy per atom, $e_g \approx 6 \times 10^{-27}$ J. These energies per atom due to gravitation must be compared with the magnitudes of other relevant energies

(b) Atomic Energies

The constituent atoms have energies associated with them. Atomic theory tells us that the maximum ionisation energy of the hydrogen atom is 13.2 eV. Now 1 eV $= 1.6 \times 10^{-19}$ J, making the ionisation energy for hydrogen $e_i = 2.1 \times 10^{-18}$ J. The ionisation energy is greater for more complex atoms.

Atoms attract each other to form molecules by electric forces, either true or virtual. The mean associated energy is approximately 1/100th of the ionisation energy, that is $e_m \approx 2 \times A \times 10^{-20}$ J $= 4 \times 10^{-19}$ J. It is this energy that holds the molecules together. These energies can be associated with an effective temperature T by $e = kT$ where $k = 1.38 \times 10^{-23}$ J/K is the Boltzmann constant. As an example, the temperature associated with molecules is always less than $T < e/k \approx 3 \times 10^4$ K. This is the temperature where melting or disintegration would occur. The ionisation energy itself provides an important temperature. For hydrogen this is $T \approx 2 \times 10^{-18}/1.38 \times 10^{-23} \approx 10^5$ K.

Hydrogen gas beyond this temperature is ionised into protons and electrons, that is, it forms *a plasma*.

(c) Internal Atomic Pressure

An atom is a three-dimensional object with a small but finite radius. It generally has a spherical shape although there can be exceptions. The positively charged nucleus attracts the negatively charged orbiting electrons and this is opposed by the electrons which have the property of resisting confinement in space beyond a certain limit. This limit is related to the Planck constant of action, h through the Uncertainty Principle of Heisenberg. No measurement can be made exactly and moving objects are less easy to measure than those standing still. It is now accepted that any measurement has an intrinsic uncertainty that must be accounted for. Let Δr be the uncertainty in the location of a particle[1] and Δp the uncertainty in its momentum. The Uncertainty Principle then stares that these separate uncertainties are related according to $\Delta p \Delta r \approx h$. This means that if the motion of an object is known without error its location must be entirely unknown and vice versa. If an electron is within an atom with characteristic dimension Δr, its momentum can only be assigned to within the uncertainty $\Delta p \approx h/\Delta r$. Applied to an atom, because the electrons have unknown locations within the atom they must be supposed to be everywhere. The electrons, therefore, form a virtual "gas" with a pressure which inflates the atom.

(d) Strength of the Atom

The electrons have the property of defending fiercely the individual volumes in which they are contained.[2] They can be pictured as moving within a "cell", the movement becoming the more violent the smaller the volume. The change of energy with change of pressure constitutes a pressure, in this case the pressure due to electrons. It is a quantum effect and is called the degenerate pressure of the electrons because it does not depend on the temperature. This (repulsive) pressure balances the attractive pressure of the attraction between electrons and protons to provide a stable volume for the atom. The electron pressure inside atoms is very large. For a hydrogen atom,

---

[1] It could be any particle and need not be an atom or atomic particle. Everyday objects are associated with a vanishingly small uncertainty.

[2] Electrons can behave as particles or as waves, depending on the circumstances. This duality of identity is characteristic of atomic particles. Protons and neutrons share the same property.

for example, the electron component of the pressure is about 5×10^{12} N/m^2 or about 5 million atmospheres. This is balanced exactly by the pressure of attraction between the electron and proton to provide equilibrium. The internal balance of more complicated atoms can involve substantially higher pressures. It is strange to think that an atom is the seat of such energies.

Atoms become unstable in the presence of external pressures of this magnitude and ultimately lose electron(s) to regain stability. They are said to suffer pressure ionisation. If the external pressure becomes high enough even complex atoms can lose all their electrons and so form a plasma of protons and electrons. The process of such a disintegration is highly complicated but is very important from an astronomical point of view. We will see that it is important in determining the behaviour of planets.

(e) Rotation

Suppose the planet to have a closely spherical figure with radius R and with a mass M. Its rotation characteristics are described by the *inertia* of the body (which is a resistance to movement), involving the inertia factor α. The inertia of the body to resist rotation about an axis, equivalent to the mass in straight-line (linear) motion, is expressed in the form $I = \alpha M R^2$. For a homogeneous sphere $\alpha = 0.4$. For a point mass, the limiting case when the radius becomes zero, is defined to be $\alpha = 0$ but that case is excluded from discussions of planets. If the sphere is rotating about an axis with angular speed ω ($= 2\pi/T$, where T is the rotation period), the energy of rotation, E_r, is given by the expression $E_r = \frac{1}{2}I\omega^2$. This can be written alternatively as $E_r = \frac{2\pi^2 \alpha M R^2}{T^2}$. We can speak of the rotation energy per atom in the rotating body and this will evidently be given by the expression $e_r = \frac{2\pi^2 \alpha R^2 A m_p}{T^2} = 3.3 \times 10^{-26} \alpha R^2 / T^2$. For a body of Jupiter-like properties, with $A = 1$, $R = 5 \times 10^7$ m, $\alpha = 1/5$ and $T \approx 10$ hrs $= 3.6 \times 10^4$ s, the formula gives $e_r \approx 10^{-46}$ J — an entirely negligible quantity in comparison with the gravitational energy.

4.2.　Packing Atoms Together

Material bodies are formed by packing atoms together in a volume of space. Each atom attracts the others with the force of gravity but there are also atomic forces to contribute as well. Remember, the total effect of gravity increases as the number of atoms is increased. There are three important cases.

(a) Few Atoms

The dominant forces here are the chemical forces (inter-atomic forces) between the atoms, which depend on the electrical attractions between the electrons and the protons inside the atoms and outside. The gravitational self-force is small and the body is an irregular accumulation of dust or small molecules. This régime will include a range of low masses and applies until the gravitational self-energy[3] approaches that of the atomic interactions which is of the order 10^{-20} J. For silicate type materials ($\rho \approx 3 \times 10^3$ kg/m^3, $A \approx 30$), for a mass $M \approx 10^{17}$ kg, $R \approx 10^4$ m the gravitational energy is of order $e_g \approx 10^{-21}$ J. This is less than the atomic interaction forces showing that gravitational energy plays no significant role in forming the structure of bodies of this general mass. It acts merely to hold the constituents loosely together should they break apart, as it might result in a collision with another body. The upper limit of this group of objects cannot be closely defined because it depends on the composition but the group will include objects such as asteroids, comets, meteorites and the smaller satellites of planets. The shape of such a body (its figure) will most likely be irregular and this will characterise the class. Each body will contain no more than a few times of 10^{46} nucleons.

(b) The Main Bodies

For greater mass gravity has an increasing influence in the packing of the atoms as the number of atoms increases. While the atomic forces are determined by the constituent atoms the gravitational force increases with the number of atoms. For low gravitational energy the atoms are held together, the tighter the more atoms are present. If there are N nucleons, each of mass m_p and volume v_a, the total mass is M and the total volume V with $M = N m_p$ and $V = N v_a$ giving $\frac{M}{V} \approx \frac{M}{R^3} =$ constant. The constant is essentially the material density, apart from a numerical factor. This is the relation that describes the planets of the Solar System. The density for each planet is, of course, a characteristic of it depending on its precise composition.

For a silicate body of mass about 10^{19} kg and a radius of order 10^4 m, the mean gravitational energy per nucleon is $e_g \approx 10^{-22}$ J. This gravitational energy is of the same order as the interaction energy, that is, $e_g \approx e_i$. The constant force of gravity will move the constituent atoms into the volume

---

[3]The gravitational self-energy will simply be called the gravitational energy from now onwards.

with the smallest surface, which is a sphere. The effect of gravity is to cause the material to "flow" over time to approach this minimum energy state. Gravity will not have a significant role for lower masses.

(c) The Extreme Size — Hydrogen

As the mass of the body increases so does the mean gravitational energy within it. For a body the size and mass of Jupiter ($M_J \approx 2 \times 10^{27}$ kg, $R_J \approx 6 \times 10^7$ m) the mean gravitational energy per nucleon is found to be $e_g \approx 4 \times 10^{-18}$ J. This is very nearly the same as the ionisation energy of the hydrogen atom. A body of this size is close to becoming a plasma body, and is close to the maximum size that a hydrogen planet can be. The central pressure is found to be about 5×10^{12} N/m^2. It is not wise to stress the numerical coincidence too strongly because the arguments have been approximate order of magnitude estimates, but, in fact, the conclusions are confirmed by more exact calculations.

The values here are mean values and the gravitational energy necessary to maintain the requisite pressure and energy throughout the planet will require a somewhat greater mass. In fact a body of some 10 M_J will be needed to show full volume ionization. The mean kinetic temperature will be of order 10^5 K but the central temperature is perhaps an order of magnitude greater. Increasing the mass only slightly greater, to about 13 M_J, yields a temperature of about 10^7 K. Naturally occurring hydrogen contains a small proportion of deuterium (the nucleus contains a proton and a neutron) and at about this temperature deuterium burning can begin. This introduces a new feature. Bodies where such deuterium burning occurs are usually called brown dwarfs. Full hydrogen burning requires a slightly higher temperature and can begin with a mass of about 10^{29} kg.

(d) Extreme Size — Silicates or Ferrous Materials

The same arguments apply for silicate compositions though conditions are rather more complicated. Nevertheless, some conclusions can be reached. The maximum ionization energy for atoms is greater than for hydrogen and the strength of the atoms is greater. Atomic stability will apply to a higher value of the gravitational self-energy and so to a higher mass than for a hydrogen body. It is difficult to be precise but the silicate range will be greater than the hydrogen range by about the number of nucleons present, which for silicates will be a factor of about 30, up to about 10^{29} kg. Pure iron bodies could have even higher masses. We have no experience of bodies of this type in the solar system. It is quite possible that they may be found eventually in exo-planet systems.

4.3. The Mass–Radius Relation

The arguments of the last Section can made firm with numerical values. It has been found that three forces between them determine the equilibrium structure of a planet. One is the force of gravity which constrains the size of the material. The second is the attraction between the protons and electrons in the atom, providing a stable atomic size. This also tends to compress the material. The third opposing force is the degenerate pressure of the electrons. The action of these three forces provides the equilibrium configuration for a planet.

Both the gravitational and electric forces have an inverse square dependence on the distance. This being the case it is possible to appeal to the virial theorem. If K is the kinetic energy (in this case of the electrons) and P is the potential energy (now of gravity and electric interactions) the virial theorem[4] states that $2K + P = 0$. Finding expressions for K and P allows a relation to be derived between the mass M and the radius R for material of atomic mass A and mass number Z. The expression for R that results is the relatively simple form

$$R = \frac{2\beta \frac{Z^{5/3}}{A^{5/3}} M^{1/3}}{\alpha_1 \frac{Z^2}{A^{4/3}} + \alpha_2 M^{2/3}} \tag{4.1}$$

where α_1, α_2 and β are three constants with the numerical values

$$\alpha_1 = 6.534 \times 10^7 \text{ m}^3/\text{kg}^{1/3}\text{s}^2, \quad \alpha_2 = 4.0 \times 10^{11} \text{ m}^4/\text{kg}^{2/3}\text{s}^2$$

and

$$\beta = 5.726 \times 10^6 \text{ m}^4/\text{kg}^{2/3}\text{s}^2.$$

These constants depend only on the atomic properties. This is the mass-radius expression. It might appear to be a complicated expression but when it is realised that it applies to all bodies whose equilibrium is controlled by these three forces the simplicity of the expression becomes apparent. It

---

[4]The general expression for a force depending on the nth power of the distance is $2K = (n+1)P$. The sign of the force is immaterial, whether attractive or repulsive. Not all particles need to have the same polarity. For gravitation $n = -2$, the theorem becomes $2K + P = 0$ as quoted in the main text.

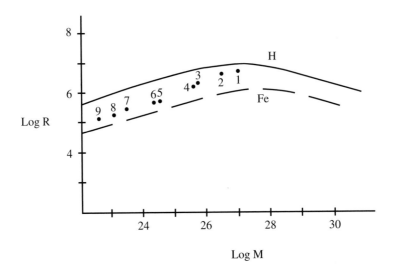

Fig. 4.1.   The plot of mass (M in kg) against radius (R in m) for a planet. The numbers refer to: 1 Jupiter; 2 Saturn; 3 Neptune; 4 Uranus; 5 Earth; 6 Venus; 7 Mars; 8 Mercury; and 9 Moon.

is also surprising that the values of the constants depend only on atomic properties.

The different behaviour at low and high M can be seen at once. For low M, that is, when $\alpha_1 \frac{Z^2}{A^{4/3}} \gg \alpha_2 M^{2/3}$, it is seen that $R \propto M^{1/3}$, the proportionality factor being linked to the material density. At the other extreme, for very large M, $\alpha_1 \frac{Z^2}{A^{4/3}} \ll \alpha_2 M^{2/3}$ and then $R \propto \frac{1}{M^{1/3}}$, which is different. At low M, the radius R increases with M while at large M, R decreases as M increases. This shows that R goes through a maximum value as M increases. This is confirmed by plotting R against M as in Fig. 4.1. Two curves are presented, one for a pure hydrogen composition and the other for pure iron. It is seen that the curve for iron is always associated with a lower radius than the corresponding curve for hydrogen.

4.4.   Maximum Size and Mass

The condition for the maximum radius, R(max), and the associated mass, $M(R\text{max})$ can be found from Eq. (4.1). It is found that:

$$R(\text{max}) = \frac{\beta}{\sqrt{\alpha_1\alpha_2}} \frac{Z^{2/3}}{A} \approx 1.12 \times 10^8 \frac{Z^{2/3}}{A} \text{ m}. \tag{4.2}$$

If the radius of Jupiter, R_J the largest planet in the Solar System, is used as a yardstick this is written instead

$$R(\text{max}) \approx 1.47 \frac{Z^{2/3}}{A} R_J .$$

The corresponding maximum mass, $M(R\text{max})$, is

$$M(R\text{max}) \approx \left(\frac{\alpha_1}{\alpha_2}\right)^{3/2} \frac{Z^2}{A} = 2.1 \times 10^{27} \frac{Z^3}{A^2} \text{ kg}$$

or, in terms of the Jupiter mass

$$M(R\text{max}) \approx 1.2 \frac{Z^3}{A^2} M_J .$$

For a hydrogen composition we set $Z = A = 1$. It seems that Jupiter is only slightly below the maximum radius for a pure hydrogen planet. Of course, Jupiter is not composed of pure hydrogen but contains helium and other elements as well. Whether this mass for Jupiter has any physical significance is not clear.

It must be stressed that the values given in this Section are not exact because they are based on some approximations, and especially the validity of the virial theorem. They are, nevertheless, reliable to a factor about 2 and are a sufficient approximation to allow important physical conclusions to be derived.

4.5.   Defining a Planetary Body

Something special happens at the maximum radius because the material becomes more compressible for masses beyond this special value. To see what it is, consider the gravitational energy and especially the gravitational energy per atom in the volume and consider a hydrogen composition. This does not restrict the validity of the arguments in principle but makes the analysis simpler.

The mean energy per atom at maximum radius, written $\varepsilon(\text{max})$, is of order $\varepsilon(\text{max}) \approx 10^{-18}$ J. This is the same order of magnitude as the energy required to ionize a hydrogen atom. For masses beyond the maximum radius the number of atoms ionized increases and the increased compressibility arises because more and more electrons become free. Gravity more easily compresses the electrons together leading to a decreasing volume for increasing mass. Apparently, the region where $MR^{-3} = \text{constant}$ refers to unionized atoms whereas the region where $MR^3 = $ constant refers to the condition where atoms are ionized. For the planets of the Solar System, where all the masses are below the value for the maximum radius, the interiors are unionized. On this basis it is natural to restrict the range of planets to that where the interior is not ionized but where gravity is significant. The upper end of the masses is defined this way but the smallest mass is less easy to specify precisely. It must depend on the composition and can be recognised by the body having a spherical shape. For a body composed of silicate materials it is found that this will require a mass of about 10^{19} kg.

For high masses, once ionization occurs the internal conditions can be changed significantly. Thermonuclear processes (see I.1) can occur giving an internal energy source and a radiation pressure to oppose gravity. Naturally occurring hydrogen is mainly H^1 with a single proton for its nucleus but a small proportion is heavy hydrogen, deuterium H^2, with a proton and neutron as the nucleus. The energy of this heavy hydrogen is greater than that of the lighter component and is more susceptible to interactions. The result is that a thermonuclear energy source is set up and the body ceases to be cold. This occurs at a mass of about 2.5×10^{28} kg, or about 13 M_J. The full complement of hydrogen becomes effective as a heat source when the mass reaches a value about 1.5×10^{29} kg or about 80 M_J. The body is now a small star.

There is no doubt that the masses where thermonuclear process occur is beyond the range of planets. The problem is where a planetary mass may end. The overall range is clear: it is between the lowest mass where gravity first becomes significant and the mass of about 13 M_J where the body ceases to be cold. There are two possibilities.

(a) Unionized Atoms

Following the example of the Solar System planets can be defined as cold bodies without internal ionization. This places the upper mass as that for a prescribed composition where the radius is a maximum. The region of the mass beyond R_{max} up to the mass where the body ceases to be cold is an, up to now, unnamed region which we might call super planet. The two regions of planet and super planet are well defined although conditions are simplest for the case of a hydrogen, or near hydrogen, composition. Silicate or ferrous bodies in the super planet region are unknown from the Solar System.

(b) General Conditions

The alternative definition is to accept any cold body within our definition as a planet. This includes the region of the maximum of the radius. This definition has the difficulty that, in the region of the maximum radius, a selected value of the radius is consistent with *two* values of the mass. This ambiguity might seem unacceptable.

Where the definition arises in these arguments it is that restricting planets to only unionized atoms that will be used.

4.6. Cosmic Bodies

It is interesting to explore briefly the effect of even higher masses. Only hydrogen/helium is considered because this composition fits the wider cosmic experience. Hydrogen becomes ionized for masses in excess of 10^{29} kg, the protons and electrons moving independently. The positive and negative electric charges cancel to provide strict electrical neutrality and the material now forms a plasma. The temperature of the compressed plasma will rise until it reaches 1.5×10^7 K. At this point thermonuclear processes begin the conversion of hydrogen to helium and the mass begins to radiate energy into space. Gravity is opposed by radiation pressure resulting in a substantial increase of the radius. This continues until the initial hydrogen has been converted to helium so that the energy source comes to an end. The radiation pressure falls and there is a corresponding catastrophic decrease of the radius. The decrease in volume means that the electrons have a smaller space for movement.

Up to this point the electrons have been assumed to behave as simple Newtonian particles but at this stage they show a quantum behaviour. Being Fermi particles each electron occupies a single cell in space[5] and resist the compression of the cell. The protons form a lattice (metallic) structure. The electron pressure within each cell now opposes gravity and maintains a stable equilibrium which is independent of the temperature and so is degenerate. The body can radiate any excess energy away but cannot radiate away the degenerate energy associated with the electrons. This is the condition of a white dwarf star.

The mean density is of the order $\rho \approx 10^9$ kg/m^3 and the pressure P within the body is given by

---

[5]Two electrons can lie together if their spins are opposite.

$$P = 0.0485 \frac{h^2}{m_e} \left(\frac{Z}{Am_p} \right)^{5/3} \rho^{5/3}$$

where m_e is the electron mass. The central pressure and density are given, respectively, by $P_c = 0.770 \frac{GM^2}{R^4}$ N/m^2 and $\rho_c = 1.43 \frac{M}{R^3}$ kg/m^3. The central density is generally of the order $\rho_c \approx 10^{12}$ kg/m^3. The mass and radius of the body were seen previously to satisfy the condition $MR^{-3} = $ constant.

The further increase of the mass increases the motion of the electrons within the cells and so their speed until they approach the speed of light. The electrons can offer no further resistance to the compression of gravity and a new regime is established. This situation was first examined by Chandrasekhar who showed that this upper mass is 1.44 M(sun) $\approx 2.8 \times 10^{30}$ kg. This Chandrasekhar mass is calculated to be given by the expression

$$M_{ch} = 0.20 \left(\frac{Z}{A} \right)^2 \left(\frac{hc}{Gm_p} \right)^{3/2} m_p \,.$$

Beyond that the electrons cannot exist and in fact form neutrons by combining individually with protons. The plasma was electrically neutral initially and the new collection of neutron particles is also electrically neutral. Neutrons are also Fermi particles, occupying their own individual cells in space, and so resist gravity on this basis. The object is now called a neutron star. It is a degenerate body with a configuration independent of the temperature. The radius now falls to a few kilometres even for a solar mass and the mean density is of the order $\approx 10^{17}$ kg/m^3. This is a nuclear density. The relation between radius and mass has the same general form as for the white dwarf but the numerical size is different. It is astonishing that the behaviour of these objects can be described by the ratios of the fundamental constants alone.

There is a limit to the resistance that even the neutrons can offer gravity and so to the mass of the neutron star. This upper mass is subject to some uncertainty but is likely to be about 4 M(sun) $\approx 10^{31}$ kg. What happens beyond this mass is very much conjectured. If there are no other forces to resist gravity the material can do little else except to collapse into a vanishingly small volume to form a black hole. On the other hand, it is likely that sub-neutron particles exist such as quarks and that these particles will be fermions able to resist gravity. A hierarchy of dense states of matter could then exist, each layer with a smaller radius than the one before, with a black hole as a limiting case which may or may not be achieved in nature.[6]

[6]There has not yet been an unequivocal detection of a black hole though there are several likely candidates.

4.7. Planets and Satellites: Planetary Bodies

The tradition is to have planets orbiting stars and satellites orbiting planets but the arguments we have developed do not involve the type of motion undertaken by the body. Whether a body is a planet or satellite, on that reckoning the behaviour of the body in the appropriate mass range will be the same. It is recognized now that such bodies may even be moving independently through space, having been ejected from some system. With this in mind it is more appropriate to speak of a planetary body rather than a planet so that all circumstances can be encompassed. Planets will generally be the more massive planetary bodies than the satellites.

The recognition of the common structures of the more massive bodies of the Solar System blurs the usual demarcations. Is Pluto really a planet or is it a larger example of a Kuiper object — but if orbiting the Sun is the criterion why aren't these bodies planets? What of the asteroids? They certainly orbit the Sun and some are massive enough to be spherical showing gravity is significant there. The new classification could be based on objects of the same general type. The terrestrial planets and the major planets are obviously two different types of planet. Pluto has always fitted oddly into that category but fits naturally into the group of Kuiper objects. The asteroids also form a homogenous whole.

Moving to higher masses, it is clear that planets and satellites form the low mass grouping of a wide range of cold bodies of comic importance. Broken only by temporary energy production, planetary bodies, super planets, white dwarfs and neutron stars form a hierarchy of related objects distinguished by mass. It is perhaps surprising that nature has adhered so clearly to a single structural template.

*

Summary

We can summarize a number of characteristics of a planetary body. The precise features will depend on the mass of the body and especially on its composition.

1. The mechanical equilibrium of the main volume is dominated by gravity.

2. The constituent atoms remain unionised.

3. Neither rotation forces nor magnetic forces play any basic part in the equilibrium. Any effects will be entirely secondary.

4. The material will slowly "flow" under gravity to reach a minimum energy configuration with a spherical surface figure in the absence of rotation. Rotation forces will cause the figure to become slightly ellipsoidal.

5. The interior will assume a configuration of virtual hydrostatic equilibrium. This will involve the layering of the interior according to the density, the heavy materials sinking towards the centre.

6. The constituents will form a heavy core region, a least dense surface crustal region with an intermediate density mantle in between. The whole structure can be encased in an atmosphere, which may be very dense or extensive.

7. The strength of gravity will be weak in the surface region and the crust will not follow the rule of hydrostatic equilibrium which apply to the rest of the volume.

8. It has not been necessary to make an explicit appeal to thermal forces to understand the mechanical equilibrium. The body will have retained heat from its formation and radioactive materials will contribute after that. The central regions will be hotter than the outside and the equilibrium will need to account for a flow of heat to the surface. The surface conditions will depend on the thermal state of any encasing atmosphere.

5

Fluid Flows and Magnetism

The structure of a planetary body has been seen to behave as a viscous fluid. The major planets and two of the terrestrial planets have measurable magnetic fields. These two aspects are linked, as we will see now.

5.1. The Fluid State

Matter exists in three physical states — solids, liquids and gases. Solid bodies show the ability to withstand shear and so maintain a fixed shape over very long time periods. They are also incompressible so the volume is maintained essentially unchanged: the volume changes only slightly under the action of external pressures, even of considerable strength. These properties result from the feature at the atomic level that the molecules composing the solid are packed close together in space. The macroscopic incompressibility is a reflection of the incompressibility of the individual atoms in their packing under normal circumstances. There are some exceptions (most notably He^3) but in general all materials become solid at low enough temperatures. Raising the temperature increases the energy of motion of the molecules and eventually the energy of atomic motion becomes comparable to the interaction energy. Then the atoms are free to move relative to the equilibrium sites and the material shows some fluidity — it has melted. There has been a small change of density (most often an increase but not always, e.g. water and germanium) but the fluid remains largely incompressible. Unlike a solid, it has no fixed shape and is now in the liquid phase. A further increase of temperature will increase the kinetic energy further until it is greater than the interaction energy. At this point the fluid has passed from liquid to gas phase. The gas has no particular volume or shape and is very compressible. Some materials (for example CO_2) pass

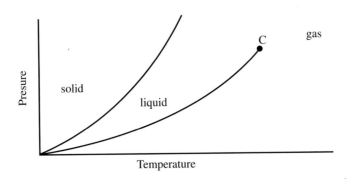

Fig. 5.1.   A typical pressure – temperature diagram linking the three phases of matter.

directly from the solid to gas phase, showing no liquid phase, and are said
to sublimate. This picture is very qualitative but sufficient for our present
purposes.

The relation between solids, liquids and gases can be portrayed on a
pressure-temperature plot., shown in Fig. 5.1. It is seen that the solid region covers the low temperatures and pressures and the liquid region covers
the higher temperature, lower pressure region. There is a point, marked C,
beyond which the material remains in the vapour phases irrespective of pressure. This is the critical point and the corresponding pressure is the critical
pressure. There is no distinction between liquid and vapour for temperatures
beyond the critical, the two becoming one. The vapour can then have very
high densities without having full liquid characteristics. The precise form of
the phase diagram differs in details from one liquid to another but the general
characteristics remain similar. We might add that water is an exceptional
substance in ways that will become clear later. One strange property is that
ice floats: the solid is less dense than the liquid at the melting point. This
is strange and unique for ordinary materials. It has the vital consequence
that lakes freeze from the top down and not from the bottom up. This is
important for life at low temperatures. It also has the consequence that ice
cubes float in a drink and do not sink. The reason for these phenomena is
seen in Fig. 5.2 which makes a comparison between the behaviour of water
and an ordinary material at atmospheric pressure. In ordinary matter fusion (or melting) is associated with an increase of volume so that the liquid
phase has a lower density than the solid. Having the greater density, the
solid sinks in the liquid. It is seen from Fig. 5.2 that water acts the other
way round. Here the volume *decreases* on fusion so the liquid has a higher

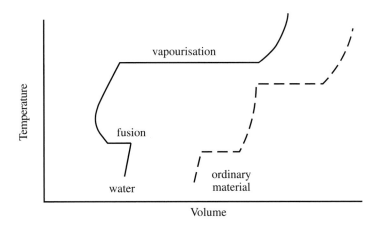

Fig. 5.2. A schematic temperature–volume diagram for water and for an ordinary material.

density than the solid. The liquid sinks causing the solid to rise and so to float. Vapourisation has the same effect for water and for ordinary liquids. Both fusion and vapourisation occur at constant temperature and involve the release of energy (the latent heat) which arises from the relaxation of interaction between the atoms and molecules of the substance. In general the atoms move a small distance apart on melting but gain freedom of movement on vapourisation. The case of water is anomalous. Now the structure of the two atoms of hydrogen and the one of oxygen forming the molecule make a spatial rearrangement on melting to form a denser packing. Consequently water becomes rather more dense than ice. The effect of pressure on water is to produce a wide range of polytropes of ice as is shown in Fig. 5.3. Ten variations are shown there, the unit of pressure being 10^9 N/m^2 which is closely 10^4 bar. Many of the polytropes are associated especially with lower temperatures. These will be particularly relevant at the lower temperatures associated with the icy satellites. The mechanical properties of these various forms of ice remain largely unknown and appeal must be made to expectations from crystallography. Ice II has a rhombohedral form and is more densely packed than ice I. The density of ice I is about the normal value 920 kg/m^3, that for ice II is likely to be about 25% greater. Ice II has no melting line but transforms directly to ice III or ice V. The densest ice is ice VII with a density of 1.5×10^3 k/gm^3 at atmospheric pressure. We need not enter into details here but such data may be necessary in understanding eventually the detailed equilibrium of the icy satellites.

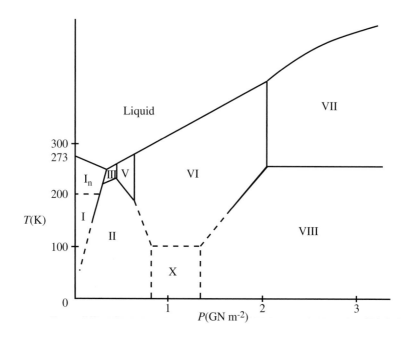

Fig. 5.3.   The pressure – temperature phase diagram for water (after Mishima and Endo).

5.2.   The Importance of Time Scales

The atoms and molecules in a liquid are in a random motion restricted to a small volume, or cell, about an equilibrium site. The fluid property of the macroscopic material arises because the equilibrium sites themselves are able to relax into new positions. The time scale appropriate to the atomic vibrations can be estimated from the Heisenberg relation, $Et \approx h$, where E is the energy, t the associated time scale and h (= 6.6×10^{-34} Js) is the Planck constant. If $E \approx 10^{-22}$ J (the interaction energy) then $t \approx 10^{-11}$ s. For longer time intervals the fluid property will be valid. This is actually a long time on the atomic times scale. The time for a beam of light to pass an atom is of order 3×10^{-19} s which is 8 orders of magnitude less so, in nuclear terms, the random movement of the atom is a rather slow activity. If a disturbance has a time scale less than 10^{-11} it can change the motion but not if it is greater, although the equilibrium position itself could be affected. Time scales are, therefore, important. For an impulsive force the liquid will behave as a solid. For a slow acting force the liquid shows normal fluidity. This distinction is of great importance in planetary science. The material of a planet may well appear solid in everyday activities but can show flow characteristics under the action of gravity over cosmic time scales.

5.3. Specifying Fluid Behaviour

Fluid motion takes place within the constraints of the conservation statements of mass, energy, and momentum. Rather than set down these various requirements in detail, which is lengthy and cumbersome, we will use a dimensional approach which will provide a set of dimensionless groupings of use later.

(a) The Buckingham Π-Theorem

This provides an invaluable method for deriving the dimensionless groupings relevant to a particular physical situation. The method centres around N (say) variables that we are required to arrange into dimensionless groupings, with n (say) separate basic dimensions describing the variables. There will be $(N - n)$ independent groupings $\Pi_1, \Pi_2, \ldots \Pi_n$ of dimensionally homogeneous entities which are free of units to describe the behaviour of the physical system. The full description of a physical system is then given in the form $F(\Pi_1, \Pi_2, \ldots \Pi_n) = 0$ where F is some function. This can be written alternatively in a form $\Pi_i = F(\sum_{j \neq i} \Pi_j)$. The dimensionless numbers Π will be present subject to a numerical factor that cannot be specified by an appeal to dimensions. This is the Π-theorem, usually ascribed to Buckingham though sometimes also to A. Vaschy who was also instrumental in its development. The approach has a very mathematical history dating back to the late 19th century and is related to group theory. We will not follow this history. The problems of establishing the theorem may be severe but its use is known to be reliable and is very easy. We will use it now to specify various fluid situations.

(b) The Practical Usefulness of the Dimensionless Π Numbers

A composite quantity without dimensions is independent of scale in space or time provided that space-time-mass transform according to the set

$$L'_i = aL_i, \quad t' = bt, \quad M' = cM$$

linking the dashed and undashed co-ordinate systems. This is a variation according to a simple scale, relating two systems of different size, speed and mass but of geometrically identical form. If the external forces can also be scaled in the same way there is here a method upon which dynamical and geometrical scaling can be based.

A relation of the type $F(\Pi_i, \Pi_j) = 0$ will be valid in either coordinate frame (L, t, M) or (L', t', M'). Information about the relation in one frame can immediately be transformed to information in the other. One frame can be in the laboratory where experiments can be conducted. The other frame could be on a large scale where the geometrically identical object is to be constructed. One example is a bridge: the design can be finalised in the laboratory prior to manufacture. Another is a planetary structure which can be studied on the laboratory scale to give insights into the actual planet. These are examples of the invariance of groups.

We might notice in passing that a ten parameter group is of central importance for Newtonian mechanics, the so-called Galileo-Newton group. This has the four sub-groups:

(1) the group of translations in space S $x_i' = x_i + c_i$

(2) the three space rotations R $x_i' = \sum a_{ij} x_j \quad (i, j = 1, 2, 3)$

(3) the time translation T $t' = t + c$

(4) moving axes M $x_i' = x_i - b_i t$

These transformations are dynamic: the earlier ones are of similitude.

It is realised that the linking of the dimensionless numbers to experimental procedures provides a very powerful method of investigating complicated physical situations where mathematical methods alone would be powerless. There are problems of exact geometrical scaling to be sure. For instance, it is not easy to scale joints and bolts so the results of the scaling experiments may only be approximate although still significant. These difficulties usually give only minor problems. The scaling can, however, never be precise and the final product is always unknown to some extent.

5.4. Isothermal Insulating Fluids

Here, the flows of insulating fluids occur without change of temperature. The movement is constricted by the conservation of mass and of momentum. There are three basic dimensions, mass, M, length, L, and time T. The quantities describing the flow are collected in Table 5.1. There are 6 independent variables so, with three basic quantities, there are 3 possible independent combinations. The selection of which three quantities are to

Table 5.1. The symbols and dimensions for quantities of isothermal flows.

Quantity	Symbol	Dimensions
Speed	U	LT^{-1}
Time	τ	T
Distance	x	L
Pressure	p	$ML^{-1}T^{-2}$
Density	ρ	ML^{-3}
Rotation	Ω	T^{-1}
Shear viscosity	η	$ML^{-1}T^{-1}$
Kinematic viscosity	$\nu = \dfrac{\eta}{\rho}$	$\dfrac{L^2}{T}$
Acceleration of gravity	g	$\dfrac{L}{T^2}$

be the basic ones is a matter of choice but once selected the rest is routine. The three independent parameters are then found by combining the three chosen basic quantities with each of the remaining quantities in turn.

As an example, let us choose ρ, v, L and the viscosity, μ. Groups of quantities ρ, v, L, μ are to be dimensionless. We express this requirement in the form $\rho^\alpha v^\beta L^\gamma \mu^\delta$ where α, β, γ, and δ are three numbers to be chosen to get rid of the dimensions. With four variables and three basic units we will need to express the grouping in terms of one variable — let us choose δ. Then we have the combination

$$\left(\frac{M}{L^3}\right)^\alpha \left(\frac{L}{T}\right)^\beta L^\gamma \left(\frac{M}{LT}\right)^\delta \tag{5.1}$$

which is to have no dimensions. This is achieved if α, β, γ and δ satisfy the conditions

for M : $\quad \alpha + \delta = 0$

for L : $\quad \beta - 3\alpha + \gamma - \delta = 0$

for T : $\quad -\beta - \delta = 0$.

There are four quantities and three equations so three variables must be expressed in terms of the fourth. It does not matter which one chooses — choose δ to be the fourth variable. Then

$$\alpha = -\delta, \quad \beta = -\delta, \quad \gamma = -\delta$$

and the relation (5.1) becomes $(\frac{\mu}{\rho L U})^\delta$. The quantity inside the brackets is the dimensionless number called the Reynolds number named after the engineer Reynolds[1]:

$$\Pi_1 = \frac{\rho U L}{\mu} \equiv \mathrm{Re}.$$

It is the ratio of quantities with the dimensions of force. One is the expression for the force of viscosity within the fluid, $F_v \approx \frac{\mu U}{L^2}$ and the other for the force of fluid inertia $F_I \approx \frac{\rho U^2}{L}$ so that $\Pi_1 = \frac{F_I}{F_V}$. Experimental scaling using this expression would generally require $UL = $ constant. Changing density and changing the material to affect the viscosity will have only a limited effect.

Other dimensionless numbers are the ratio of the expressions for other forces and are derived in an analogous way. Instead of linking the viscosity with the basic quantities let us link the time, for a time dependent flow. Then we require (ρ, v, L, t) to be dimensionless. This gives the dimensionless Strouhal number

$$\Pi_2 = \frac{U\tau}{L} \equiv \mathrm{St}$$

for time dependent advection. Scaling will require $\frac{U}{L} = $ constant. Running cameras fast or slow will require the time to scale as $\tau = \frac{L}{U}$. This can be important in visualising geophysical phenomena.

Rotation effects can be included in the same way. By requiring (ρ, U, L, Ω) to be dimensionless, where Ω is the angular velocity we obtain the dimensionless Rossby number

$$\Pi_3 = \frac{\Omega L}{U} \equiv R_o.$$

Scaling now is according to $\frac{L}{U} = $ constant but the rotation rate can be decreased by increasing $\frac{L}{U}$. The pressure can also be accounted for through the quartet (ρ, v, L, p). This gives the dimensionless number

$$\Pi_4 = \frac{pU^2}{p}.$$

This number relates the pressure head of a flowing fluid to the conditions of the flow. It is controlled by the equation of state of the material and so is not a number that can be chosen arbitrarily once the fluid has been assigned.

---

[1]He studied the viscous flow of fluids, especially in pipes.

Other numbers can be constructed from these. For instance, the Rossby number can be written alternatively

$$\Pi_3 = \frac{\Omega L}{U} = \left(\frac{\Omega L^2}{\nu}\right)\left(\frac{\nu}{LU}\right) = \frac{\left(\frac{\Omega L^2}{\nu}\right)}{\Pi_1}$$

giving the Taylor number

$$\Pi_5 = \frac{\Omega L^2}{\nu} \equiv \mathrm{Ta}$$

which relates the forces of rotation to the viscous forces. For a given fluid, scaling is such that $\Omega \approx \frac{1}{L^2}$.

The flows generally occur under gravity and the ratio of the gravitational to inertial forces is then of interest. If g is the local acceleration of gravity, this ratio is expressed in terms of the number

$$\Pi_6 = \frac{gL}{U^2} \equiv \mathrm{Fr},$$

often called the Froude number. It is a number particularly associated with wave motions. Scaling is according to $\frac{L}{U^2} = $ constant.

It will be realised that many of these scaling laws are inconsistent with one another. For instance, Reynolds scaling (appropriate to viscous dissipation) requires that $UL = $ constant whereas Froude scaling requires $\frac{L}{U^2} = $ constant. This raises problems in the design of engineering structures and compromises have to be made. It is here that design becomes also an art and requires experience.

5.5. Thermal Insulating Fluid Flows

Temperature variations can be included if account is taken of the conservation of energy. In terms of dimensions it is necessary to introduce the fourth variable temperature Θ to give the four basic variables M, L, T, Θ. This, then, leads to the addition of the heat quantities shown in Table 5.2. Thermal problems then involve expressing the heat flow, q, in terms of the various controlling variables. Explicitly

$$q = F(\rho, \beta\Delta\theta, \lambda, \eta, C_P, L, U).$$

This provides a set of six Π numbers by following the same method as before. These are as follows:

$$\Pi_7 = \frac{g\beta\Delta\theta\rho^2 L^3}{\eta^2}$$

Table 5.2.   The symbols and dimensions of quantities for thermal fluid flows.

Quantity	Symbol	Dimensions
Heat	H	$\dfrac{ML^2}{T^2}$
Heat transfer rate	Q	$\dfrac{ML^2}{T^3}$
Heat current density	q	$\dfrac{M}{T^3}$
Heat transfer coefficient	$\mathcal{H}$	$\dfrac{M}{T^3\Theta}$
Specific heat	C_p	$\dfrac{L^2}{T^2\Theta}$
Thermal conductivity	λ	$\dfrac{ML}{T^3\Theta}$
Thermal diffusivity	$\kappa = \dfrac{\lambda}{\rho C_p}$	$\dfrac{L^2}{T}$
Coef. of thermal expansion	$\beta = \dfrac{\Delta\theta}{\theta_o}$	$\dfrac{M}{L^3\Theta}$
Acceleration of gravity	g	$\dfrac{L}{T^2}$

and usually called the Grashof number, Gr. It relates the buoyancy force to the inertial force. Introducing the specific heat of the material, Cp, gives the Prandtl number

$$\Pi_8 = \frac{\eta C_p}{\lambda} = \left(\frac{\eta}{\rho}\right)\frac{\rho C_p}{\lambda} = \frac{\nu}{\kappa} \equiv \mathrm{Pr}\,.$$

It allows a comparison between the kinematic viscosity, $\nu = \frac{\eta}{\rho}$, with the thermal diffusivity, $\kappa = \frac{\lambda}{\rho C_p}$. It is the only dimensionless grouping that involves only properties of the fluid and not of the flow. Planetary problems involve material with high viscosity so Pr is very large there (perhaps of the order 10^4). Laboratory problems are dominated by a strong thermal conductivity giving a small Pr there, (≤ 1).

The Grashof number Π_7 can be rearranged into a more significant form as follows

$$\frac{g\beta\rho^2 L^3\Delta\theta}{\eta^2} = \left(\frac{g\beta L^3\Delta\theta}{\kappa\nu}\right)\left(\frac{\kappa}{\nu}\right) = \frac{\Pi_9}{\mathrm{Pr}}\,.$$

The first ratio is very important in describing natural convection and is called the Rayleigh number, Ra after Lord Rayleigh who first recognised its significance. Then

$$\Pi_9 = \frac{g\beta L^3 \Delta\theta}{\kappa\nu} = \text{Ra}.$$

Neither Ra nor Gr involve the flow velocity and are useful in describing free convection (under gravity). Other numbers include the flow velocity and can be used to describe thermal conditions with forced convection. One such number is the Péclet number, Pe

$$\Pi_{10} = \frac{C_p \rho U L}{\lambda} \equiv \text{Pe}$$

relating convected to conducted heat flow. A further related number is the Nusselt number Nu involving the flow of heat per unit area per unit time, q, (that is the heat flux)

$$\Pi_{11} = \frac{gL}{\lambda\Delta\theta} = \text{Nu}.$$

The heat flux q across a surface is expressed empirically as being directly proportional to the temperature difference, $\Delta\theta$, between the surface and the environment. This relation can be written as $q = \mathcal{H}\Delta\theta$ where $\mathcal{H}$ is an empirical quantity called the heat transfer coefficient. This must be found experimentally for each situation. Introducing this coefficient, the Nusselt number takes the form

$$\text{Nu} = \mathcal{H}\frac{L}{\lambda}.$$

It is sometimes important in forced convection problems to include the heat flow in the dimensionless expression. This can be arranged by forming the Stanton number, St. Beginning with the Nusselt number we write successively

$$\mathcal{H}\frac{L}{\lambda} = \mathcal{H}\frac{1}{U\rho C_p}\frac{UL}{\nu}\frac{\nu}{\lambda} = \Pi_{12} \times \text{Re} \times \text{Pr}$$

where

$$\Pi_{12} = \mathcal{H}\frac{1}{U\rho C_p} \equiv \text{St}.$$

Using these numbers, heat transfer problems can be expressed quite generally in the form Nu $= f(\text{Re}, \text{Pe})$ or St $= f(\text{Re}, \text{Pe})$ for forced convection and Nu $= F(\text{Gr}, \text{Pr})$ or St $= F(\text{Gr}, \text{Pr})$ for free convection where f and F

represent functions, usually to be found empirically. Sometimes it is more
convenient to replace Gr by Ra because Ra involves the viscosity whereas
Gr does not. The usual procedure is to assume a relation of the form Nu =
A Re$^\alpha$ Pe$^\beta$, for example, and use experimental rigs to determine the constants A, α and β. Thus, holding Re constant, experiment allows an estimate
to be made of β. Taking the logarithm we have the expression for Nu in the
form

$$\log \mathrm{Nu} = \alpha \log(A\ \mathrm{Re}) + \beta \log \mathrm{Pe}\,.$$

This can be plotted as a straight line showing log Nu against log Pe with slope
β allowing β to be found. The intercept on the Pe axis allows $\alpha \log(A\ \mathrm{Re})$
to be found. Alternatively, keeping Pe constant, α can be found. It is then
possible in principle to deduce an empirical value of the constant A. The
values obtained this way will be only approximate but probably sufficient
for many practical purposes.

5.6.   Natural Convection: Volcanic Activities

The inside of a planet is hot. This is partly due to the compression of
the material with depth due to gravity which provides a static temperature
profile where heat does not flow from one region to another because it is
a condition of thermodynamic equilibrium. This temperature distribution
shows an adiabatic temperature gradient. There are also heat sources due
to the decay of radioactive elements and to the heat of formation of the
body itself which are not adiabatic. For smaller bodies orbiting larger ones,
ther gravitational compression/expansion effects will also raise the temperature due to the non-elastic dissipation of heat. A temperature gradient
above the adiabatic value will provide instabilities. A packet of material
will have a temperature above the local equilibrium value and so will have
a lower density. This density difference in the presence of gravity provides
a buoyancy force which causes the packet to rise to a region of similar density. Heat is carried upwards, along the gradient of gravity, and in this way
heat rises from the centre to the outside of the body. This is the process of
convection.

Movement of the fluid (perhaps plastic) material is opposed by viscosity of the fluid. The buoyancy force must overcome the viscous force for
motion to take place. Once movement occurs heat transfer occurs as well
reducing the temperature of a packet of material and so reducing the buoyancy force to the stage where the viscous force dominates again. The heat

builds up and the buoyancy force takes over, and convection begins again. In this way the viscosity controls the internal motions, the heat transfer is by natural convection. The controlling dimensionless numbers are then Ra, Re, Pr so that there will be a describing expression of the form $\mathcal{F}(\text{Ra}, \text{Re}, \text{Pr}) = 0$. There can be effects of rotation in which case the Taylor number, Ta, will often be the appropriate number to introduce into the arguments. Direct calculations of the processes using the well known mathematical equations of fluid mechanics are not possible because the fluid is too complicated.[2]

The precise thermal behaviour of the surface region depends on many factors but all modify the single process of releasing heat into space. A liquid surface will bubble and release heat in a simple way but a solid surface is more complicated. Hot magma will form underneath the surface layer and heat will certainly be conducted to the surface by simple conduction through the surface rock. If there is more heat than can be carried away by that process pressure will build up and the magma will break through the surface at weak spots. This breaking through will generally be severe and the ejection event is called a volcano. Sub-surface material in the form of rocks and lava will be ejected into the air at high temperature accompanied by the associated gases, which will be CO_2 and sulphur compounds, especially for silicate materials. Whether the volcanoes are widely spread (as for Venus), or constrained to certain regions or lines (as for Earth and for Mars) depend on the crustal thickness and strength. Continual eruptions will build up ash and lava material to form a substantial mountain but the height to which it can rise must depend on the strength of the crust and on the chemical nature of the lava. For instance, thick lava (such as rhyolite) produces steep sided volcanic domes because it is not able to travel very far. In contrast, basalt is a runny lava when hot and flows easily to produce low angled shield volcanoes. There is also a contrast between a very viscous material which leads to explosive eruptions. Volcanology (the study of volcanoes) has advanced considerably since data have come from other bodies in the Solar System.

5.7. Boundary Conditions

The boundary of a fluid shows special conditions.

---

[2]Fluid mechanics can only deal analytically with so-called simple liquids where the stress and strain are proportional to each other. Geophysical materials are far more complicated and not described in simple terms even now.

(i) Constant Temperature Fluid against a Rigid Surface

George Stokes in the mid-19th century realised that a fluid remains at rest when in contact with a rigid surface. On the other hand, the mainstream certainly flows past with a definite speed away from the surface. There must, therefore, be a region near the surface containing a gradient of speed from zero at the surface to the maximum of the main flow a distance away. This is called the velocity boundary layer. The flow within it will be laminar at lower speeds but turbulent at higher speeds. The surface has the form of a parabola, thickening as the square root of the distance downstream from the leading edge of the surface. The layer also increases with the square root of the shear viscosity of the fluid. The boundary layer thickness can, therefore, becomes quite large when involved with a molten magma. It is usually quite small for a gas: as an example, that surrounding an aeroplane in flight is no thicker than a cigarette paper. Prandtl showed (in the early 20th century) that the effect of viscosity is largest close to the surface.

(ii) Constant Temperature Curved Surface

The effect of a curvature of the surface is to hold the boundary layer with the curvature at first but when the layer breaks away (separates) from the surface in turbulent flow as the speed increases. Separation is delayed by surface roughness, a wrinkled surface holding the boundary layer to it for a greater distance downstream than a smooth surface. For a projectile passing through a fluid the main viscous resistance to the flow occurs in the downstream region (the wake) so the conditions at which separation of the boundary layer occurs are most important.

(iii) Thermal Flows

The situation is more complicated if the temperature of the fluid is different from that of the rigid barrier. In geophysical applications the surface temperature is usually higher than that of the fluid (though for engineering problems the situation is often the other way around). Now heat is carried downstream by the mainstream flow but in the boundary layer the heat flow is mainly by simple thermal conduction. The range for which conduction dominates forms the thermal boundary layer. The thickness is proportional to the inverse square root of the Prandtl number. It achieves high values in some geophysical applications. Conditions within the thermal boundary layer are controlled by those within the velocity boundary layer. Cooling

Table 5.3.   Symbols and dimensions of various electrical quantities.

Quantity	Symbol	Dimensions
Magnetic field	H	ampere/metre
Electric field	E	volt/metre
Electrical conductivity	σ	
Permittivity	ε	coulomb/m^2
Magnetic permeability	μ	henry/m
Electric charge	q	coulomb
Electric current	j	ampere

by fluid flow is not a very efficient mechanism. For higher temperatures the most effective cooling mechanism is radiation.

5.8.   Electrically Conducting Fluids

Most planetary materials are able to conduct electricity, at least to a limited extent, and magnetic fields are widespread. These electric-magnetic effects must be accounted for. This involves both the motion of a fluid as well as the rules of electricity and magnetism. Again appeal can be made to dimensionless numbers through the Π-theorem. We will quote results rather than repeat the standard procedures again. The appropriate quantities are listed in Table 5.3.

A fluid able to conduct electricity, whether a conducting gas or a liquid, potentially constitutes an electric current. If this is permeated by a magnetic field it is acted on by a force, called the Lorentz force, F_L, which acts on the moving fluid *and* on the permeating magnetic field. If the angle between the direction of the fluid current, j, and that of the axis of the magnetic field of strength B is $\angle\theta$ then $F_L = \rho j B \sin\theta$, where ρ is the fluid density. This mutual relation between the behaviour of the fluid and the behaviour of the magnetic field is called magnetohydrodynamics. It links magnetism and fluid motions. It is entirely non-relativistic which means that the magnetic effects are important but not the electrical.[3] This new situation can be described by using its own dimensionless numbers.

[3]It is non-relativistic because the Maxwell's displacement current is neglected which means that time-dependent electric forces play no part. Electromagnetic waves are not involved.

The neglect of the displacement current can be expressed by the condition

$$\frac{\varepsilon \omega}{4\pi\sigma} \ll 1$$

where ω is the inverse of the characteristic time.

Ohm's law is assumed to apply to the fluid in the form

$$j = \sigma E + \frac{U x B}{c} \sin\theta$$

where θ is the angle between the fluid moving with speed U and the direction of the magnetic field. It is found that this relation can be expressed in terms of $B = \mu H$ alone in the form

$$\frac{H}{\tau} = \frac{U H \sin\theta}{L} + \left(\frac{c^2}{4\pi\mu\sigma}\right)\frac{H}{L^2}. \tag{5.2}$$

This relation links the change in the magnetic field in time with two separate effects. Consider the first case of the fluid at rest so the only term is the second term on the right-hand side. This gives a characteristic time τ defined by

$$\tau = \frac{4\pi\mu\sigma L^2}{c^2}. \tag{5.3}$$

and represents the time of decay of the stationary field. A full analysis shows that the field diffuses into the fluid. If energy is not injected into the field, via an electric current, the field will decay with this characteristic time. Theory shows that the field at any time τ is linked to the initial field H_o by the relation

$$H = H_o \exp\left\{-\frac{1}{\tau}\right\}. \tag{5.4}$$

The time for the decay of the field seen in (5.3) depends on the product σL^2. The decay time becomes very large if either σ or L is very large, or both are. In the laboratory neither quantity is very large and consequently magnetic field decay is very rapid. In astronomy, L is certainly very large and σ is often large as well. The magnetic field for the case when $\sigma L^2 \gg 1$ is said to be frozen into the fluid since it cannot decay spontaneously.

The coefficient of the second term in (5.2) is called the magnetic viscosity, η_M — it plays the same role in the magnetic case as the viscosity in an insulating fluid. Explicitly

$$\eta_M = \frac{c^2}{4\pi\mu\sigma}. \tag{5.5}$$

It becomes smaller the larger the electrical conductivity of the fluid. Now we can move to the second case when the fluid is in motion and both terms are present in (5.2). Now the magnetic field is carried along by the fluid opposing the diffusion through the fluid. The ratio of the quantities on the right-hand side of (5.2) defines the magnetic Reynolds number $\mathrm{Re}_M = \frac{UL}{\eta_M} = \frac{4\pi\mu\sigma UL}{c^2}$.

5.9. Application to Planetary Magnetic Fields

The arguments of the last section have been used in attempts to understand intrinsic planetary magnetic fields. The arguments are centred on (5.2). It is argued that a planetary field will result from the motion of an electrically conducting fluid region of appropriate geometry within the planet. It is supposed that fluid kinetic or thermal energy inside is converted to magnetic energy which is then magnified by the mechanism described by (5.2). Because the field must escape the region of its source, it must not be frozen into the fluid so ReM must not be too high.

The effect of the permeating magnetic field on the motion of the conducting fluid is balanced by the effect of the conducting fluid on the permeating magnetic field. An equipartition of energy will be set up for equilibrium and the whole will form a simple consistent dynamical system. For planetary sciences, this is the condition to describe the intrinsic magnetic field of the planetary body. The mathematical description combines the description of the fluid (technically through the Navier-Stokes equations) with the equations of magnetism. This is called the magnetic dynamo theory. The mathematical problem set by this equipartition is far too difficult to solve directly, at least at the present time, and simplifications must be made if progress is to be made. This is done by assuming the flow pattern for the fluid and attempting to find the magnetic consequences.

It is natural at first to suppose that there is a relation between the flow pattern and the forces of rotation so that a symmetry can be expected between the axis of the fluid flow and the axis of the magnetic field. Unfortunately this was shown in the 1930s to be not possible,[4] a result that has haunted this study ever since. Modern work is undertaken in numerical terms using the most advanced computer systems and enormous effort has been put into solving the problem but it must be admitted that very little

---

[4]This is the famous theorem of T. G. Cowling published in 1934. It shows that there cannot be a precise symmetry between the fluid and magnetic vectors if the magnetic field is to be self-sustained.

progress of a fundamental nature has been made so far.[5] Some of the phenomena can be understood qualitatively, for instance field reversals due to instabilities, but no quantitative understanding which can lead to the prediction of events has yet been achieved. The description of the planetary magnetic fields must still be made empirically. Planetary intrinsic magnetic fields described in these terms have met great difficulties by the arrangements of the fields discovered in Saturn, Uranus and Neptune. These are very difficult observations to explain in terms of the dynamo theory.

*

Summary

1. The approach to the description of fluid flow using dimensional analysis is described using the Π-theorem. This is a method of general application and its application is considered in a little more detail.

2. A set of dimensionless numbers is derived to describe the flow of an isothermal fluid flow and for thermal flows.

3. The method is extended to cover fluids that conduct electric currents. This leads to a set of dimensionless numbers associated with magneto-hydrodynamics.

4. The application of these arguments to planetary problems is outlined. In particular, the appearance of volcanoes due to internal heat flow to the surface and the presence of a planetary intrinsic magnetic field are both considered.

---

[5]Sir Fred Hoyle once remarked that if many people have been spending many hours using a common method to solve a problem without success then it may be that they are using the wrong approach.

Part II

GENERAL FEATURES OF THE SOLAR SYSTEM

The members of the Solar System are reviewed together with their compositions.

6

The Larger Members of the Solar System

The Solar System includes a wide range of objects from the large to the very small. The planets, satellites and asteroids are the larger and more massive members and their general properties are considered now.

6.1. The Sun

The Sun is the biggest body in the Solar System with a mass of 1.99×10^{30} kg. The mean radius is 696,00 km (which is 108 Earth radii) giving it a mean density of 1.41×10^3 kg/m^3. The surface temperature is 5,800 K making it spectroscopically a dG2 star (the d is for dwarf,[1] a designation showing it is on the main sequence in the Hertzberg-Russell (H-R) diagram). The luminosity is 3.9×10^{28} watts. The Sun is probably about 5×10^9 years old and will have about the same future before its fuel begins to become exhausted. This places it in its middle age. The source of energy at its centre is the hydrogen-helium cycle in which hydrogen is transmuted to helium (see IV.22). The central temperature is about 1.5×10^7 K. The Sun is a broadly average star, very similar to some 95% of the stars in the Universe. The properties of stars are considered in IV.

6.2. The Planets

The Sun is accompanied by a range of objects as it moves through space of which the most massive are the planets. There are nine accredited members but only eight are serious ones. The ninth (Pluto) is now generally included

---

[1]It will become a giant when it rearranges itself following the exhaustion of its thermo-nuclear fuel. It will then have left the main sequence of the Hertzsprung-Russell (H-R) diagram.

Table 6.1.   The orbital characteristics of the planets including Pluto.
(1 AU $= 1.49 \times 10^{11}$ m)

Planet	Mean distance (AU)	Orbital eccentricity	Orbital inclination	Period (years)
Mercury	0.387	0.2056	$7°0'$	0.2409
Venus	0.723	0.0068	$3°24'$	0.6152
Earth	1.000	0.017		1.000
Mars	1.524	0.093	$1°51'$	1.8809
Jupiter	5.203	0.048	$1°18'$	11.8623
Saturn	9.539	0.056	$2°29'$	29.458
Uranus	19.19	0.047	$0°46'$	84.01
Neptune	30.07	0.0086	$1°47'$	164.79
Pluto	39.46	0.249	$17°19'$	248.54

with the Kuiper objects (see Sec. 6.8) and so is not classed as a planet though
it is often still included with them.

(a) The orbital characteristics

The details of the orbits of the planets about the Sun[2] are collected in
Table 6.1. Several features can be noticed. First, all the planets lie essentially in the equatorial plane of the Sun. The plane of the orbit of the
Earth itself is called *the ecliptic* which is also the apparent orbit of the
Sun against the constellations. Deviations from the common plane are generally less than $3°.5$ with the exceptions of the innermost member (Mercury)
and the outermost member (Pluto). As far as is known there is no planetary orbit nearer the Sun than Mercury nor outside the orbit of Pluto.
Second, the system of planets extends out to about 40 A.U., which is
5.96×10^9 km. This is not the full extent of the Solar System because,
as will be seen later, there are many small objects and influences beyond
this distance. The times to orbit the Sun vary from less than 100 (Earth)
days to nearly 250 (Earth) years. For the outer members Neptune will have
made a little more than one orbit since its discovery (in 1855) whereas Pluto
will have moved relatively little since its discovery in 1930. Of the others,
Uranus was discovered in 1781 while all the remainder were well known
in antiquity.

[2]Remember, the Sun and planets actually orbit about the centre of mass of the Solar
System but this is nearly the same thing.

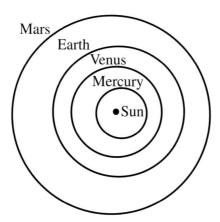

Fig. 6.1.   The orbits of the four terrestrial planets about the Sun as centre. Notice each orbit appears circular by itself but it is in fact an ellipse.

Kepler showed that the planetary orbits are ellipses and a property of much interest in later arguments will be the eccentricity of the orbit, e. This is the degree to which the shape departs from a circle. It was seen in Sec. 2.1 that the semi-minor axis, b, of an ellipse is related to the semi-major axis a by the expression

$$b^2 = a^2(1 - e^2), \quad \text{that is}, \quad e = \left[1 - \left(\frac{b}{a}\right)^2\right]^{1/2}.$$

An eccentricity of $e = 0.1$ gives the ratio $b/a = 0.995$ meaning b is smaller than a by only 0.5%. It is seen in Table 6.1 that, with the exceptions of Mercury and Pluto, all the eccentricities are less than this. Mars is the greatest but the values for the other planets are half this difference or less. This means that the orbits are circular to a high degree of accuracy. This feature is seen in Fig. 6.1 which shows a representation of the orbits of the terrestrial planets (Mercury, Venus, Earth and Mars) about the Sun as centre. It is seen that the orbits appear circular although they are not lined up in a simple way about the Sun as the centre of the force.

(b) The physical quantities

The masses, equatorial radii, the mean densities and the surface gravities of the planets are listed in Table 6.2. It is immediately evident that the planets fall naturally into two groups, the four large and massive planets of relatively low mean densities (the major planets) enclosing the four smaller

Table 6.2. The physical characteristics of the planets. The flattening f is defined by $f = (R_{eq} - R_p)/R_{eq}$ where R_{eq} is the equatorial radius and R_p the polar radius.

Planet	Mass (10^{24} kg)	Mass (Earth = 1)	Equatorial radius (10^3 km)	Equatorial radius (Earth = 1)	Mean density (10^3 kg/m^3)	Surface gravity (m/s^2)	Flattening f
Mercury	0.3302	0.0553	2.440	0.383	5.427	3.70	0
Venus	4.869	0.815	6.052	0.949	5.204	8.87	< 0.002
Earth	5.974	1.000	6.378	1.000	5.520	9.78	0.0034
Mars	0.6419	0.107	3.393	0.532	3.933	3.69	0.0065
Jupiter	1899	317.8	71.49	11.21	1.326	23.12	0.0649
Saturn	568.5	95.16	60.27	9.449	0.687	8.96	0.0980
Uranus	86.6	14.54	25.56	4.01	1.27	8.69	0.0229
Neptune	102.4	17.15	24.76	3.88	1.64	11.0	0.0171
Pluto	0.014	0.0025	1.44	0.226	2.03	0.45	—

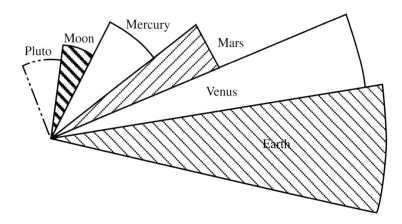

Fig. 6.2.   The relative sizes of the terrestrial planets. The Moon and Pluto are included for comparison.

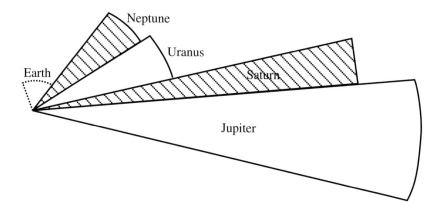

Fig. 6.3.   The relative sizes of the major planets. The Earth is included for comparison.

and less massive planets (the terrestrial planets) but with much higher mean densities. One planet, Jupiter, has a greater mass and greater volume than the rest of the planets added together. The relative radii are shown in Fig. 6.2 for the terrestrial planets and Fig. 6.3 for the giant or major planets. Pluto remains an anomaly but is included for comparison.

Two properties are immediately clear from Figs. 6.2 and 6.3. One is that the two groups each contain four planets: apparently the major planets have not come singly. Secondly, each group of four contains two groups of similar planets. For the terrestrial planets, Earth and Venus are very similar in size and density. Mars and Mercury are not dissimilar in size though not in density. For the major planets, Jupiter and Saturn are similar in size and density while Uranus and Neptune also form a pair of similar mass, size and therefore density. It will be seen later that the two sets of planets, terrestrial and major, are in fact separated by a region of very small bodies (the asteroid belt). A region of small bodies also lies outside the orbit of Neptune, as was first hypothesised by Edgeworth and a little later by Kuiper, a belt exists beyond Neptune containing icy objects similar to Pluto. Beyond this, Oort believed there is a cloud of small ice bodies forming the Oort cloud. More will be said of both belts later.

6.3. Satellites

All the planets with the exceptions of Mercury and Venus, have companion bodies of lower masses called *satellites*. While Earth has one (the Moon) and Mars has two small ones (Phobos and Deimos) the major planets have many, perhaps as many as 90. The satellites form several groups of objects, some large and some small and can be divided into three general categories: principal satellites, secondary satellites and small satellites. The physical data for the principal and secondary groups are included in the tables that follow. Table 6.3 includes the principal satellites Their relative sizes are shown in Figs. 6.4 and 6.5. The secondary satellites are treated in Table 6.4 and Fig. 6.5. The smaller satellites are associated with the major planets and especially with Jupiter and Saturn. Many of them are small, irregular bodies and we need not give details of them here.

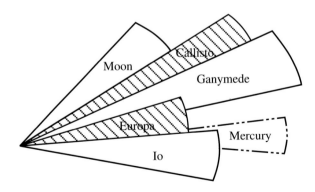

Fig. 6.4.   The four primary Galilean satellites of Jupiter. Data for the Moon and Mercury are included for comparison.

Table 6.3. The physical data for the principal satellites. R_p is the radius of the parent planet.

Planet	Satellite	Radius, R (10^3 km)	Mass (10^{23} kg)	Mean density (10^3 kg/m^3)	Mean radius of orbit, R (10^5 km)	(R/R_p)
Earth ($R_p = 6.38 \times 10^3$ km)	Moon	1.738	0.735	3.34	3.84	60.09
Jupiter ($R_p = 7.19 \times 10^4$ km)	Io	1.816	0.892	3.55	4.13	5.74
	Europa	1.563	0.487	3.04	6.71	9.33
	Ganymede	2.638	1.490	1.93	10.7	14.88
	Callisto	2.410	1.064	1.81	18.8	26.15
Saturn ($R_p = 6 \times 10^4$ km)	Titan	2.56	1.36	1.90	12.22	20.37
Neptune ($R_p = 2.475 \times 10^4$ km)	Triton	1.353	0.215	2.05	3.55	14.34

Table 6.4.   The physical data for the secondary satellites.

Planet	Satellite	Radius (10^2 km)	Mass (10^{20} km)	Mean density (kg/m^3)	Mean radius of orbit, R (10^5 km)	R/R_p
Mars ($R_p = 3.4 \times 10^3$ km)	Phobos	$19.2 \times 21.4 \times 27$ km	0.96×10^{-5}	1.90	0.094	2.76
	Deimos	$11 \times 12 \times 15$ km	2.0×10^{-5}	2.10	0.235	6.91
Saturn ($R_p = 6 \times 10^4$ km)	Mimas	1.95	0.376	1.20	1.85	3.08
	Encaledus	2.50	0.740	1.10	2.38	3.97
	Tethys	5.25	6.26	1.00	2.95	4.92
	Dione	5.60	10.5	1.49	3.77	6.28
	Rhea	7.65	22.8	1.24	5.27	8.78
	Hyperion	1.85×113	16.0	1.00	1.48	24.68
	Iapetus	7.20	19.3	1.20	3.56	59.33
Uranus ($R_p = 2.6 \times 10^4$ km)	Miranda	2.4×2.33	0.659	1.20	1.85	5.00
	Ariel	5.81×5.78	13.5	1.67	1.91	7.39
	Umbriel	5.85	11.7	1.40	2.95	10.27
	Titania	7.90	35.3	1.71	4.38	16.85
	Oberon	7.60	30.1	1.63	5.83	22.54
Neptune ($R_p = 2.47 \times 10^4$ km)	Nereid	2.35	0.17	1.30	55.62	225.2
Pluto ($R_p = 1.15 \times 10^3$ km)	Charon	6.25	16.0	1.7(?)	0.196	

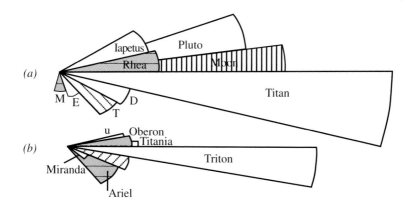

Fig. 6.5. The primary and secondary satellites of (a) Saturn, including the Moon and Pluto for comparison and (b) Uranus and Neptune. (M = Mimas, E = Enceladus, T = Tethys, D = Dione, U = Umbriel and Ne = Nereid)

6.4. Planetary Rings

That "objects" surround Saturn has been known since Galileo first pointed his new telescope at the planet in 1611. The true explanation of their structure as rings of particles was given by Huygens in 1659. This structure was regarded as unique to Saturn for about three centuries until high powered telescopes on Earth together with the observations of the Voyager space probes found and confirmed that each of the major planets is associated with a ring structure of some kind.[3] The four ring systems proved to be very different, with that of Saturn being the most substantial and probably durable. The details of the rings for Jupiter and Saturn are given in Table 6.5. It is seen that those for Saturn extend out to some 8 Saturn radii.

The Saturn rings carry a surprising amount of matter. The A and C rings have masses similar to some of the secondary satellites while the B rings has up to ten times that mass. It has been found that small satellites are associated with the rings of Saturn, acting as "shepherds" to provide stability for them. Even the Main rings of Jupiter and the G ring of Saturn have masses comparable to those of the satellites of Mars.

It is not known how permanent the rings are. Collisions between the particles in the Saturn ring will presumably reduce the particle sizes

[3]Nine narrow rings were observed around Uranus in 1977 when the planet was seen to occult a star. Jupiter's gossamer rings were discovered by Voyager 1 in 1979 during a flyby. A few years later the rings around Neptune were discovered from Earth, again by the occultation of a star.

Table 6.5.　The characteristics of the rings of Jupiter and Saturn.

Planet	Name of ring	Distance from centre/R_P	Radial width (km)	Mass (kg)	Albedo
Jupiter	Halo	1.4–1.7	22,800	–	0.05
	Main	1.72–1.81	6,400	10^{13}	0.05
	Gossamer	1.81–3.2	99,000	–	0.05
Saturn	D	1.11–1.24	7,150	–	–
	C	1.24–1.52	17,500	1.1×10^{21}	0.12–0.30
	Maxwell gap	1.45	270		
	B	1.52–1.95	25,500	2.8×10^{22}	0.5–0.6
	Cassini division	1.95–2.02	4,700	5.7×10^{20}	0.2–0.4
	A	2.02–2.27	14,600	6.2×10^{21}	0.4–0.6
	Enke gap	2.214	325		
	Keeler gap	2.263	35		
	F	2.324	30–500	–	0.6
	G	2.72–2.85	8,000	1×10^{17}	–
	E	3–8(?)	300,000(?)	–	–

Table 6.6.　The characteristics of the rings of Uranus and Neptune. (The masses and the albedos are yet unknown.)

Planet	Name of ring	Distance from centre/R_P	Radial width (km)
Uranus	1986 U2R	1.49(?)	2,500(?)
	6	1.597	1–3
	5	1.612	2–3
	4	1.625	2–3
	Alpha	1.707	7–12
	Beta	1.743	8–12
	Eta	1.801	0–2
	Gamma	1.818	1–4
	Delta	1.843	3–9
	Lambda	1.909	1–2
	Epsilon	1.952	20–100/
Neptune	Galle	1.68	15(?)
	LeVerrier	2.14	110
	Lassell	2.15–2.4	4,000
	Arago	2.31	< 100
	Adams	2.53	< 50

with time. The tenuous rings of the other planets are probably made of materials ejected from the accompanying satellites. The ejections arise from bombardment of the surfaces by ions from the primary planet and from the solar wind and so will continue into the future. The material in the rings of Jupiter, Uranus and Neptune will be transient but renewed by further ejections from the satellites. It is not known what their future forms may be like.

6.5. Angular Momentum

The planets, and the Sun, are in rotation and the planets in orbits around the Sun so the System is the seat of considerable rotational, or angular, momentum. If M is the mass of a body (supposed to be a point mass) in circular motion at a distance R_0 about a centre with speed V, the angular momentum of the motion, $L = MVR_0$. It is of interest to determine the values of L for the orbits of the various bodies. The rotational momenta for the individual planets are also of interest. The angular momentum for planetary rotation will be written as l and given by the expression $l = I\omega$ where I is the moment of inertia of the body of mass M and radius R_P, and ω is the angular velocity of rotation of the body. It is a result of mechanics that I is to be written $I = \alpha_P M R_P{}^2$. Here α_P is the dimensionless inertia factor and is a measure of the distribution of the material inside. For a homogeneous body $\alpha_P = 0.4$ while for a point mass the limiting value $\alpha_P = 0$ applies. These limiting values contain the range of conditions in between, the inertia factor becoming smaller the more concentrated the matter in the body. But ω can be written $\omega = v/R_P$ where v is the rotation speed. This gives finally $l = \alpha_P M v R_P$. These rotational properties are collected in Table 6.7 for the orbital and rotational motions of the planets. The values of α_P are also included. Data for the Sun are also included. These involve the rotation of the Sun only, the contribution of the small rotation of the Sun about the centre of mass of the whole System being negligible.

It is seen that the angular momenta of the major planets are some two orders of magnitude greater than that of the Sun. It seems that the rotation of the Sun involves no more than 1% of the angular momentum of the total System. This contrasts strongly with the mass of the System. Here the Sun includes 99.9% of the total mass. The reason for this dichotomy must be explained if the origin of the Solar System is to be understood.

Table 6.7.   Giving values of the various angular momenta associated with the planets. α_P is the inertia factor, V the orbital velocity, L the orbital angular momentum, v the rotational velocity and l the rotational angular momentum. Data for the Sun are also given.

Planet	α_P	Mean orbital V (km/s)	Orbital L (kg m^2/s)	Mean rotational v (km/s)	Rotational l (kg m^2/s)
Mercury	–	47.87	9.15×10^{37}	–	–
Venus	0.340(?)	35.02	1.85×10^{40}	–	–
Earth	0.331	29.97	2.66×10^{40}	0.465	5.87×10^{33}
Mars	0.376	24.14	3.53×10^{39}	0.241	1.99×10^{32}
Jupiter	0.264	13.07	1.93×10^{43}	12.7	4.59×10^{39}
Saturn	0.21	9.67	7.82×10^{42}	3.68	2.64×10^{37}
Uranus	0.23	6.83	1.69×10^{42}	2.94	1.53×10^{35}
Neptune	0.29	5.48	2.51×10^{42}	2.73	2.02×10^{35}
Pluto	–	4.75	3.91×10^{38}	–	–
Sun	≈ 0.05	≈ 0	≈ 0	1.88	1.3×10^{41}

Table 6.8.   The obliquity angles (that is the tilt of the equator to the orbit) and the angle between the rotation and magnetic axes are called magnetic axis. Venus and Uranus are assigned retrograde rotations, marked (r). (1 gauss $= 10^{-5}$ T)

Planet	Obliquity (degrees)	Magnetic axis (degrees)	Surface magnetic field (gauss)	Magnetic moment (tesler m^3)
Mercury	0.1	+14	0.003	
Venus	177.4(r)	–	$< 2 \times 10^{-5}$	
Earth	23.45	+10.8	0.31	
Mars	25.19	–	< 0.0003	
Jupiter	3.12	−9.6	4.28	
Saturn	26.73	−0.0	0.22	
Uranus	97.86(r)	−59	0.23	
Neptune	29.56	−47	0.14	

6.6.   Magnetism and Rotation

The planets are in rotation. Several of them also have an intrinsic magnetic field.

These data are listed in Table 6.8. Positive rotation is defined as anti-clockwise motion looking down from the north pole. On this basis the planets Venus, Uranus together with Pluto have retrograde rotations.

*

Summary

1. The Sun is the principal member of the Solar System. It is accompanied by 8 planets and Pluto.

2. Four planets form the terrestrial group of planets near the Sun with high density but small radius.

3. Four planets with low density but large mass form the major planets.

4. Each group has similar planets in pairs — Jupiter–Saturn and Uranus–Neptune for the major planets and Earth–Venus for the terrestrial planets.

5. Pluto has been included as a planet until recently but more recently it is recognised as a member of a large number of small icy bodies orbiting the Sun beyond the radius of Neptune. These are collectively called the Edgeworth–Kuiper objects.

6. The major planets have many satellites which (with one exception) include a large proportion of water-ice. The terrestrial planets have but three satellites, two small ones for Mars but the Moon for Earth.

7. The satellites fall into three categories: primary satellites comparable in size and mass to the Moon, or larger; secondary satellites that are spherical; and the third group that are small and irregular.

8. The major planets each have a systems of rings in the equatorial plane. Those for Saturn are the most substantial.

9. The planets have intrinsic magnetic fields with the exceptions of Venus and Mars.

7

Smaller Members:
Asteroids, Comets and Meteorites

The Solar System contains a very large number of very small bodies of various kinds. It is believed they are the remnants of the formation of the Solar System. These are our concern now.

7.1. Asteroids

There has been interest over the centuries in accounting for the particular values of the semi-major axes of the planets and the aim was to develop a theoretical formula to represent this. One long standing empirical formula has become known as Bode's law, or sometimes Bode's series. It was, in fact, recognised originally by Wolf as far back as 1741 and developed by Titius in 1772. It was given a mathematical formulation by Bode in 1778. To get the sequence, start with the series of numbers beginning with 0 and 3 and then double every member of the series from 3 onwards — this gives 0, 3, 6, 12, 24, 48, 96, 192, 384, 768, and so on. Next, add 4 to each number in the series to obtain 4, 7, 10, 16, 28, 52, 100, 196, 388, 772, and divide each number by 10. This gives the Bode series

$$0.4, 0.7, 1.0, 1.6, 2.8, 5.2, 10, 19.6, 38.8, 154.4 \ldots .$$

These numbers are surprisingly in agreement within a few percent of the mean distances of the planets from the Sun in AU which are

$$0.387, 0.723, 1.0, 1.524, 5.203, 9.539, 19.19, 30.07, 39.46 .$$

The sequence clearly fails for Neptune and totally fails for Pluto. It is surprising that it is so accurate up to and including Uranus.

At the time Bode proposed this series no planets were known beyond 10 AU and there was no planet at the predicted distance of 2.8 AU. This series was very influential in the 18th and early 19th centuries. It was gratifying that the new planet Uranus fitted the series beautifully and searches were made for the missing planet at 2.8 AU which was confidently expected to be there. Sure enough, on 1 January 1801, Piazzi at Palermo Observatory discovered a new object which moved in an orbit about the Sun with a calculated distance of 2.76 AU, corresponding to the previous gap in the Bode series. The body was called Ceres. It turned out to be too small to be a planet and was, in fact, the first of a new class of object called Asteroids, orbiting the Sun between the orbits of Mars and Jupiter. Ceres was given

Table 7.1. The 18 asteroids with radii in excess of 110 km, together with three very small asteroids. The second column gives the order in which the members were discovered. The sixth column lists the eccentricities e of the orbits and the last column lists the inclination of the plane orbits to the ecliptic.

Name	Number	Radius (km)	Mean orbit size (10^6 km)	Orbital period (yrs)	e for orbit	Orbital inclination
Ceres	1	467	413.9	4.6	0.077	10.58
Pallas	2	263	414.5	4.61	0.232	34.82
Vesta	4	255	353.4	3.63	0.090	7.13
Hygiea	10	204	470.3	5.55	0.120	3.84
Davida	511	163	475.4	5.65	0.182	15.94
Interamnia	704	158	458.1	5.36	0.146	17.32
Europa	52	151	463.3	5.46	0.101	7.47
Juno	3	134	399.4	4.36	0.258	13.00
Sylvia	87	130	521.5	6.55	0.082	10.84
Eunomia	15	128	395.5	4.30	0.187	11.75
Euphrosyne	31	128	472.1	5.58	0.228	26.34
Psyche	16	127	437.1	4.99	0.138	3.09
Cybele	65	120	513.0	6.36	0.104	3.55
Bamberga	324	115	401.4	4.40	0.337	11.10
Patienta	451	113	458.2	5.36	0.075	15.23
Doris	48	111	465.5	5.49	0.073	6.55
Herculina	532	111	414.7	4.63	0.175	16.33
Camilla	107	111	521.8	6.54	0.080	10.02
Icarus	–	≈ 2	160.9	1.12	0.827	22.9
Hildago	–	15	350.0	3.60	0.657	42.5
Eros	433	≈ 25	217.5	1.76	0.223	10.8

the number 1: there are now known to be in excess of 25,000 of these bodies. The details of those with a radius greater than 110 km are given in Table 7.1.

The vast majority of asteroids are small objects with radii in the range of tens of km or less. Many of these smaller bodies have resulted from the collision of larger bodies and their trajectories have been determined by collision dynamics. This process was first recognised in the early 1900s by the Japanese astronomer Kiyotsugu Hirayama and the associated families of asteroids are called Hirayama families. It has been possible to link some present members together to gain knowledge of the original body but this has not proved possible in general.

All known asteroids move in the same direction as the planets, that is their orbits are said to be prograde. Although the majority of the orbits lie within 25° of the ecliptic some inclinations are greater — for instance Icarus makes an angle of 42.5°. The majority of orbits lie between those of Mars and Jupiter but not all do.

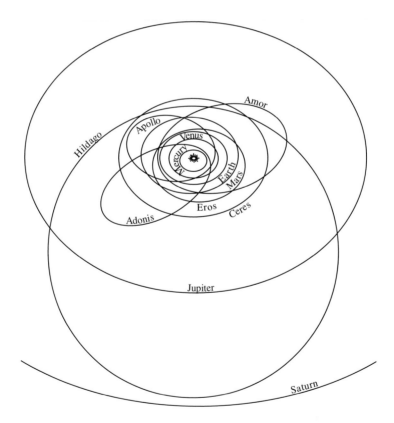

Fig. 7.1.   The orbits of asteroids that cross those of Venus, Earth, Mars and Jupiter.

The orbits of asteroids spread widely in the Solar System. This is seen in Fig. 7.1 which shows the orbits involving also Venus, Earth, Jupiter and Saturn. Some 116 objects have been found to have perihelia beyond the Jupiter orbit but semi-major axes inside the orbit of Neptune. These have been called Centaur objects. Bodies beyond Neptune, often referred to as trans-Neptunian objects, will be considered in the next Section.

A representation of the spread of the orbits of all the asteroids is given in Fig. 7.2.

There it is seen that there are a significant number of orbits inside the Mars orbit and outside the Jupiter orbit. Two members are shown moving in essentially the Jupiter orbit. These are the Trojan asteroids which move at specific and constant distances from Jupiter, at the two Lagrangian points. It is also clear that there are many asteroids nearer the Sun than the Earth. These, with semi-major axes less than 1.0 AU, form the so-called Aten asteroids. The details of this group are only approximately known and many, if not most, of them remain to be discovered but some 152 are known at the present time. Being nearer the Sun they are not easy to distinguish in the solar glare and is likely to remain an obscure group for some time to come. There is a separate group of bodies with perihelion distance less than 1.3 AU and these are often called the Amors. Some 866 have been listed to date.

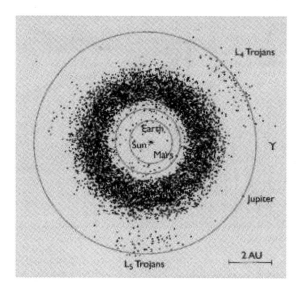

Fig. 7.2.   The distribution of asteroid orbits out to the distance of Jupiter (NASA).

Some asteroids have orbits which cross that of the Earth and so can pose a threat to Earth. The Apollo asteroids, named after the asteroid Apollo which was discovered in 1932, have perihelion distances less than 1.0 AU. With diameters of a few kilometres they are difficult to detect but their effects could be disastrous if they struck the Earth. Some will have done this in the past as existing meteorite craters bear witness. A special watch has recently been mounted to detect and monitor these objects of which 903 have been listed. There are 414 potentially hazardous asteroids (PHA) uncovered so far that are under continual observation. A small perturbation of any orbit would cause grave distress. There is, incidentally, no firm answer yet to the question of what to do if an Earth colliding asteroid were to be detected. Fortunately, such collisions are very rare. Substantial impacts can be expected every few hundred thousand years and major ones every few millions of years.

Figure 7.2 might suggest that the distribution of asteroids is essentially continuous but that is not quite true. In 1866, Daniel Kirkwood realised that there are two prominent gaps in the rings corresponding to one-third and one-half the period of Jupiter. This must surely indicate resonance

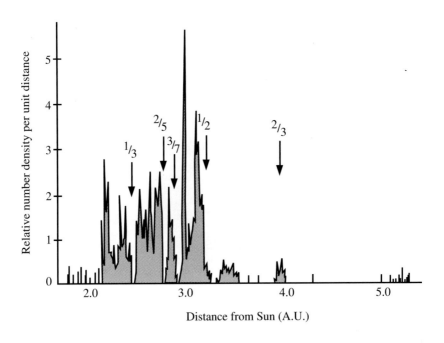

Fig. 7.3. The distribution of asteroids has gaps corresponding to resonances of period with the planet Jupiter.

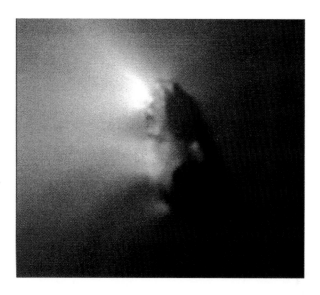

Fig. 7.4. A photograph of comet Halley taken in 1986. The Sun is to the left (ESA).

phenomena. Other gaps were found at $2/5^{\text{th}}$ and $3/7^{\text{th}}$ the Jupiter period. A further gap is found at $2/3^{\text{rd}}$ the Jupiter period and others have been listed. This more complex structure is shown in Fig. 7.3. The gaps are now very clear as is the spread from the inner to the outer Solar System. The gaps are maintained even though collisions between asteroids will introduce new smaller members and will change the orbital characteristics of the whole.

7.2. Comets and Meteor Showers

Comets have been a prominent object in the sky since antiquity. Perhaps the best known is Comet Halley which has, in the event, been observed for several thousand years. It is well known now for two special reasons. First as the foreboding of doom[1] for King Harold at Hastings in 1066 (but what about Prince William of Normandy who won the battle?). Second and more important, it was the comet whose return was successfully predicted by William Halley using Isaac Newton's new theory of universal gravitation. Unfortunately, Halley didn't live to see his prediction confirmed but the return gave enormous support, and acceptance, for Newton's theory. In 1986 the comet was brought into the daily world when ESA's Giotto space

[1]Presumably he would not have been so concerned if he could have seen Fig. 7.4.

Table 7.2. Eleven better known short-period comets with associated meteor showers. Halley was known to the Chinese at least as far back as −239 BC. New comets are being discovered even today.

Comet	Designation	Perihelion distance (km)	Orbit period (yrs)	Orbit eccentricity	Orbit inclination (°)	Meteor showers
Biela	3 D	1.28×10^8	6.62	0.756	12.55	Andromedids
Encke	2 P	4.96×10^7	3.29	0.849	11.91	Taurids
Giacobini-Zinner	21 P	1.491×10^8	6.61	0.706	31.86	Draconids
Halley	1 P	8.88×10^7	75.8	0.967	162.22	Eta Aquarids, Orionids
Swift-Tuttle	109 P	1.44×10^8	135	0.963	113.47	Perseids
Tempel-Tuttle	55 P	2.36×10^8	33.3	0.906	162.49	Leonids
Tuttle	8 P	1.58×10^8	13.6	0.819	54.94	Ursids
Arend-Rowland	C/1956 R1	0.538×10^7	?	1.000	119.95	–
Hale-Bopp	C/1995 O1	1.36×10^8	?	0.995	89.43	–
Hyakutake	C/1996 B2	3.44×10^7	?	0.999	124.95	–
IRAS-Araki-Alcock	C/1983 H1	1.36×10^8	?	0.990	73.25	–
Thatcher	1861 1	–	410	–	–	Lyrid
Phaethon	3200	–	1.4	–	–	Geminid

probe approached it to within 60 kilometres, returning extraordinary data to Earth. A photograph of the encounter is shown in Fig. 7.4. The comet was found to be a chunk of material with dimensions of the order of a few tens of kilometres. This appears to be typical of all comets.[2] Several comets appear in the sky during a typical year. Some are a return of known comets while others are new. The origin of new members will be considered in Sec. 7.3.

Comets are divided into two classes — short period and long period. In very broad terms, comets with an orbital period of less than about 200 years are called short-period while the rest are long period. About 100 short-period comets are known. While it may be believed that many more exist, the difficulty always is observing them. They remain passive objects beyond about 5 AU from the Sun and become visible only nearer than this. We are able to see easily only those members that come this close to the Sun.

Comets follow elliptic orbits about the Sun, like planets, although the energy is often small and the elliptic form can only be properly confirmed if they return. It is seen from Table 7.2 that both the orbital eccentricities are generally large and that the inclinations to the ecliptic spread over a wide range. Again, the semi-major axes vary enormously leading to a wide range of orbital periods. They can approach quite close to the Sun and then move deeply to the edge of the Solar System. When close to the Sun, the core (nucleus) is heated strongly by the Sun releasing volatile materials in the form of a tail.[3] This release does not occur away from the Sun. After a sufficient number of orbits (perhaps between a few hundred and a few thousand) the reservoir of volatile components will be exhausted, the comet becoming an inert dark mass. It would probably appear then much like a small asteroid.

Comets generally eject small particles in their orbits and these can enter the Earth's atmosphere. There they are slowed by friction becoming temporarily luminous. These flashes of light can be seen in the sky every night and are called meteors. When the orbit of a comet crosses the orbit of the Earth the concentration of meteor grains can be very high and the appearance of meteors has the form of a "shower". These interactions take place at specific times of the year and appear to emerge from particular constellations from which they take their name. Some of the links between comets and meteor showers are collected in Table 7.2. The comets interact

---

[2]The NASA automatic craft Stardust made a similar encounter with the comet Wild 2 in January 2004.

[3]There are usually two separate tails, one composed of dust and the other composed of ionized atomic particles. This latter tail interacts with the solar wind.

gravitationally with the planets so their orbits are always subject to some change. In general, the planet takes energy from the comet making its orbit flatter. The orbital directions can be prograde or retrograde, and approximately equal numbers of each direction have been observed.

7.3. Meteorites

The smaller meteors are burned up in the Earth's atmosphere and drop to the surface as dust. Solid remnants of the larger meteors, on the other hand, can reach the surface, causing a crater: these remnants are called meteorites. The extent to which the surface of the Earth has been subjected to such bombardment has only become clear recently from photographs taken in Earth orbit. The effects of such collisions can be very catastrophic as will be seen later.

7.4. The Edgeworth–Kuiper Belt

It is unlikely that all the material for planetary formation was exhausted with the appearance of Neptune and Pluto. There could well be a considerable quantity beyond the orbit of Neptune that did not condense and which still remains in the form of an extended second "asteroid" belt of pristine material from the formation of the Solar System. That such an outer region should exist was proposed by Edgeworth (1947) and shortly afterwards by Kuiper (1952). Although originally hypothetical, a number of Kuiper belts objects (as they have become called) have been discovered since the first one was observed in 1992. It was given the designation 1992 QB1 and later named Quaoar (after a native American tribe that originally lived in the Los Angeles basin). Its radius is estimated to be about 500 km and its surface temperature about 50 K. It appears to follow a closely circular orbit and to lie generally in the plane of the ecliptic. Spectra have recently been taken of the surface at near infra-red wavelengths and these show the general characteristics of water. There is an indication that the ice is in a crystalline form (in which case the water molecules are arranged in a lattice form) and not the amorphous (structureless, random) arrangement that might have been expected at the low temperature of the surface. Crystalline ice is thermodynamically stable only above 110 K and this indicates that the surface has a varied history yet to be unravelled. Some 60 or 70 such bodies are known and more are being found in this fast growing area of study. It is believed now that Pluto and its satellite Charon are, in fact, Kuiper objects. These objects have a resonance effects with Neptune. For instance, for Pluto (and Charon) are in a 3:2 resonance with Neptune, Pluto making 2 orbits round the Sun to Neptune's 3. It is very likely that Neptune's satellites Triton and Nereid, and Saturn's satellite Phoebe are captured Kuiper objects resulting from close encounters in the past.

It is believed that there could be as many as 70,000 trans-Neptunian objects each with radius in excess of 100 km, composed primarily of ice and silicate materials. They lie in a belt which extends from Neptune (about 30 AU) out to about 50 AU. This could provide a collective mass comparable to that of the Moon. The objects in this cloud have random motions in their orbits and collide, fragments being ejected from the belt. The eccentricities are likely to be small leading to small semi-major axes. It is proposed that the Edgeworth–Kuiper belt is the origin of the short period comets.

7.5.   The Oort Cloud

New comets appear at a constant rate of a little less than 1 per year and it is necessary to find where these come from. A significant step was taken by Jan Oort in 1948 who proposed a reservoir of potential comets beyond the normal scale of the Solar System, including the Edgeworth–Kuiper belt.

New comets can come from one (or I suppose all) of only three sources. They could appear from outside the Solar System. This would imply that the solar neighbourhood is full of small objects but there is no other evidence for this. Even if this were true, objects entering the System from outside would most likely have sufficient kinetic energy to provide a positive total energy throughout the System. They would avoid capture by gravity — the path would be hyperbolic past the Sun and not an elliptical orbit of the type shown by captive comets. Another possibility is that they are manufactured within the Solar System but there is no evidence for this. This leaves the third possibility of a reservoir of ancient bodies beyond the planets "left over" from the formation of the Solar System 4.5×10^9 years ago. These bodies will not have been replenished since. For stability, the bodies must orbit the Sun forming a mass cloud enclosing the System. This is the cloud proposed by Oort. The comets from the Oort cloud will have a large eccentricity and large semi-major axis. This means their orbital period will be large and so will form long period comets.

For this to be true the number of objects must be large. One per year would require 4.5×10^9 objects to start with but more if the supply is not to be exhausted now. The number can be constrained further by remembering characteristics of the observed new comets. There are two crucial ones. First, the orbits have large eccentricities and second the orbits can be prograde or retrograde equally. The first would suggest the cloud to have large eccentricity. The second implies leakage from a group of bodies in some form of dynamical equilibrium. The leakage would occur due to interactions between the particles of the cloud and other influences. There would also be a leakage out of the Solar System, these bodies not being seen from Earth. Oort estimated that the leakage rate of one per year from a random cloud of particles would be rather less than 1% of the total which would imply the presence of 10^{11} or 10^{12} members originally. The rate of ejection of particles suggests the material density of the cloud is low and therefore it must be extensive. Estimates suggest it may spread out as far as 2 light years, which is half way to the nearest star! This conclusion is consistent with the random sense of the orbits since they wold arise from a system of bodies in equilibrium.

It is necessary to specify some mechanism that could cause members to be ejected. Oort suggested several. One is passing stars. This is a rare event but would be sufficient to disturb the equilibrium. The second is a perturbation due to a giant molecular cloud which typically has a mass of 10^5 solar masses ($\approx 10^{35}$ kg). There is alternatively the possibility of planetary mass objects within the cloud. It is difficult to know what matter might be in the volume around the Solar System. Evidence from other stars suggests that an encompassing dust cloud is not uncommon. This dust emits infra-red (temperature) radiation which can be detected out from the star. A photograph of β Pectoris actually shows a cloud extending out to perhaps 900 AU around the star. This suggests also that interstellar space must contain many small objects ejected from the parent stars.

The situation has become complicated by the discovery of a body, designated 2003B12 and called Sedna unofficially, orbiting the Sun in an extremely elliptical orbit. It is about 90 AU from the Sun at the present time. Its perihelion distance is likely to be about 75 AU but at aphelion it may be as far away as 900 AU. Sedna is likely to have a radius of some 850 km and a surface temperature of perhaps 33 K at its closest approach to the Sun. The orbital time is probably 10,500 years. Sedna presents several problems. First, its orbital distance is rather closer to the Sun than might be expected for Oort cloud objects. Could there be many such inner Oort cloud bodies? If so, what is their spread in size? The second problem involves the surface colour. It would be expected to show a white surface compatible with being composed very largely of water ice. Surprisingly, the surface of Sedna shows a red colour not dissimilar to that of Mars but rather less strikingly so. Can this mean that Sedna has a surface ferrous component? The outer Solar System is becoming an interesting but confusing place. It is likely to provide surprises in the future.

*

Summary

1. The Solar System contains a wide range of small objects. Neglected until recently, these bodies are now the focus of much interest in astronomy.

2. The earliest to be recognised are the asteroids, a range of bodies with radii less than 500 km and most with broadly circular orbits between those of Mars and Jupiter.

3. More recently, a range of bodies first proposed to exist by Edgewooth and independently by Kuiper, have been found beyond the orbit of Neptune. It is realised now that Pluto and its satellite Charon are members of this group. They are composed of silicate materials but with a very high proportion of water-ice (probably in excess of 80% by mass).

4. Asteroid collisions produce small bodies called meteorites. These pervade the Solar System and some have struck the Earth in the past and some will do the same in the future.

5. Comets are a strong feature of the Solar System. There are two types, short period and long period comets.

6. The source of the short period comets is thought to be the Edgeworth–Kuiper belt.

7. To account for the long period comets, Oort proposed the existence of a further belt of icy particles surrounding the outer reaches of the Solar System. This has become known as the Oort cloud. This has not been found directly in the same way that the Edgeworth–Kuiper belt has been found.

8

The Material of the Solar System

What is the Solar System made of? How is the material distributed? That is our topic now.

8.1. The Solar/Cosmic Abundance of the Elements

Observations show a universal distribution of the chemical elements. The observations are of limited extent (for instance, they cannot be made directly inside a body) but their general occurrence is significant. The broad distribution is shown in Fig. 8.1. There are some strange apparent anomalies but generally the curve shows that the lightest elements are the most common.

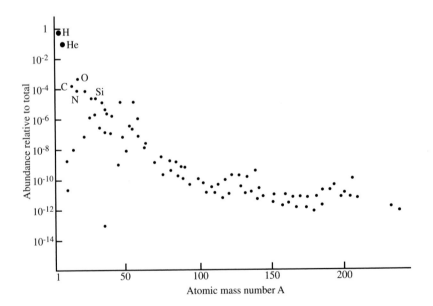

Fig. 8.1. The cosmic distribution of the chemical elements.

Table 8.1. The cosmic distribution of the most abundant 14 chemical elements.

Element	Relative abundances by number (silicon = 1)	Relative abundances by mass
Hydrogen	3.18×10^4	0.9800
Helium	2.21×10^3	
Oxygen	2.21×10	
Carbon	1.18×10	0.0133
Nitrogen	3.64	
Neon	3.44	0.0017
Magnesium	1.06	
Silicon	1	
Aluminium	0.85	
Iron	0.83	
Sulphur	0.50	0.00365
Calcium	0.072	
Sodium	0.06	
Nickel	0.048	

The most common is hydrogen with helium (a chemically inert gas) the next. Perhaps as much as 98% of the cosmos by mass is composed of these two elements. A more detailed list is given in Table 8.1. The 14 elements listed there account for 99.865% of the whole: all the other elements are present in no higher a proportion than 0.135%, all put together. It is clear that the molecules formed from the 14 elements can be expected to appear universally. Many of the chemical elements familiar in everyday life (such as uranium) lie in this 0.135% of the very rare elements.

8.2. The Formation of Molecules

One or two important molecules can be guessed at once. The first and second most abundant chemically active elements are hydrogen and oxygen (helium is chemically inactive). An immediate outcome must be water and we can expect this to be a common molecule. The occurrence of carbon and nitrogen suggests the presence of such molecules as NH_3 and CH_4. More complicated molecules are based on oxygen. These, in fact, form the silicate materials to be considered later. Iron is surprisingly abundant. It forms oxides with oxygen (water) but it can be expected to be more abundant than can be absorbed in this way, suggesting the occurrence of free iron.

These expectations can be given a chemical basis through their electronic configurations and their affinities for different types of crystalline bonds.

There are three divisions.

(a) Lithophilic elements. This comes from the Greek for stone and these elements tend to be associated with oxygen in oxides and silicates. Such of the more common elements are K, Na, Ca, Mg, Rb, Sr, Ba Li, the rare earths, Mn, Be, Al, and Cr.

(b) Chalcophilic elements (from the Greek for copper). These elements tend to appear as sulphides and include Cu, Zn, Sn, Pb, Co, Mo, Hg, Cl, Bi and Ag.

(c) Siderophilic elements (from the Greek for iron). These are generally metallic elements such as Fe, Ni, Pt, Pd, Os, Ir, Au, and As.

The basis of these divisions is the position of the elements in the Periodic Table. There are several criteria but the most important is probably the electro-negativity, E, a term introduced by Pauling in the late 1950s. It is a measure of the ability of an atom to attract electrons and so form an electrically negatively charged anion. This ability is highest for the halogen group of elements (for instance H, F) and lowest for the metallic elements that tend to lose electrons and so become cations. Other elements fall in between. It is moderately high for oxygen and high for sulphur. Elements are allotted a number from 0 to 4 representing the strength of the chemical bonds that are formed. The lithophilic elements tend to have $E < 1.6$ on the scale while the chalcophilic lie in the range $1.6 < E < 2.0$. The siderophilic elements have E in the range $2.0 < E < 2.4$.

Two elements having an appreciable difference in the electro-negativities, such as Na and Cl forming NaCl, have a strong electrostatic ionic bond between them. This applies to the liophilic elements with $E < 1.6$. Elements with comparable values of E tend to share electrons and so form covalent bonding. Metallic elements have a large number of shared electrons in a tight lattice structure giving rise to the typical properties of high electrical and heat conductivities. The behaviour of a particular element can vary to some extent according to the relative quantities of different elements in a mixture. For instance, in terms of E number iron should be chalcophilic but in sufficiently high proportions the iron left over will show the features of a siderophilic metal. These matters will be found important later in determining the inner compositions of the planets.

Table 8.2.   The measured abundance of a range of elements in the Earth's crust.

Element	p.p.m $\equiv$ gm/10^3 kg	Element	p.p.m $\equiv$ gm/10^3 kg
Oxygen	4.55×10^5	Lithium	18
Silicon	2.72×10^5	Lead	13
Aluminium	8.3×10^4	Boron	9.0
Iron	6.2×10^4	Thorium	8.1
Calcium	4.66×10^4	Rare earths	6–2
Magnesium	2.76×10^4	Caesium	2.6
Sodium	2.27×10^4	Bromine	2.5
Potassium	1.84×10^4	Uranium	2.3
Titanium	6.32×10^3	Tin	2.1
Hydrogen	1.52×10^3	Arsenic	1.8
Phosphorus	1.12×10^3	Tungsten	1.2
Manganese	1.06×10^3	Iodine	0.46
Barium	390	Antimony	0.2
Strontium	384	Cadmium	0.16
Sulphur	340	silver	0.08
Carbon	180	Mercury	0.08
Chlorine	126	Platinum	0.01
Chronium	122	Gold	0.004
Nickel	99	Helium	0.003
Zinc	76	Radon	1.7×10^{-10}
Copper	68	Radium	Trace
Cobalt	29	Neptunium	Trace
Nitrogen	19	Plutonium	Trace

8.3.   The Compositions of Terrestrial Materials

The composition of some materials can be found by chemical analysis in the laboratory for the surface of the Earth and for meteorites. The two can best be considered separately.

(a) The Surface of the Earth

The outer region of the Earth (called the crust) can be measured directly. The average results are collected in Table 8.2. It is seen that oxygen is the most abundant element so that the materials are essentially silicates.

These elements do not exist in this simple form. Instead, they form a series of minerals although some ferrous elements remain in metallic form.[1]

[1] This is true for some other metals as well.

An assessment of the general composition of the crust is collected in Table 8.3. It is found that the crust is composed of two separate regions, one continental (land) and the other oceanic (the seas).

(b) Meteorites

Meteorites, having struck the Earth from space, can be picked up on the surface of the Earth so it is possible to gain some knowledge of the material

Table 8.3. Some minerals found in the crust of the Earth. There is a distinction between the crust of the continents and that of the oceans.

Mineral	Continental crust %	Oceanic crust %
Quartz (SiO_2)	60.1	49.9
Brookite (Ti_2)	1.1	1.5
Alumina (Al_2O_3)	15.6	17.3
Haematite (Fe_2O_3)	3.1	2.0
Ferrous oxide (FeO)	3.9	6.9
Ferrous sulphide (FeS)	trace	trace
Iron (Fe)	trace	trace-
Nickel (Ni)	trace	trace
Cobalt (Co)	trace	trace
Magnesium oxide (MgO)	3.6	7.3
Lime (CaO)	5.2	11.9
Sodium oxide (Na_2O)	3.9	2.8
Potassium oxide (K_2O)	3.2	0.2
Phossphoric anhydride (P_2O_5)	0.3	0.2

Table 8.4. Some minerals common to the Earth and to meteorites.

Mineral	Composition
Kamacite	(Fe, Ni)(4–7% Ni)
Taemite	(Fe, Ni)(30–60% Ni)
Troilite	FeS
Olivine	$(Mg, Fe)_2SiO_4$
Orthopyroxine	$(Mg, Fe)SiO_2$
Pigeonite	$(Ca, Mg, Fe)SiO_3$
Diopsite	$Ca(Mg, Fe)SiO_6$
Plagiooclase	$(Na, Ca)(Al, Si)_4O_8$

away from the Earth. It is found that, in general, meteorites contain the same minerals known on Earth. Some of these are listed in Table 8.4.

One of the most interesting and important meteorites is the Allende meteorite which resulted from the breakup of a meteor over the Mexican village of Pueblito de Allende on February 8, 1969. It has a mean material density of about 3670 kg/m^3, which is rather higher than the crustal density of the Earth of 2,800 kg/m^3. The measured age of the Allende specimen is 4.57×10^9 years which is generally accepted to be the age of the Solar System itself. The material here is, apparently, a sample of that from which the Solar System was formed. It has an unfractionated chemical composition and is typical of chondritic meteorites. The wide range of chemical elements present can be compared with the adundance from the solar atmosphere. The comparison involving 69 chemical elements is shown in Fig. 8.2.

Apart from providing a picture of the original materials from which the Solar System was formed, the Allende meteorite has a further piece of

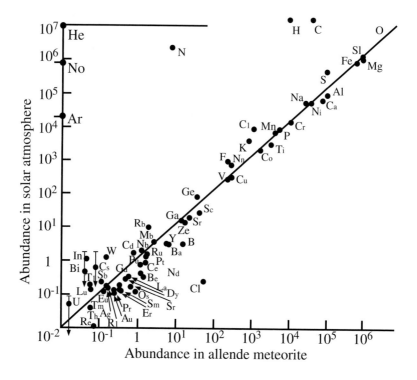

Fig. 8.2. The comparison of the abundance of 69 elements in the Allende meteorite with these elements found in the solar atmosphere (after J.A. Wood).

information to offer. It shows traces of the nucleus ^{25}Mg which is the decay product of ^{26}Al by β^+ decay The decay has a half life of only 7.5×10^5 years and this is the significance. There is no natural abundance known now but there is a 10% abundance of ^{25}Mg on Earth. The nucleide will have resulted from a supernova explosion and the Solar System will need to have formed within 7.5×10^5 years afterwards to contain the initial aluminium. This indicates that a supernova was involved with the formation of the System.

8.4. The Moon

A large quantity of lunar surface material was returned by the American Apollo lunar missions and a smaller quantity was returned using a Russian automatic probe. Laboratory measurements have provided the chemical

Table 8.5. The ratio of the abundances Moon/Earth for a range of volatile chemical elements.

Element	Abundances Moon/Earth	Element	Abundances Moon/Earth
Bismuth	1.15×10^{-3}	Potassium	6.5×10^{-2}
Thallium	4×10^{-3}	Germanium	6.9×10^{-2}
Calcium	7×10^{-3}	Sodium	8×10^{-2}
Zinc	8.5×10^{-3}	Lead	9×10^{-2}
Gold	9×10^{-3}	Copper	1.05×10^{-1}
Silver	1×10^{-2}	Gallium	3×10^{-1}
Rubidium	3.5×10^{-2}	Manganese	1.8
Indium	3.8×10^{-2}	Sulphur	4.4

Table 8.6. The ratio of the abundances Moon/Earth for a range of refractory elements.

Element	Abundances Moon/Earth	Element	Abundance Moon/Earth
Iridium	1.1×10^{-1}	Magnesium	1.9
Nickel	4×10^{-1}	Titanium	2
Aluminium	8×10^{-1}	Thorium	4.2
Silicon	9.5×10^{-1}	Uranium	5
Calcium	1.6	Barium	6
Iron	1.9	Chromium	10.5

composition of these rocks. The comparison between Moon and Earth abundances for volatile elements is shown in Table 8.5. The corresponding data for refractory elements is shown in Table 8.6. It is clear from the Tables that the Moon is richer in refractory elements than Earth but the Earth is richer in volatile elements. This presumably will have implications for the formation of the Moon and the Earth. The minerals on the Moon are very similar to those found on Earth with the exception that lunar material is almost entirely devoid of water. This means there are no clays or iron oxides (rust).

The maria rocks (see later) were basalt flows, the rocks having formed from the cooling of a melt. The lunar basalts are silicates composed almost entirely of the same minerals as the Earth. The most common mineral is pyroxene (Ca, Mg, Fe) as yellow-brown crystals a few centimetres long. Olivine (Mg, Fe), occurring occasionally, is in pale green crystals. Feldspar, or plagioclase, is a (Ca, Al) silicate which forms thin colourless crystals. Ilmenite, an opaque black-bladed (Fe, Ti) oxide, is also common. Spinel (Mg, Fe, Cr, Al) oxide also occurs in small amounts in the lunar lavas. There are some new lunar minerals. One is pyroxferroite which is a pyroxene enriched especially by iron. Another is armolcolite which is a (Fe, Ti, Mg) oxide not unlike ilmenite.

Another are tiny orange-red crystals called tranquillityite. This is a (Fe, Ti, Si) oxide but unusually enriched with (Zn, Yt, and U), all rare elements. These minerals are all consistent with a cooling from a molten lava in a low oxygen atmosphere.

Table 8.7. The relative abundances of oxides on the Venus surface at the two sites of Venera 13 and Venera 14.

Oxide	Relatve abundances (% mass)	
	Venera 13	Venera 14
SiO_2	45	49
Al_2O_3	16	18
MgO	10	8
FeO	9	9
CaO	7	10
K_2O	4	0.2
TiO_2	1.5	1.2
MnO	0.2	0.16

8.5.   Venus

The surface of Venus has been studied very thoroughly by orbiting spacecraft using radar techniques (see II.13) but two direct observations of the surface were obtained by Russian automatic craft, Venera 13 and 14. In spite of high pressures and temperatures, data of the composition were found and these are collected in Table 8.7. It is seen that the analysis of each site is essentially the same as the other, with small variations. The oxides are essentially the same as those found on Earth.

8.6.   The Material of the Solar System

The materials that arise from the cosmic abundance provide a set of specific molecules with associated material densities. The mean densities of the planets reflect the material composition. It is possible, therefore, to make an informed guess of the main constituents of the various components of the solar system. For example, the abundance is consistent with the appearance of a range of silicate materials — oxides — and it is known that these also form the main bulk of the material of the surface of the Earth. The mean densities of silicates are in the range 2,800–3,500 kg/m^3 so the Earth's density of about 5,500 kg/m^3 must require a substantial proportion, say x, of the heavy ferrous materials, of which iron is the most abundant

Table 8.8.   The possible compositions of the planets on the basis of their mean material densities.

Planet	Mean density $(10^3$ kg/m$^3)$	Possible material composition
Mercury	5.42	Free ferrous (comprising some 75% of total mass) with silicates
Venus	5.25	Silicates with ferrous comprising some 28% of total mass
Earth	5.52	Silicates with ferrous comprising some 32% of total mass
Mars	3.94	Silicates with some free ferrous materials
Jupiter	1.314	Almost entirely He and he with some heavier elements above cosmic abundance
Saturn	0.690	Mainly H and He with some elements above the cosmic abundance
Uranus	1.19	Mainly H and He but some H$_2$O and other materials
Neptune	1.66	Similar to Uranus

with a density of about 7,900 kg/m^3. This would make $x \approx 34.5\%$, a value remarkably close to the actual value of about 32%. In fact, the compression of the material has been neglected among other lesser factors necessary to achieve an accurate assessment. In this spirit, a brief survey of the planets is set out in Table 8.8.

A comparable list for the satellites and other small bodies is given in Table 8.9. It must be remembered in all these tables that these are preliminary estimates and are subject to detailed changes.

Table 8.9. Possible compositions of small bodies on the basis of material mean densities.

Small body	Mean density (10^3 kg/m^3)	Possible material composition
Earth — Moon	3.35	Silicates with small free ferrous component if any
Jupiter — Io	3.55	As for Moon with slightly more free iron
Europa	3.04	As for Moon/Io but with water ice
Ganymede	1.93	Silicates with a proportion of water ice
Callisto	1.81	As for Ganymede but with more water ice
Saturn — Titan	1.90	Silicates with water ice
Neptune — Triton	2.0	Silicates with water ice
Icy satellites	1.0–1.4	Mainly water ice with some silicate materials
Comets	2.0 (?)	Silicates with water, solid CO_2 and volatiles
Kuiper bodies		Silicates with variable quantities of water ice
Oort bodies	2.0 (?)	As for comets

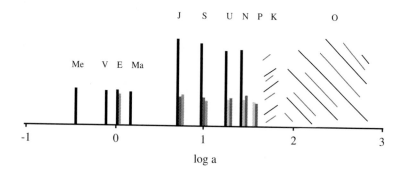

Fig. 8.3. The locations of the components of the Solar System on a logarithmic scale. Ma — Mars; V — Venus; E — Earth; Ma — Mars; J — Jupiter; S — Saturn; U — Uranus; N — Neptune; P — Pluto; K — Kuiper region; O — Oort region. Black — hydrogen/helium; dark shading — silicates; light shading — water.

8.7. Material in Orbit

The material of the Solar System is distributed in orbital arrangements and we give a representation of this now. The extension is quite large, perhaps as much as 1000 AU $\approx 1.5 \times 10^{14}$ m. For comparison, 1 light year $= 9.4 \times 10^{12}$ m so the region may well spread out to 1.6 light years (ly). The nearest star is about 4.2 ly away so the gap between the two systems is not so large that stray Oort particles could not transfer from one system to the other.

    The orbital distribution of the material is shown schematically in Fig. 8.3. The shadings refer to chemical elements — black for hydrogen, dark shading for silicates and light shading for water either as liquid or solid. The two differing regions of four planets each are clearly evident as is the encasing regions of Kuiper particles and Oort particles.

*

Summary

1. The abundances of the chemical elements in the solar system follow a pattern called the cosmic abundance of the elements. The Sun follows the same abundance which is sometimes called alternatively the solar abundance.

2. Hydrogen and helium account for some 98% of the cosmic abundance by mass, while oxygen, carbon and nitrogen account for about 1.33%. The remaining elements account for the remaining 0.67%.

3. Most of the chemical elements fall into one of three groupings: lithophilic elements which tend to be associated with oxygen in silicates and oxides; chalcophilic which tend to appear as sulphides; and siderophilic elements which generally remain as metals.

4. The basis of the division appears to be the ability of the nucleus of the atom to attract electrons for negatively charged anions.

5. The material compositions of the surface regions of the Earth and of meteorites appear similar though not identical.

6. The composition of lunar rocks is very different, mainly because there is a complete lack of water.

7. The surface composition of Venus has been tested at two sites by Russian Venera automatic craft. It is very closely the same as that for Earth.

8. The surface composition of Mars also follows broadly the same form as for the other terrestrial planets.

9. The composition of the Solar System is very largely hydrogen/helium, silicates/sulphates, ferrous metals and water. Most of the hydrogen/helium is concentrated in the major planets and in the Sun. These bodies are orbited by silicate and silicate/water-ice bodies to give a characteristic distribution of material about the Sun.

9

Finding the Ages of Rocks: Geochronology

Radioactive Dating

The history of the civilisation collected in the Old Testament has been traced back by theologians 4004 years, and this was supposed in the 18th century to be the date the Earth was formed. Other civilisations were discovered later to make the age of the Earth shorter still. The early geologists were unhappy with so short a time scale within which to set the geological record. This constraint was eased towards the end of the 19th century when Lord Rayleigh found an age of the Earth of a few tens of millions of years, based on calculations of the heat content and heat loss by the Earth. Early in the 20th century, Rutherford and Blackwood found an age of some 4,000 million years using the radioactive decay of uranium salts in rocks. Geologists now were happy with the time scale available to them. This latter age determination has since been confirmed and refined to give the accepted age of the Earth now as about 4,500 million years to within a few percent. How can the ages of rocks and planets be deduced this way? How does the radioactive decay method work? What are its possibilities for development? This is our concern now.

9.1. Atoms and Radioactive Decay

All matter is composed of atoms and the majority have remained unaltered since the birth of the planet. The longevity of an atom is central to the methods for determining the age of a particular collection of atoms which form a rock.

9.1.1. Some comments on atomic structure

The central nucleus is composed of protons (p) and neutrons (n), collectively called nucleons. Their masses are very closely similar (the neutron mass is slightly greater than that of the proton) and can be taken to be the same for our purposes with little loss of accuracy. The nucleon mass is $m_p = 1.67 \times 10^{-27}$ kg. Enclosing the nucleus is a collection of electrons, each with mass 9.1×10^{-31} kg. The electron carries a negative electrical charge of value 1.602×10^{-19} coulomb while the proton carries a positive charge of the same magnitude. The atom is electrically neutral so the number of extra nuclear negatively charged electrons must equal the number of positively charged protons in the nucleus. This number is called the atomic number, usually denoted by Z. The number of nucleons is denoted by A so the number of neutrons in the nucleus is $N = (A - Z)$. $= Z$,: The structure of a particular atom X can be specified by $_Z^A X$, or even $^A X$ because specifying the chemical material X also specifies Z. The chemical behaviour of an element is determined by the number of extra-nuclear electrons Z, the natural elements forming a sequence with Z number varying from 1 to 92. Atoms with higher Z numbers have been made artificially using nuclear reactors but they are unstable with very short life-times.

9.1.2. Atomic transformations: Isotopes

Atoms may have the same Z number but different A number and are called isotopes of the element. They will have identical chemical properties because they share a common Z number. Although they cannot be separated by chemical means they have different masses and can be separated this way using a mass spectrometer. (We might notice that the presence of isotopes also implies the existence of atoms with the same A but different Z. These will not concern us explicitly now.) Every element has isotopes but the number varies widely from one atom to the next and not all are stable. One isotope of an element is the most abundant in nature for that element, above the others. This corresponds to a particularly stable arrangement of nucleons within the nucleus. The neutron has no electric charge so the forces holding the nucleus together cannot be electrical — they are called nuclear forces but the details will not concern us now.

The lightest atom is that of hydrogen with one extra-nuclear electron ($Z = 1$) but there are two naturally occurring isotopes, one with a single proton as a nucleus $_1^1 H$ and the other with a neutron as well and called heavy hydrogen or deuterium $_1^2 H$. These isotopes occur naturally in the ratio $_1^1 H : _1^2 H = 6250 : 1$. This implies that ordinary water contains a heavy variety in the same proportion. There is a third, artificial, isotope with two nuclear neutrons called tritium $_1^3 H$. This is not stable and spontaneously ejects a neutron from the nucleus to become deuterium

$$\ce{^3_1H} \Rightarrow \ce{^2_1H} + n \,.$$

The spontaneous ejection of nuclear particles is called natural radioactivity and is widespread in the atomic table. The effect is to change the initial element into a different element with a different Z number.

At the other extreme of the mass scale, the heaviest natural element is uranium with $Z = 92$. There are three isotopes with A numbers 238, 235 and 234 with the proportions for occurrence $^{238}_{92}U(99.275\%)$, $^{235}_{92}U(0.720\%)$ and (0.005%). In these nuclei 92 protons are accompanied by, respectively, 146, 143 and 142 neutrons. Two other isotopes have been made artificially, using nuclear reactors, $^{236}_{92}U$ and $^{233}_{92}U$. The last one is used in nuclear technology together with $^{235}_{92}U$. All the isotopes are unstable, the nuclei ejecting atomic particles over a period of time. This decay is conventionally describes in terms of the time for one-half an original mass to decay and called the half-life of the element. The half lifes of the uranium isotopes are found to be $^{238}_{92}U(4.51 \times 10^9$ year), $^{235}_{92}U(7.00 \times 10^8$ year), $^{234}_{92}U(2.47 \times 10^5$ year), $^{236}_{92}(2.34 \times 10^7$ year) and $^{233}_{92}U(1.62 \times 10^5$ year). These times are long: $^{238}_{92}U$ has a half life very similar to the age of the Earth. The artificial nuclei $^{236}_{92}U$ and $^{233}_{92}U$ also have very long half lifes and once created cannot be eliminated easily.

9.1.3.   Radioactive series

The decay of the uranium isotopes is not a single event but is followed by further decays. The initial element is called the parent and the nucleus after the decay is called the daughter nucleus. If the daughter itself gives rise to a further daughter nucleus there is a decay chain. There may be many intermediate steps but eventually the chain ends with a stable nucleus. In the case of uranium each isotope gives rise to its own chain and the end of each one is a different isotope of lead. This is shown in Table 9.1 which also includes the number of ejected particles to complete the chain. In fact each decay to the daughter is achieved by the emission of either a helium nucleus ($A = 4$, $Z = 2$) and called an α-particle, or by the emission of an electron, called a β-particle. In losing an α-particle the nucleus loses two units of positive charge and four units of mass. In losing an electron it loses one unit of negative electricity (or gains one unit of positive charge) but the mass is unaltered. The ejections are accompanied by high energy electromagnetic radiation, or more specifically the emission of a γ-photon.

There are radioactive series other than uranium and thorium (half life 1.41×10^{10} years) with end points other than lead. For example, the artificial element neptunium, $^{237}_{93}Ne$ (half life 2.14×10^6 years) forms a series with end

Table 9.1. The decay chains for the two naturally occurring isotopes of uranium and that for thorium. γ photons are also ejected at each stage.

Parent element		End of decay chain	No. of ejected particles	
			α	β^-
$^{238}_{92}$U	$\Rightarrow$	$^{206}_{82}$Pb	8	6
$^{235}_{92}$U	$\Rightarrow$	$^{207}_{82}$Pb	7	4
$^{232}_{90}$Th	$\Rightarrow$	$^{208}_{82}$Pb	6	6

point the stable isotope of bismuth, $^{209}_{82}$Bi, achieved after 14 emissions of atomic particles.

9.1.4. Single radioactive decay

Radioactive decay is a statistical phenomena. From an initial sample of N radioactive atoms, the number of atoms, δN, that will decay in the element of time δt is directly proportional to N. Explicitly

$$\delta N \propto N \delta t, \quad \delta N = -\lambda N \delta t \tag{9.1}$$

where λ is a decay constant characteristic of a particular element. The negative sign is present because the initial number of atoms is decreased by the decay. The ratio $\frac{\delta N}{N}$ is called the activity.

The number of atoms that are lost during the time interval τ is obtained by integrating (9.1) over this time interval. Then

$$\frac{dN}{dt} = -\lambda N$$

so integrating over the time from 0 to τ gives

$$N_\tau = N_0 e^{-\lambda \tau} \tag{9.2}$$

where N_0 is the initial number of atoms, or parent atoms. N_τ are the daughter atoms. The number of parent atoms lost during the interval is $(N_0 - N_\tau)$ and this is also the number, D_τ, of daughter atoms produced because the total number of atoms remains constant.

An expression for the half life, $T_{1/2}$, of the decay follows from (9.2). This is the time for one-half the initial atoms to decay by the radioactive process so that $N_\tau = 1/2 N_0$. Then, using the natural logarithm

$$T_{1/2} = \frac{\ln 2}{\lambda} = \frac{0.693}{\lambda}. \tag{9.3}$$

It is seen that a large decay constant corresponds to a short half-life and vice-versa. For example, the half life of ^{238}U is 4.51×10^9 years $= 1.42 \times 10^{17}$ s giving $\lambda = 4\,9 \times 10^{-18}$ s^{-1}. This ensures that the exponential factor in (9.2) differs little from unity over a long time interval.

9.1.5. A radioactive chain

The decay of the parent atom may produce a daughter atom which is itself radioactive. And the next daughter may be radioactive as well. The result is a chain of disintegrations, each depending on the one before. The number of atoms at each stage is the algebraic sum of the gain of new members by disintegration of the atom before and the loss of members due to disintegration. The chain will stop when the atoms produced at a step cease to be radioactive. The details of a chain of n steps are shown below with D being the final number of daughter atoms.

$$\frac{dN_1}{dt} = -\lambda_1 N_1 \Rightarrow \frac{dN_2}{dt} = \lambda_1 N_1 - \lambda_2 N_2$$

$$\Rightarrow \cdots\cdots \tag{9.4}$$

$$\frac{dN_m}{dt} = \lambda_{m-1} N_{m-1} - \lambda_m N_m \Rightarrow \frac{dD}{dt} = \lambda_n N_n \,.$$

The series involves a set of decay constants. The set is sometimes called the Bateman Relations. If the time for each step is known the time to follow the full chain can be calculated. The chain might refer to material in a rock giving a method for dating it. The set is difficult to solve analytically and is best achieved using computers.

9.2. Nuclear Reactions

The transformation from a parent atom to a daughter atom involves the emission of atomic particles and electromagnetic radiation by the parent. This may occur naturally (natural radioactivity) as has been seen for uranium, or can be induced artificially by making radioactive nuclei in nuclear reactors. Either way, the particles emitted belong to the same group.

The discovery of radioactivity by Becquerel in 1896 involving the uranium series showed the presence of three distinct types of rays. They were called α, β, and γ respectively because their forms were then unknown. It was soon found that α and β rays are atomic particles while γ rays are electromagnetic radiation. More precisely, α-particles are helium nuclei with $Z = 2$ and $A = 4$; β-particles are electrically charged with a mass of 9.1×10^{-31} kg and are either negatively charged electrons or positively charged positrons; and the γ rays are very high energy electromagnetic radiation. The loss of an α-particle causes the nucleus to lose 4 units of mass and 2 units of positive charge. The loss of a β-particle does not sensibly change the mass but causes the loss or gain of one unit of electric charge. The loss of γ rays reduces the energy but does not change the mass or charge.

The nuclear rearrangements are now known to be more complicated and result essentially from the transformation of protons to neutrons and neutrons to protons in the nucleus. Thus

$$n \Rightarrow p + \beta^- + \bar{v}$$

$$p \Rightarrow n + \beta^+ + v. \tag{9.5}$$

Here β^- denotes an electron, β^+ denotes a positron, v denotes a massless neutrino but with energy and momentum and $\bar{v}$ denotes an antineutrino. The first process is more common, the nucleus gaining a positive charge. An example of geophysical interest is the natural conversion

$$^{14}_{6}C \Rightarrow {}^{14}_{7}N + \beta^- .$$

There is no γ radiation. The half-life for the decay is 5730 years making it of archaeological interest. The second decay process is called decay by positron emission. It is most often the result of the artificial production of a nucleus. One example is the artificial nucleus ^{58}Co with the decay scheme

$$^{58}_{27}\text{Co} \Rightarrow {}^{58}_{28}\text{Ni} + \beta^+ .$$

The half-life is 71.3 days. It is of use in medicine. Again, there is no radiation emitted during the process. There is an enormous range in the half lives of the elements.

Whilst the transformation of proton to neutron can take place only within the nucleus, the inverse transformation can take place outside as well. In fact a neutron is itself a radioactive particle with a half life of about 13 minutes. Neither form of β-particle is sufficiently energetic to exist within a nucleus but can be captured and absorbed from outside. This might arise from the capture of a high-energy electron from outside the atom or from the capture of an orbiting valence electron. Then

$$p + \beta^- \Rightarrow n + v. \tag{9.6}$$

9.3. An Elementary Method for Dating Rocks

The formula (9.2) can be used to determine the age of a rock sample in the simplest case of a single radioactive event. The number of parent atoms is unknown but the number at time τ, N_τ, and the number of daughter atoms then, D_τ, can be measured. Then $N_0 = N_\tau + D_\tau$. Then, (9.2) becomes alternatively

$$N_0 - D_\tau = N_0 \exp\{-\lambda\tau\}$$

so that

$$D_\tau = N_0[1 - \exp\{-\lambda\tau\}]$$
$$= N_\tau[\exp\{+\lambda\tau\} - 1]. \qquad (9.7)$$

This is converted immediately into the form

$$\tau = \frac{1}{\lambda}\ln\left[1 + \frac{D_\tau}{N_\tau}\right]. \qquad (9.8)$$

It is required to measure D_τ and N_τ and to find λ. This is essentially how, in 1907, Rutherford and Boltwood first obtained ages far in excess of those predicted by Lord Rayleigh at that time. The first experiments were based on the decay of uranium yielding α-particles (that is helium nuclei) and measuring the quantity of helium gas present in the rocks, together with the quantity of uranium found at the same time. Unfortunately rocks are permeable to helium so this method gives too short an age. Later, Rutherford and Boltwood used lead as the daughter, so achieving a more accurate value.

There are, however, problems apart from the assumption that no daughter atoms have been lost. It is necessary for daughter atoms to result from only one source of parent atoms. It assumes also that the sample has been left undisturbed since its formation. There could have been chemical reactions between the daughters and other minerals. The daughter might be a gas and the rock could be porous so that some quantity of the daughter could have been lost. Accurate age determinations requires improved methods. For a chain it is necessary to solve the set of expressions (9.4). The process can be simplified in a chain where one stage requires overwhelmingly more time to complete than any of the others. In that case the chain effectively reduces to a single stage.

9.4. The Closure Temperature

Age measurements are very much more complicated than might seem from the previous example for several reasons. One is that a component of the decay may be a gas and there is the possibility of a partial loss through the porosity of the rock. The loss depends on the temperature of the mineral and there is a temperature, called the closure temperature, above which gases cannot be retained. Age measurements using gaseous daughter atoms refer, therefore, to the time when the temperature fell below the particular closure temperature. The closure temperatures vary enormously from one mineral to another. As a general guide, they fall within the range $300°C$ to $1,000°C$.

9.5. Selecting a Particular Decay Mode

Accurate age determinations require several age determinations for a given specimen and in each case a similar magnitude for the number of parent and daughter atoms so that the ratio is near unity. It is also necessary in some cases to be able to distinguish the behaviour of a particular isotope especially for meteorites and other studies. Again, the stability of an element in a particular mineral is very variable between the minerals and may not even be present in the mineral that is to be dated. A variety of decay schemes have been developed to cover the wide range of cases met in practice. We give one here as an example.

9.5.1. The rubidium–strontium method

There are two isotopes of rubidium, ^{85}Ru and ^{87}Ru. Strontium has four isotopes, ^{84}Sr, ^{86}Sr, ^{87}Sr, and ^{88}Sr in the ratios 0.6%, 10%, 7%, and 83%. The decay process $^{87}_{37}$Rb $\rightarrow$ $^{87}_{38}$Sr is the one of interest now. Using (9.7) we write

$$[^{87}\text{Sr}]_{\text{now}} = [^{87}\text{Rb}]_{\text{now}}[\exp\{\lambda\tau\} - 1].\tag{9.9}$$

Since the decay constant is known this expression could be used to determine the age if the two abundances ^{87}Sr and ^{87}Rb are known. This age would presume that there was no ^{87}Sr present initially but this cannot be certain. Instead it is necessary to rearrange (9.9) to show the initial strontium content $[^{87}\text{Sr}]_0$ giving

$$[^{87}\text{Sr}]_{\text{now}} = [^{87}\text{Sr}]_0 + [^{87}\text{Rb}]_{\text{now}}[\exp\{\lambda\tau\} - 1].\tag{9.10}$$

To proceed this expression must be normalised with a quantity which does not change with time. ^{86}Sr is not a product of radioactive decay so its present amount should have remained unaltered from the earliest times. This allows us to write $[^{86}\text{Sr}]_{\text{now}} = [^{86}\text{Sr}]_0$. For this purpose, divide (9.10) by $[^{86}\text{Sr}]_{\text{now}}$.

$$\frac{[^{87}\text{Sr}]_{\text{now}}}{[^{86}\text{Sr}]_{\text{now}}} = \frac{[^{87}\text{Sr}]_0}{[^{86}\text{Sr}]_{\text{now}}} + \frac{[^{87}\text{Rb}]_{\text{now}}}{[^{86}\text{Sr}]_{\text{now}}}[\exp(\lambda\tau) - 1]$$

$$= \frac{[^{87}\text{Sr}]_0}{[^{86}\text{Sr}]_0} + \frac{[^{87}\text{Rb}]_{\text{now}}}{[^{86}\text{Sr}]_{\text{now}}}[\exp(\lambda\tau) - 1]. \qquad (9.11)$$

The $\frac{\text{Sr}}{\text{Br}}$ ratio will vary from one region of a mineral to another due to the idiosyncrasies of the melting process although each region will decay at the same rate. Consequently a plot of $\frac{[^{87}\text{Sr}]_{\text{now}}}{[^{86}\text{Sr}]_{\text{now}}}$ against $\frac{[^{87}\text{Rb}]_{\text{now}}}{[^{86}\text{Sr}]_{\text{now}}}$ for different mineral samples within the whole will provide a straight line with slope $[\exp\{\lambda\tau\} - 1]$. With λ known, τ follows immediately. This procedure is known as the whole rock isochron now for rubidium and strontium. The plot will not pass through the origin if strontium was there originally. The intercept on the Sr axis is then given by the factor $\frac{[^{87}\text{Sr}]_0}{[^{86}\text{Sr}]_0}$ giving the initial quantity $[^{87}\text{Sr}]_0$: the ratio is called the initial ratio for strontium.

The measurements for the oldest meteorites give an age of about 4,550 million years and an initial ratio of 0.699, called the Basic Achondrite Best Initial or BEBI for short. The oldest known terrestrial rocks have an age of about 3,800 million years and an initial ratio of 0.700. Modern basalt rocks show a full rock value of the strontium ratio of 0.704. If some portions of a sample have been heated differently from other portions, the corresponding initial ratios will be different. Such measurements can provide information about the thermal history of a specimen.

9.5.2. Other decay schemes

The strontium-rubidium approach is not the only isochron available. The long half-life of rubidium (4.88×10^{10} years) makes it valuable for older rocks but either a rock may not contain rubidium or the rubidium abundance may be very small. Some alternative isochrons are listed now but without detail since the method is essentially the same as for rubidium and strontium.

(i) Samarium–neodymium

These elements are both silvery white metals with rare earth elements, of the lanthanide group. They both have industrial uses. Neodymium has seven natural isotopes of which $^{144}_{60}\text{Nd}$ (23.80%) has a half life of 2.4×10^{15} years and so is effectively a stable nucleus. Samarium has nine isotopes and one of them, $^{147}_{62}\text{Sm}$, decays to $^{143}_{60}\text{Nd}$. The decay chain analogous to (9.11) is set up with $[^{144}_{66}\text{Nd}]_{\text{now}} \approx [^{144}_{60}\text{Nd}]_0$ as the normaliser being the present value. The isochron is constructed as before. High accuracy has been achieved in measurements and, although ideal for dating older samples, the method is also useful for dating younger rocks.

(ii) Lutetium–hafnium

The hard rare earth metal lutetium has two isotopes, one is stable and the other is $^{176}_{71}\text{Lu}$ with a half life of 2.2×10^{10} years. This is virtually stable for work with bodies with an age of one-tenth the value, and can be used as a normaliser for isochron analysis. Hafnium, a silvery ductile metal, is largely impervious to weathering. It has 6 naturally occurring isotopes of which $^{176}_{72}\text{Hf}$ is the decay product of $^{176}_{71}\text{Lu}$ by β^- decay. This isochron has proved useful for the study of magmas.

(iii) Rhenium–osmium

Rhenium is a silvery metal with two naturally occurring isotopes, one stable and the other, $^{187}_{75}\text{Re}$, decays by β^- decay to $^{187}_{76}\text{Os}$ which is stable. The half life of $^{187}_{75}\text{Re}$ is 4.3×10^{10} years and acts as a normaliser for the isochron. This process again is very suitable for the analysis of magmas.

(iv) Thorium–lead

The decay scheme now is based on $^{232}_{90}\text{Th} \rightarrow {}^{208}_{82}\text{Pb}$. The isochron is constructed with the non-radiogenic isotope ^{204}Pb as the normaliser. Thorium and lead are more usually retained by rocks than is uranium making this a valuable tool for studying older rocks.

(v) Uranium–lead: The concordant diagram

The decay chain $^{238}_{92}\text{U} \rightarrow {}^{206}_{82}\text{Pb}$ is used with the decay chain $^{235}_{92}\text{U} \rightarrow {}^{207}_{82}\text{Pb}$ using ^{204}Pb as the normaliser. It is known that $\frac{[^{235}\text{U}]_{\text{now}}}{[^{238}\text{U}]_{\text{now}}} = \frac{1}{137.88} = 7.2526 \times 10^{-3}$.

Any periods of high temperature will affect the quantity of lead present, lead being very volatile. This will lead to a predicted age which is too short. A test of whether this has or has not happened in a given specimen can be judged by plotting the ratio $\frac{^{206}Pb}{^{238}U}$ against the ratio $\frac{^{207}Pb}{^{235}U}$ for rocks of different ages. These ratios can be specified with some accuracy. Measured points that lie on the curve have not lost lead: those that lie below it have. The curve is called the concordant diagram or concordia. Points that deviate from the line are called discordant points. The time since the lead was lost can be deduced from the diagram but the matter is too technical to pursue here.

(vi) Potassium–argon

^{40}K has a half life of 1.28×10^9 years decaying by β^- decay to ^{40}Ar which is a stable isotope accounting for 99.600 of the naturally occurring gas. An isochron can be constructed by using ^{36}Ar as the normaliser. It is presumed that the present ratio of the two stable argon isotopes $\frac{[^{40}Ar]}{[^{35}Ar]} = 295.55$ is also the initial ratio. The method is suitable for younger rocks.

9.6. Dating Using Nuclear Reactors

Nuclear reactors allow naturally stable nuclei to be made unstable by bombardment by neutrons. The advantage is that stable nuclei which may be abundant in rocks can be brought into the measurement. One example is using the stable isotope of potassium ^{39}K. It is the most abundant isotope of potassium (93.3% of a naturally occurring sample) but can be converted to the radiogenic isotope of argon ^{39}Ar by collisions with 2 MeV (fast) neutrons. ^{39}Ar does not occur naturally and is transformed back to ^{39}K by β^- decay with a half-life of 269 years. Then

$$^{39}K + n \rightarrow {}^{39}Ar + p, \quad {}^{39}Ar \rightarrow {}^{39}K + \beta^- .$$

The number of argon atoms produced in the reactor during a given time interval is proportional to the number of initial potassium atoms so that $[^{39}Ar]_{now} = \sigma[^{39}K]_{now}$. σ is a constant which depends on the details of the interaction between neutrons and potassium atoms and the time of irradiation. It can be found by conducting the experiment using a specimen of known age, and to this extent the method is comparative. An isochron can be established using $[^{39}Ar]$ as the normaliser.

An entirely different alternative method is based on the damage caused in a crystal by the passage of fast, heavy, atomic nuclei. The radioactive decay within a mineral releases particles that collide with other mineral particles, transferring energy at each stage and causing damage until it is brought to rest. One such nucleus is ^{238}U which disintegrates spontaneously into two particles of essentially equal mass, together with two or three high energy neutrons. Damage is caused by the particles which can be observed under the microscope after etching the sample. The number of disintegrations observed in a sample can be interpreted in terms of the age of the sample using the reactor data. It is necessary for older rocks to account for other radioactive materials, of shorter half-lifes, that may have been present earlier. For the oldest rocks this could involve plutonium. The stability of the fission tracks depends on the random movement of the neighbours not being too large and this depends on the temperature. Although the method has its skills and uncertainties it can allow the thermal history of a specimen to be traced.

*

Summary

1. The radioactive transformations between atoms provides an accurate way of measuring the ages of rocks.

2. The radioactivity may be naturally occurring or can be induced artificially by irradiation in a nuclear reactor.

3. The decay is defined in terms of a decay constant which is related to the half-life of the atom. This is the time for half an original assemblage of atoms to disintegrate.

4. The decay process may begin with a parent atom which decays into another (daughter) atom which itself may decay further. The result is a radioactive chain with a set of decay constants.

5. A rock specimen may be pervious to atoms of certain gases. If these are daughter atoms of a decay process in the rock, measuring their number at any time will provide a false (low) estimate of the age of the rock. Each mineral has a temperature above which gases cannot be retained. This is called the closure temperature.

6. The decay mode to be chosen in any particular case is chosen to provide the maximum accuracy. The choice depends not only on the more abundant radioactive element in the rock but also on the age of the rock.

7. Seven day modes are treated in the text.

8. If the more abundant element is stable, and so not radioactive, the decay mode method can still be applied using a nuclear reactor. The stable atom can be destabilized using fast neutrons. The approach usually requires normalization by using a rock of known age and so is a comparative method.

9. It is also possible to date a rock specimen by the study of the radioactive damage within it. A sample of the rock is cut and etched to reveal the track system. The method has various uncertainties.

10

Surfaces and Interiors

Before reviewing the principal observed features of the planets it is convenient to consider the various ways that such information has been obtained. It will be necessary to consider the two groups of members of the Solar System separately. The techniques currently being used were developed originally for studies of the Earth which shows strong rigidity and this is by far the best known member of the System. The methods developed there are available for application to the other terrestrial planets[1] in the future. The giant fluid planets must be treated separately, as explained in the last Section. It is important, however, to realise that actually living on a planet's surface does not offer any decisive advantage in finding out about it below and above the surface. The interior is largely impenetrable to direct observation and the atmosphere cannot be viewed in any comprehensive way without advanced technical means. The one region readily accessible is the surface.

10.1. The Surface Figure

We live on the surface of the Earth but the surfaces of other bodies are known only indirectly. It would be tempting to say that the surface is the part you see immediately but a dense atmosphere may enclose the true surface (for example, Venus) making it necessary to specify the wavelength involved in this statement. The visible wavelengths are relevant but other wavelengths can also give important information. In particular, the form and nature of the surface can be assessed from orbiting spacecraft carrying sensitive spectrometers or by using radar frequencies.

[1]And also to the icy satellites and the asteroids.

10.1.1. The shape of a planet

The most important studies are made by walking over the surface and observing the features in detail. This was how the surface of the Earth was first explored. This allows the present geology to be documented and, through fossil evidence that may be there either animate or inanimate, the details of the past evolution can be deduced. The study is a long and painstaking task which, as far as the Earth is concerned, has not yet been achieved on land and has only just begun in the oceans. The first steps have been taken on Moon by human astronauts and on Mars by lander vehicles. Two parachute landings of space vehicles have been made on Venus. This is the definitive approach which is difficult to achieve though presumably one day the surface will be explored in detail.[2]

One of the prime objects in the early exploration of the Earth was to find its shape, or figure. This is the object of the study of Geodesy. Observations of the planets using telescopes made it clear that all the terrestrial planets are very closely spherical and that the departures from it can be usefully accounted for by ascribing the bodies a slightly oblate spheroid shape which is the figure generated by rotating an ellipse about its minor axis. The precise figure is specified by the flattening f. If R_e is the equatorial radius and R_p the polar radius, the flattening is defined by

$$f = \frac{R_e - R_p}{R_e}.$$
(10.1)

To the first order in the flattening the radius r is written

$$r = R_e(1 - f \sin^2 \lambda),$$
(10.2)

where λ is the latitude. We see that at the equator ($\lambda = 0$) $r = R_e$ while at the pole ($\lambda = \pi/2$), $r = R_e(1 - f) = R_p$. For the terrestrial planets the flattening is of the order 3×10^{-3} or less. The flattening will arise primarily from rotation. The gravitational acceleration on the surface of a sphere rotating with angular velocity ω, written g_{rot}, will be *less* than that on a non-rotating sphere. The effect of rotation will vary with latitude and so, therefore, will the measured acceleration of gravity. Explicitly, the dependence is taken to be

$$g_{\text{rot}} = g - \omega^2 R_e \cos^2 \lambda.$$
(10.3)

Alternatively, the acceleration can be calculated to give a reference gravity formula of the form

---

[2]This is the strongest reason for manned exploration of the Solar System.

$$g_{\text{rot}}(\lambda) = g_e(1 + \alpha \sin^2 \lambda + \beta \sin^4 \lambda). \tag{10.4}$$

Here α and β are two constants. This form was adopted by the International Association of Geodesy in 1967 for the Earth with defined values of the three constants (g_e, α, β). For the Earth, rather more than 50% of the variation is due to rotation and rather less than 50% to the ellipsoidal shape. Gravity observations are expressed as deviations from the standard form (10.4). The acceleration of gravity, g, is greater the smaller the radius because the surface is nearer the centre then. This means that g is slightly larger in the polar regions than at the equator.

Planetary bodies are neither perfectly spherical nor are they perfect oblate spheroids. Elevated regions and valleys on all the planets show variations of many kilometres. In terms of the radius of the planet, the surface is extraordinarily flat — typically deviations of a few parts in ten thousand — but there are deviations on a human scale which need quantifying. This is done on Earth by introducing the geoid. This is a surface of gravitational equipotential. The gravitational potential is constant along such a surface so the gradient of the potential is perpendicular to the surface. This means that a plumb line will always hang perpendicular to the equipotential surface under the action of both gravity and the rotation of the planet. The equipotential surfaces for other planets will, when defined, presumably be given different names for the surface of constant potential. The study of equipotentials has been a particularly successful activity of mathematicians in the past and especially by Clairaut and Sir George Darwin, among many others.

The development of geodesy on Earth involved careful surveying on the continents but the oceans define the geiod there. Up to 50 years ago the continental geoid was very much better known than that for the oceans. Measurements at sea are made difficult by the motion of the ship, an inconvenience not met with on land. The oblate spheroid of (10.2) is an approximation to the geoid and for Earth is called the reference spheroid or reference ellipsoid. It is a mathematical construction modelling the gravity field for a model rotating Earth. The surface gravity on the spheroid is to be calculated from the gravity formula (10.4). The introduction of orbiting satellites in 1957 transformed geodesy because it allowed worldwide coverage over short time periods. Radar studies have allowed the Earth geoid over land and sea to be determined with incredible accuracy. The marine geoid is defined to about 10 cm and theory and observation now agree to within ± 50 cm. These studies are available for application to the other planets including the fluid major planets. The accuracy that can be achieved will act as a detailed constraint on theories of the interiors of these bodies.

The use of satellites has allowed very accurate measurements to be made of the gravitational field. The gravitational acceleration g can be expressed in terms of a gravitational potential U by $g = -\text{grad } U$. For any point distance r from the centre of the planet with latitude ϑ, let us write the potential in the form

$$U = g_e R \left[\frac{R_e}{r} + J \left(\frac{R_e}{r} \right)^3 \left(\frac{1}{3} - \cos^2 \theta \right) \right] + O(J^2) \qquad (10.5)$$

with g_e being the equatorial gravity, R_e the equatorial radius of the planet and where J is a parameter to be found. Incidentally, the symbol J is used here for the coefficient in honour of Sir Harold Jeffreys who contributed so greatly to these matters in the 1930s. It is clear from (10.5) that the second term in the bracket is associated with the deviations from spherical form. The parameter J is, therefore, an indicator of the flattening, f, of the body. We can notice that the rotation characteristics of the planet do not appear in (10.5) because the orbiting craft does not partake of the rotation of the planet.

A full analysis of the motion of the orbiting craft shows that the plane of its orbit precesses about the rotation axis of the planet with an angle measured by Ω, say. The axis of spin of the satellite also precesses with angle α, say. The orbit has a constant angular momentum. It is found that the rate of presession varies directly as J (and so with the oblateness) but also through α and the radius of the orbit, r. The time variation of Ω, $d\Omega/dt$, can be measured with an accuracy of better than 0.05% so this offers an accurate way of measuring J. The analysis applies equally to circular or elliptic orbits. While the best values of J can be expected to be obtained from the study of artificial orbiters the method can also be used studying the motion of natural satellites of a planet.

10.1.2.   Radar studies

In some cases the observed disk of the object is a rather featureless circle with the surface covered in cloud. This is the case for Venus and also for Titon, the satellite of Saturn. Another problem is a surface which is fully visible but not so far visited by a lander: this is the case of the innermost planet Mercury. It is also the case of the outermost planetary type body Pluto.

10.2.   The Interior

It was seen in I.4 that the inside of a planetary body will generally consist of three separate regions, a crust, a mantle and a core. Although the mantle and core are often chemically distinct, it is convenient to treat them together as the interior, and to treat the crust separately.

The detailed study of the full interior is made by seismology. This uses earthquakes as a source for energy. Earthquakes have been known and feared since antiquity but it was the Lisbon earthquake of 1784 that ushered in the new study. This was a very large event which was felt, over a time span, as far away as England. Ships anchored in the Solent felt the swell produced in the water and even the surfaces of ponds in the South of England were visibly disturbed. Effect were noted as far away as Poland. John Michel of London studied these linked phenomena and realised that the time sequences at different points could be reconciled if the energy from the source had spread in the form of waves. He realised that, however unlikely for the solid Earth, the waves must be involved with an elasticity of the material. He therefore proposed that the Earth responds to the passage of energy in the same way as an elastic solid. Notice that this does not imply that the material is an elastic solid but only that it behaves like one if subjected to strong stimuli of short duration.

10.2.1.   Density with depth

In the surface region the gravitational interaction is not as strong as the interaction between atoms so that chemical forces are the controlling factor. The mantle lies below this region and here conditions of a mean hydrostatic equilibrium can be assumed. Accordingly, the change in pressure from the top to the bottom of a very thin sheet of material is due entirely to the weight of material above. Put another way, the pressure increases with depth due to weight of the material above. If r is the distance measured from the centre of the planet and the planet is not rotating, the change of pressure, $dP(r)$ at level r, over the thin section dr can be written

$$-dP(r) = \rho(r)g(r)\,dr \tag{10.6}$$

where $\rho(r)$ is the mean material density at the thin slice and $g(r)$ is the acceleration of gravity there. This can be written in terms of the total mass $M(r)$ within the sphere of radius r. Since

$$g(r) = \frac{GM(r)}{r^2} \quad \text{and} \quad \rho(r) = \frac{3M(r)}{4\pi r^3}$$

we can write (10.6) alternatively as

$$\frac{dP}{dR} = -\frac{3GM^2(r)}{4\pi r^5} = \frac{4\pi}{3} G\rho^2(r)r \,. \tag{10.7}$$

This expression can be integrated if the density profile is known, so that formally

$$P(r) = \int \frac{4\pi G}{3} \rho^2(r)r \, dr \,.$$

This integral can be evaluated over the whole body for the trivial case of a uniform density, ρ_0, to give the central pressure

$$P_c = \frac{2\pi G}{3} \rho_0^2 R^2$$

where R is the total radius. This is not realistic as a property because the material will be compressible and this has not been accounted for.[3] It does, however, give a lower limit which could be useful in order of magnitude estimates. A meaningful calculation of the pressure with depth can be made only if the density profile is known. This is made by taking explicit account of the compressibility of the material.

For a compressible material, a positive increment in the pressure will cause the volume of unit mass to decrease by amount $-\Delta V$, say. The isothermal bulk modulus, κ, of the material is defined as the change in pressure, ΔP, with change of volume, ΔV, per unit volume V. Explicitly

$$\kappa = -\frac{\Delta P}{\frac{\Delta V}{V}} \,. \tag{10.8a}$$

For unit mass of material, $V = 1/\rho$ so that $\Delta V = -\frac{1}{\rho^2}\Delta\rho$. Passing to the infinitesimal form, (9.7) becomes instead

$$\kappa = \rho \frac{dP}{d\rho} \,. \tag{10.8b}$$

We can now write successively $\frac{dP}{d\rho} = \frac{\kappa}{\rho}$ and $\frac{dP}{d\rho} = \frac{dP}{dr}\frac{dr}{d\rho}$. This means

$$\frac{d\rho}{dr} = \frac{\rho}{\kappa}\frac{dP}{dr} \,.$$

[3]Early studies at the end of the 19th century encountered difficulties because the compressibility of the material was not taken into account.

We can use (10.7) to introduce $\frac{dP}{dr}$ allowing an expression to be constructed for the change of density with the depth. Then

$$\frac{d\rho}{dr} = -\frac{4\pi}{3}\frac{G\rho^3(r)r}{\kappa(r)} .$$
(10.9)

Since $\kappa > 0$ always, the negative sign shows that the density increases with decreasing r, that is with increasing depth. This expression can be used to determine the density with depth by iteration, starting from a specified upper value and moving downwards, provided the details of the bulk modulus with depth are known alternatively.

The simplicity of these various expressions should not obscure their deeper meaning. The analysis applies to material which responds like an elastic solid to a disturbance of short duration but responds like a fluid (albeit a strongly viscous fluid) to the action of a disturbance over an extended time interval. Expressed another way, the material of the planet is not a Newtonian fluid nor is it an ideal solid.[4]

10.2.2. The isothermal bulk modulus: Seismology

It is possible to use (10.9) to determine the change of density with depth if the isothermal bulk modulus can be specified. One way of achieving this is to remember the assumption of Michel and suppose the energy contained in an earthquake disturbance is propagated throughout the volume in the form of elastic waves. For simplicity, the material is supposed to be ideally elastic. The material is compressible and has rigidity so two distinct types of wave can propagate independently in the main volume. These are a longitudinal (compressional push-pull) sound wave with no rotation and a transverse wave with rotation but without compression. In seismology the longitudinal wave is called a *P*-wave and the transverse wave is called an *S*-wave. *P* can stand for primary or principal while S can stand for secondary. These two wave patterns are shown in Fig. 10.1.

The standard wave equation without dissipation has the form

$$c^2\frac{\partial^2 \boldsymbol{a}}{\partial x^2} = \frac{\partial^2 \boldsymbol{a}}{\partial t^2}$$
(10.10)

where the amplitude $\boldsymbol{a}$ is propagated in the x-direction ($+$ or $-$) with speed c. For a general disturbance there can be a y- and a z- component as well.

[4]This leads to difficulty in specifying the state of a terrestrial planet accurately. It strictly cannot be described as the Solid Earth, or Mars or whatever. Condensed is not adequate since a liquid is condensed.

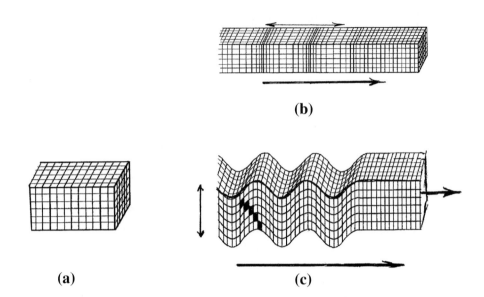

Fig. 10.1.   (a) the undisturbed medium; (b) the passage of a longitudinal P-wave; and (c) the passage of a transverse shear S-wave.

The wave proceeds without change of amplitude. This standard form is applied both to the longitudinal P-wave, with a scalar amplitude, and to the transverse S-wave where there are three component amplitudes. Of course, the wave speeds are different in each case. The longitudinal P-wave can propagate in gases, solids and liquids and involves the bulk modulus κ and the shear modulus μ. The relation between wave speeds and the elastic constants can be derived from the theory of elasticity. We need give only the results of these complicated derivations here. For P-waves it is fond that

$$c \equiv V_P \quad \text{and} \quad \rho V_P^2 = \kappa + \frac{4\mu}{3} \tag{10.11a}$$

while S-waves, showing shear at constant volume, involve only the shear modulus so that now

$$c \equiv V_S \quad \text{and} \quad \rho V_S^2 = \mu. \tag{10.11b}$$

An expression for the bulk modulus is found by combining (10.11a) and (10.11b): this is

$$\kappa = \rho \left[V_P^2 - \frac{4}{3} V_S^2 \right]. \tag{10.12}$$

All the quantities appearing here will, in general, vary with the depth in the body. If it is possible to measure the wave speeds V_P and V_S then a knowledge of the density will allow the local value of the bulk modulus to be determined. This information is then available for use with (9.4) to provide an equation for determining the density gradient, and so the density

$$\frac{d\rho}{dt} = -\frac{4\pi}{3}\frac{G\rho^2(r)r}{[V_P^2 - \frac{4}{3}V_S^2]},\qquad(10.13)$$

where r is the distance outwards from the centre. The wave speed are measured from seismology.

The density is found by iteration assuming the wave speeds are known throughout the volume. Starting from a specified density $\rho(o)$ near the surface (10.13) is used to determine the change of density for a small increment of depth dr to give the density $\rho(r - dr)$ at the next lowest level $(r - dr)$. The procedure is then repeated for the next increment, and for the next, until the lowest depth of the region is found. This is the principle of the method of Williamson and Adams for finding the density profile with depth within a solid planet. It has been widely used for the study of the interior of the Earth and is available for the future exploration of the interiors of the other terrestrial planets. The method has achieved a very high degree of sophistication and has proved to be a very sensitive tool. For the Earth it is conventional to begin the calculations at a depth of 33 km, specifying the initial density $\rho(o) = \rho\,(33\ \text{km}) = 3300\ \text{kg/m}^3$.

10.2.3. Travel times

The application of the arguments just developed involves determining the times of travel of the waves from an "earthquake" (there isn't yet a name for the waves in other planets) from the source to the point of measurement a known path-distance away. The problem, then, is to plot the paths taken by the waves. The method involves the steady correction of approximate information. It is necessary to know the wave speed in a variety of minerals under a range of conditions.

It is assumed that the density will increase with depth, however slightly. The wave will then be concave towards the surface because the speed increases with increasing density. The time of the initial disturbance and the time of receipt of the wave packets at a point on the surface are supposed known. It is also necessary to know the location of the initial disturbance. The density of the material together with the shear and bulk moduli are known approximately because the composition can be presumed known at least approximately. An approximate path for the wave can be found and a travel time can be deduced. This is corrected by successive approximations to provide an accurate value. Rather than have every observatory doing this, the travel times collected by many observatories are collected together in the form of tables for agreed general tables future use. Sir Harold Jeffreys and his onee time student K. E. Bullen were particularly prominent in constructed travel time tables.

10.2.4. The quality factor, Q

Seismic waves do not actually travel without loss of energy to the surroundings because the material is not perfectly elastic. This transfer of energy is shown by the decrease of amplitude of the wave as it travels. The degree to which the wave is damped in this way is usually described by the so-called quality factor, Q.

Q is defined as the ratio of the energy, E, stored in the wave during one period or wavelength, divided by the energy lost to the medium during the cycle, $\frac{dE}{dt}$. Explicitly,

$$Q = \frac{2\pi E}{T} \frac{1}{\frac{dE}{dt}} \quad \text{or} \quad \frac{dE}{E} = \frac{2\pi}{QT} dt \qquad (10.14)$$

where T is the time period of the wave. This expression can be integrated over time to give

$$E = E_0 \exp\left\{-\frac{2\pi t}{QT}\right\} \qquad (10.15)$$

with E_0 being the initial value of the energy. It is seen that the dissipation is less the higher the Q value. Indeed, if $Q = 0$ there is no possibility of a wave at all; as $Q \to \infty$ so the dissipation becomes smaller. The process would in practice be observed in terms of a wave amplitude, A. This is the square root of the energy so then

$$A = A_0 \exp\left\{-\frac{\pi t}{QT}\right\} = A_0 \exp\left\{-\frac{\omega t}{2Q}\right\}, \qquad (10.16)$$

where the second form introduces the angular frequency ω. The same analysis can be repeated but concentrating now on the spatial characteristics to give, for propagation with wavelength λ in the x-direction

$$A = A_0 \exp\left\{-\frac{\pi x}{\lambda Q}\right\}.$$

The values of Q derived from these formulae are for one type of wave only, either P or S. In general it is found that, for a given Q value, the P waves travel with less dissipation than the corresponding S-waves and the $Q_P \approx Q_S$. Q values for silicate vary between 10^2 and 10^4, dependent on composition and physical conditions.

10.2.5. Seismic tomography

A modern development of seismic studies to the Earth must be mentioned because it will eventually be available for use in all the terrestrial planets. This is seismic tomography which attempts to view the interior of the planet is in a way comparable to the body scan familiar in medical applications. The process is only applicable if an enormous number of reliable travel times over the main volume of the planet are available. To achieve high accuracy it is necessary to have data from a very densely spread body of observatories. The procedure is highly mathematical and only the first steps have been taken in its application to the Earth but already its worth is being demonstrated.

The aim is to recognise small variations of the wave speeds over very small volumes, ideally over very local volumes. The speed will change partly due to variations of the composition, affecting the mean elastic constants, and partly due to local temperature variations. Isolating one from the other will be a highly complex task requiring a large number of data for contiguous regions. It is necessary to have established an agreed mean seismic model of the planet so the local variations can be recognised against this framework. Huge quantities of seismic data are required to establish the framework model and such a database is now available for Earth. The current preliminary applications to the interior of the Earth can show variations in the wave speeds by a few percent but over quite large areas. This work of refining the results and restricting the volume under discussion is for the future but eventually is likely to be able to act as a fine thermometer.

10.2.6. Rotation included

The formulae are slightly modified if the body is rotating about a fixed axis with angular velocity ω. The pressure gradient must include a contribution of the effects of rotation which gives the contribution $\frac{2}{3}\rho\omega r$. Therefore, (10.9) is changed to the form

$$\frac{dP}{dr} = -\frac{4\pi G\rho^2(r)r}{3} + \frac{2}{3}\rho\omega r = -\rho r\left[\frac{4\pi}{2}G\rho(r) - \frac{2}{3}\omega\right]. \qquad (10.17)$$

Then, as before

$$\frac{dP}{dr} = \frac{dP}{d\rho}\frac{d\rho}{dr} = \frac{\kappa}{\rho}\frac{d\rho}{dr},$$

giving the density gradient in the body but now including the rotation. The effect of rotation will be small for planets of the Solar System, and will cause a flattening of the poles. This elliptic shape will be reflected in the figure at all depths.

It would seem from (10.17) that the stable rotation is possible only so long as $\frac{4\pi}{3}G\rho(r) > \frac{2}{3}\omega^2$ giving a positive pressure gradient downwards. With $\omega = \frac{2\pi}{T}$, T being the period of rotation, this means $T > \sqrt{\frac{2\pi}{G\rho}} \approx 7.5 \times 10^3$ s $\approx$ 2 hours. Correspondingly, $\omega < 8.4 \times 10^{-4}$ s^{-1}. This elementary condition is fully met in the planets of the Solar System.

10.3.   The Near-Surface Interior Region

The arguments developed so far are applicable to the interior but the surface regions can also be investigated in a similar way.

10.3.1.   Surface waves

Seismic waves can be guided along the surface layers of the planet rather by analogy with waves in a wave guide. Another analogy is with waves of length λ on the ocean where the disturbance decreases with depth until it is small at about $\lambda/2$ below the surface. The waves do not penetrate into the interior. They result from near surface earthquake disturbances on Earth and could also arise from meteor impact on other planets.

There are two types of surface wave (or arguably three if the so-called Stonely waves are included). One is named after Lord Rayleigh who predicted it in 1887. These Rayleigh waves are sometimes descriptively called ground roll waves in exploration seismology. They show a vertical particle motion in the plane containing the propagation direction. In having a retrograde elliptical motion they are analogous to the ocean waves at sea. The amplitude decays exponentially with depth, the longest wavelengths penetrating furthest, as for sea waves. The second type of wave was predicted by A. E. H. Love in 1911 and is rather more complicated. These Love waves involve a horizontal motion transverse to the direction of propagation. They occur in a layered structure and in the special case when the shear velocity in the overlying layer is greater than that in the layer below. Both the

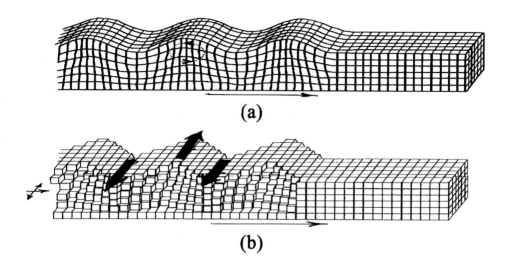

Fig. 10.2. The surface waves. (a) Rayleigh wave being a vertical elliptical motion in the plane containing the direction of motion; (b) Love wave involving horizontal motion.

Rayleigh and Love waves are dispersive, that is their velocities depending on the wavelength. The wave profile changes with the distance, the phase travelling with a phase velocity characte ristic of a particular frequency. The wave energy travels with the greater group velocity which again depends on the frequency of the wave. Usually on Earth the Love group velocities are greater than the Rayleigh group velocities. The effects of these waves at the surface are greater than the effects of the P- and S-waves. This makes then dangerous for the surface conditions but does not allow them to provide information about the interior of a planet. The two wave forms are shown in Fig. 10.2. The third type of energy propagation mentioned above, the Stonely waves, involve the energy propagation along the boundary between two media. They are not dispersive.

10.4. Free Body Oscillations

Every body has a fundamental frequency accompanied by a set of harmonic frequencies and will vibrate if the body is subjected to an appropriate impulse which excites those frequencies. A drum will vibrate and a car will vibrate especially at certain speeds. In exactly the same way a planet with rigidity will vibrate if subjected to an impulsive energy source (for instance, an earthquake). The suggestion that an elastic planet might show whole body oscillations was apparently first examined in detail by A. E. H. Love in 1911, in relation to the Earth. The feature remained a speculative theory until 1960 May 22 when the oscillations excited by the strong Chilean earthquake on that day induced oscillations of an amplitude that could be measured.

Theory predicts two independent types of free oscillations, one toroidal or torsional oscillations, denoted by T, and spheroidal oscillations, S. These have analogies with the Love and Rayleigh waves of seismology: the free oscillations represent the combinations of the wave modes. The T-oscillations are perpendicular to the radius vector and so lie within surfaces of concentric spheres within the body. The simplest S-oscillation is a purely radial motion but the general oscillation has also a tangential component. Both T- and S-oscillations have overtones (harmonics) as in a musical instrument so the general oscillation profile can be very complicated. For Earth the maximum periods are of order 2500 s (40 minutes) for the T mode and about 3200 s (about 55 minutes) for the S mode. The times for seismic waves are lower by a factor of about 20. The simplest modes are portrayed in Fig. 10.3.

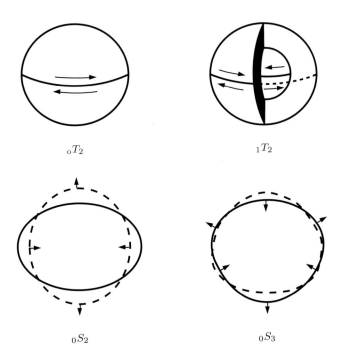

$$_0T_2 \qquad\qquad\qquad _1T_2$$

$$_0S_2 \qquad\qquad\qquad _0S_3$$

Fig. 10.3. The normal modes for T and S oscillations. In $_0T_2$ the two hemispheres oscillate in opposite directions. In $_1T_0$ the other region oscillates as for T_2 but the inner hemispheres oscillate in the opposite directions. The mode $_0S_2$ shows oscillations involving two axes while the mode $_0S_3$ involves three axes.

10.5. Empirical Equations of State

Efforts have been made in the past to replace the numerical specification of the bulk modulus that results from observation by a theoretical expression. In this way (10.9) could be used directly. This would have obvious advantages in developing the theory if an accurate expression could be found. The disadvantage of this theoretical approach is the extreme complexity and divergence of planetary materials. Nevertheless, even a very approximate expression would be useful if it were simple. An approach of this type that has proved useful in certain problems is that of F. D. Murnaghan.

The starting point is the expression (10.8) for the bulk modulus, κ. It is known that κ depends on the pressure so it is possible to represent its behaviour over at least a restricted range of pressure by a power series expansion in terms of the pressure

$$\kappa = \kappa_0 + BP + CP^2 + DP^3 + \cdots .$$

where κ_0 is the bulk modulus for zero pressure and B, C, D, ... are coefficients that will depend directly on the chemical composition and so on the pressure and the temperature. Using (9.8b) we can write

$$\rho \frac{\partial P}{\partial \rho} + \kappa_0 + BP + CP^2 + DP^3 + \cdots . \tag{10.18}$$

This expression can be integrated to give P as a function of ρ, that is to yield an essentially isothermal equation of state for the material. There are three cases that can be usefully included here.

(i) The unrealistic case of a constant bulk modulus, so that $B = C = D = \cdots = 0$ and (10.18) gives the simple form for the density:

$$\rho = \rho_0 \exp\left\{\frac{P - P_0}{\kappa_0}\right\}.$$

(ii) For a linear dependence on the pressure we have (10.18) in the form

$$\frac{d\rho}{\rho} = \frac{dP}{\kappa_0 + BP} .$$

If the lower pressure is zero (say, atmospheric), this leads to the expressions for the density and pressure

$$P = \frac{\kappa_0}{B}\left[\frac{\rho}{\rho_0} - 1\right], \quad \rho = \rho_0\left[1 + \frac{B}{\kappa_0}P\right]^{1/B} . \tag{10.19}$$

These expressions were applied to planetary problems by Murnaghan. The bulk modulus itself can be found directly from (9.24) and (9.8b). The result is

$$\kappa = \left(\frac{\rho}{\rho_0}\right)^B \kappa_0 \quad \text{or} \quad \frac{\kappa}{\rho^B} = \text{constant}. \tag{10.20}$$

For terrestrial materials the best value of B is in the range $3 \le B \le 5$. (10.20) gives a not unreasonable representation of hydrogen and helium with $B \approx 2$.

(iii) Finally the quadratic dependence on pressure has been included in calculations by R. A. Lyttleton especially in the calculations for Mars. In this case, (10.18) can be written as

$$\frac{d\rho}{\rho} = \frac{dP}{\kappa_0 + BP + CP^2}. \tag{10.21}$$

The form of the expression for the pressure depends on the relative magnitudes of κ_0, B and C: for terrestrial type materials we take $\kappa_0 > B > C$. Then it is found that

$$P = \frac{2\kappa_0(\rho^a - \rho_0^a)}{(B+a)\rho_0^a - (B-a)\rho^a}$$

where $a = (B^2 - 4\kappa_0 C)^{1/2}$. This equation of state is altogether more complicated for calculations.

It might be thought that equations of state such as these are not of use now that giant computers are available to use laboratory equation of state data directly. This is not entirely correct. Certainly, the computer power is important but it is also important to have a relatively simple procedure for planning larger calculations and for checking that errors have not crept into large computations. It is always important to have a general idea of the structure of the physical system under investigation.

10.6.   Fluid Bodies

The seismic methods do not apply if the body lacks all elasticity. This means these methods cannot be used, even in principle, for the study of fluid bodies. Other methods are less comprehensive.

(a) Surface figure

The fluid body in equilibrium will assume a spherical surface figure, this providing the minimum surface area for a given volume. Rotation will add the Coriolis force which extends the radius of the equatorial regions but decreases that of the poles. The figure is now an oblate spheroid or a slightly flattened sphere with full symmetry about the rotation axis. The degree of flattening increases with speed of rotation. For the members of the Solar System, this flattening is small but this need not be the case for other planetary systems.

(b) The interior

One direct contact with the interior is through the surrounding gravitational field. Another is through the distribution of magnetic energy over the different polar components of the field. In particular, the stronger the non-dipole components, the nearer the source of the field is to the surface. There is no quantitative consensus of the precise origin of a planetary magnetic field so the use of the field for the exploration of the interior is severely limited.

The fluid interior can be expected to have established a mean mechanical equilibrium and the increase of pressure and density with increasing depth can be calculated for various model interiors. The composition of the interior can be constrained by linking the measured surface abundances with those of the other major planets and of the Sun. The measured mean density of the planet will allow the much larger central density to be judged in the light of the modelled mantle above it. The composition will be likely to be similar to terrestrial silicates and ferrous materials. The high pressures at the centre of a Jupiter-like planet will ensure that the rigidity is low and the material is markedly plastic. It is clear that the interior of a large fluid planet is less well known than a rigid Earth-like planet.

A fluid body undergoes vibrations due to processes inside the volume and these cause whole body oscillations. The amplitudes will be small but are as yet unknown and have not yet been detected.[5] Improved measuring techniques using unmanned space vehicles can be expected to allows these oscillations to be detected and used eventually to study the interiors of heavy fluid planets.

---

[5]The full body oscillations of the Sun are well known and widely studied to yield information about the interior. The amplitudes of the solar oscillations are much larger than those expected of a fluid planet.

*

Summary

1. The study of the shape of the planet and its surface features is the study of geodesy.

2. The way the figure of the planet and its gravity are represented is explained.

3. The method of exploring the interior of a rigid planet using earthquake waves is called seismology. The method is explained in a little more detail.

4. The study of the interior is made using longitudinal P-waves and transverse S-waves. The surface layers are explored using the different Rayleigh and Love waves.

5. The P- and S-waves are not dispersive but the surface waves are. This means their propagation speed are different for different wave frequencies.

6. Waves lose energy as they propagate and this loss is represented by the ratio of the energy stored in the wave divided by the loss of energy during one cycle. This is expressed in terms of the Q-factor.

7. Recent developments of seismology are towards the recognition of different conditions of temperature and density within very restricted regions of the interior. This is seismic tomography.

8. Whole body oscillations of the Earth, first predicted about 100 years ago, were first detected in the Chilean earthquake of 1960 May 22. They are now detected in very strong earthquake events. They are related directly to the P- and S- wave modes and their study allows information to be gained about the conditions inside.

9. Efforts have been made to develop a theoretical equation of state for direct use in the theory. This has not been successful but an empirical expression has been developed for use in some cases. This work is described.

10. The problem of exploring the interior of a fluid body is considered.

11

The Solid Earth

Fig. 11.1.   The Earth from Space. (NASA: Apollo 17)

The vision is of blue water, brown land, white cloud and white ice.
The continent of Africa with Arabia is in the centre, flanked
by the Indian and Atlantic oceans. The ice of Antarctica
can be seen clearly at the southern pole.

11.1. General Parameters

The Earth is probably the most extensively studied body in the Universe and is usually accepted to be the prototype for any planet which has developed technologically capable life forms. It has several central advantages of which the most obvious are:

(i) The three phases of water can exist simultaneously on the planet. This provides a general control mechanism for damping out rapid temperature fluctuations on the surface.

(ii) It carries a large reservoir of water in its oceans which outweighs its continents as shown in Table 11.1. The precise ratio of these values depends on the amount of ice locked up at the poles. The Earth is moving out of an ice period at the present time so the quantity of ice is decreasing. The change in water level worldwide with all the ice melted would be an increase of a few hundred metres, not enough to affect these figures significantly. It would be enough to affect the population. The most densely populated areas are on and near the coasts where the elevation above sea level is generally small. The Earth is in an essentially circular stable orbit about the Sun as heat source so that its climate is stable.

Table 11.1.   The physical properties of the continents and the oceans.

	Area (km^2)	% Earth	Mass (kg)	Mean height/depth (m)	Max height/depth (m)
Land	1.49×10^8	29.2	–	840	8,840 (Mt. Everest)
Ocean	3.61×10^8	70.8	1.42×10^{21}	3,800	10,550 (Mariana trench)

(iii) The surface region contains a range of minerals for ultimate technological development.

(iv) Some would add that the Earth has a Moon (the largest satellite relative to its primary one in the Solar System, with the exception of Pluto/Charon) able to raise tides and so to stir gently the water of the oceans.

(v) The Earth is not quite a sphere with an equatorial radius $R_e = 6378.160$ km and a polar radius $R_P = 6356.775$ km giving an ellipticity (polar flattening) of $0.003353 = 1/298.24$. Its volume is 1.083×10^{21} m^3. It has an inertia factor of mean value 0.331 showing that its mass is concentrated towards the centre and rather more so than could be expected by the simple compression of the constituent materials.

The Earth spins about a fixed axis with angular velocity 7.292 rad/s. The obliquity of the orbit (the angle its rotation axis makes with normal to the plane of its orbit round the Sun) is 23°.45.

The external gravitational field has a potential V of the form

$$V = -\frac{GM}{R}\left[1 - \sum_{n=2} J_n \left(\frac{R_E}{R}\right)^n P_n(\cos\theta) + \cdots\right]$$

$$+ \text{terms involving longitude} \tag{11.1}$$

where R is the distance from the centre, θ is the co-latitude ($\pi/2$–latitude) and P_n is the Legendre polynomial of degree n. R_E is the radius of the Earth. The five J coefficients have the values

$$J_2 = 1.0827 \times 10^{-3}, \quad J_4 = -1.6 \times 10^{-6}, \quad J_6 = 6 \times 10^{-7}$$

$$J_3 = -2.5 \times 10^{-6}, \quad J_5 = -2 \times 10^{-7}. \tag{11.2}$$

The heat budget of the Earth is largely that received from the sun but also that which flows through the surface from the interior. The mean heat received from the Sun is that from a body of temperature 5,800 K at a mean distance of 1.49×10^8 km. This provides a heat flux of 1.4×10^3 W of which 1/3 is reflected back into space (forming the albedo) and 2/3 is absorbed by the Earth. With atmosphere, this maintains a temperature of 280 K but in the absence of the atmosphere it would be about 270 K, or slightly below the freezing point of water. It is interesting that the presence of the atmosphere allows life to exist on Earth. By comparison, the measured mean heat flow through the surface of the Earth is 6.3×10^{-2} W, which is a factor larger than 10^{-4} of the solar value. It is believed that the Earth is cooling from a hotter past but it is not proven although radioactivity is declining and this would be the most likely source of new heat. The solar heating is not constant. Variations in the orbit about the Sun, of the direction of the spin axis, the albedo and variations of the solar constant will all affect the quantity of solar radiation and also the temperature at the surface of the Earth. This is a complex subject which has just begun to be treated in serious detail.

The Earth is the source of a weak intrinsic magnetic field which is largely of dipole form. The molecules in the upper atmosphere are ionised by the ultraviolet radiation from the Sun forming an ionosphere. This also is associated with an even weaker magnetic field than the intrinsic field. These features will be investigated more carefully and expanded now.

11.2. The Interior Seismic Structure

This has been investigated over the last century using seismic waves and is now well established. The values are accurate to better than 1%. The raw plot is of seismic wave speeds and this is shown in Fig. 11.2 which shows both the P-wave and S-wave speeds. It is clear that there are six regions, all covered by the P-waves but only three by the S-waves. The first is a region from the surface down to about 30 km and called the crust. The exact depth is variable with location on the surface but seismic analyses conventionally begin at a depth of 33 km everywhere. The second region is between 33 km and about 400 km depth called the upper mantle. Below this is a transition region between the depths of 400 to 600 km which lies on top of the lower mantle covering the depths between 600 and 2898 km. These regions see an increase in the P-wave velocity V_P typically from about 8.2 km/s at the 33 m depth to 13.5 km/s at the lower depth. The corresponding S-wave speeds are lower, typically 4.63 km/s $< V_S <$ 7.30 km/s. Both P- and S- waves propagate through the upper and lower mantle regions. The region stops discontinuously and the wave speeds decrease discontinuously as well.

At a depth of 2898 km V_P drops to about 8.10 km/s but $V_S = 0$. Indeed, S-waves do not propagate below this depth at all, which is called the outer core and is the fifth region defined by the waves. The sixth region begins

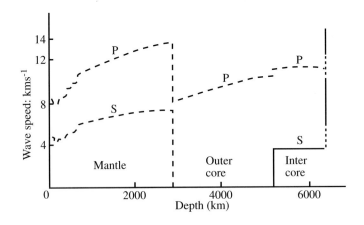

Fig. 11.2. The wave speeds for P- and for S-waves within the Earth.

at a depth of 4500 km and is called the inner core. V_P increases by about 4.7% discontinuously at the outer/inner core boundary surface and increases slightly to the centre of the Earth. Although $V_S = 0$ in the outer core indirect evidence shows that $V_S \neq 0$ within the inner core itself. This is interpreted as showing that the outer core cannot sustain shear, and so is essentially liquid in the normal sense, whereas the material of the inner core is able to sustain shear and so is solid within normal classifications. More light is cast on these regions when the seismic data are converted to density values. The result is included in Fig. 11.2. It is seen that the density increases from the surface to centre with variations in the upper mantle region, a discontinuous increase at the lower mantle/outer core boundary and a slight increase at the outer/inner core boundary. The density range is $\rho(33 \text{ km}) = 3{,}320 \text{ kg/m}^3$ and $\rho(\text{centre}) = 17{,}000 \text{ kg/m}^3$. The pressure is also plotted in Fig. 11.3: this is a continuous curve with values within the range $P(33) = 9 \times 10^8 \text{ N/m}^2$ to $P(\text{centre}) = 3.89 \times 10^{11} \text{ N/m}^2$ (which is about 10^6 atmospheres). The acceleration due to gravity is also shown in the figure which is seen to remain sensibly constant down to the mantle/core boundary below which it steadily decreases to zero at the centre.

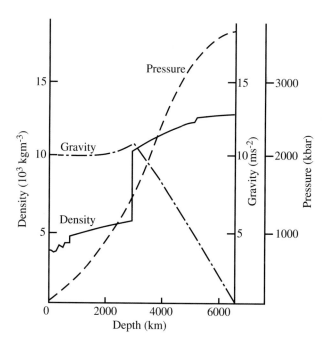

Fig. 11.3. The density, pressure and the acceleration due to gravity within the Earth deduced from measured values of seismic wave speeds.

Table 11.2.　The masses of the various divisions of the Earth's interior.

Region	Depth range (km)	Mass, M (10^{24} kg)	Relative mass M/M_E
Continental crust	0–50	0.022	0.0037
Oceanic crust	0–10	0.006	0.001
Upper mantle	10–400	0.615	0.103
Transition region	400–650	0.448	0.075
Lower mantle	650–2,890	2.939	0.492
Outer core	2,890–5,150	1.840	0.308
Inner core	5,150–6,370	0.102	0.017
Total	0–6,370	5.97(2)	1.00

The density regime can be used to estimate the masses of the various divisions within the Earth. Such a list is given in Table 11.2. The relative masses of the different regions is interesting. For instance, the lower mantle is the most massive but this region and the outer core between them account for 80% of the Earth's mass. The inner core accounts for 1.02×10^{23} kg which is comparable to the mass of the Galilean satellite Callisto. The density of the Earth is, however, some 7 times that of Callisto, reflecting a very different composition. Again, the mass of the Moon is about 3 times the mass of the combined crust while the mass of Pluto is about half that of the crust. The mass of the upper mantle is very closely that of Mars (and the continental land area is very closely the same as the total surface area of Mars). These comparisons are interesting but probably not significant.

With the density distribution known the elastic moduli themselves can be isolated. These are plotted in Fig. 11.4. The bulk modulus κ acts throughout the volume but the shear modulus μ becomes zero at the mantle/ core boundary. Not only is there a phase change at this boundary: the discontinuous change in κ at the boundary suggests a change of material which is confirmed by the density data. The crustal region is seen to have special properties.

11.2.1.　Seismic tomography

The data used to construct Fig. 11.2 have been augmented to reveal a deeper structure to the Earth. Several graphs of this type have been drawn on the basis of a great number of seismic data and the agreement between them is very good, probably better than 1%. An amalgam of the data have formed

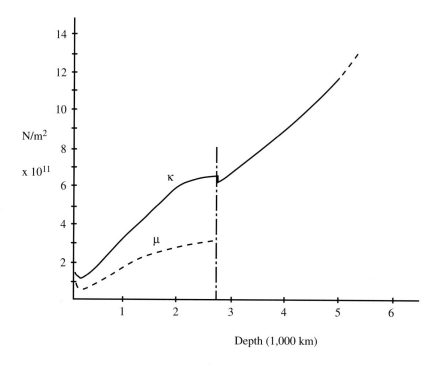

Fig. 11.4. The elastic moduli for the material of the Earth as function of the depth κ represents the bulk modulus and μ the shear modulus.

the preliminary Reference Earth Model (PREM) against which variations in the data for specific regions can be compared. Such a comparison is called a seismic tomograph of the new science of seismic tomography.

Huge quantities of data are now available and huge quantities are required for this development. It aims to isolate differences in the *P*- and *S*-wave speeds of as little as a few percent against the PREM. The variations can be due to differing temperatures in the locality or to chemical variations (affecting the wave speeds) or to both. It is usually presumed that the effect of temperature on the wave speeds is more important than the effects of chemical variation. The method is in its infancy. It requires the manipulation of very large quantities of data and has become possible with the recent development of computer technology.

11.3. An Active Structure

The Earth is not a static body. The (small) outflow of heat and the larger outflow of hot lava from volcanoes show that the internal temperature is high. Unfortunately, the internal equilibrium of a planet does not depend upon the temperature (see I.4) so it is very difficult to obtain data about the heat content. Again, earthquake activity shows the Earth's interior is susceptible to rearrangement. The behaviour of materials in the interior is particularly dominated by the size of the planet and especially through the pressure. The material is in a mean hydrostatic equilibrium and the pressure forces dense agglomerations of material which tend to sink. It is this that probably dictates the thickness of the crust itself.

11.3.1. Mantle material

Material from the top of the upper mantle can be examined in eroded mountain belts. The major components are found to be the two minerals, olivine and pyroxene, both magnesium, silicon, oxygen minerals, conatining some iron. Explicitly, olivine is $(Mg,Fe)_2SiO_4$ and pyroxene is $(Mg,Fe)SiO_3$ with densities rather less than 3,000 kg/m^3. The iron is no more than a minor constituent. It abundance in the cosmic abundance suggests it is located elsewhere inside — and this will be found to be the inner core. The most abundant material from volcanoes is basalt, again a silicate but now containing not only magnesium, silicon, oxygen and iron but also aluminium and calcium. The combination has a lower density than olivine and pyroxene and so tends to remain on the surface. The ocean floor is covered with basalt and some accumulations break surface to form islands, of which examples are Iceland and Hawaii. The widespread occurrence of basalt is associated with a widespread occurrence of volcanoes.

In the lower mantle the pressure becomes sufficiently high for olivine to suffer a transformation to different minerals, periclase (MgO) and a high pressure form of estatite. The densities are higher than the low pressure forms. Seismic evidence suggests the presence of FeO, showing iron to be a growing constituent with depth.

11.3.2. Mid-ocean ridges and ocean trenches

A major mountain chain containing a vast network of volcanoes exists under the oceans. The chain is some 40,000 km long and girdles the Earth. At places, the peaks are 1 km high or more. They were discovered only some 40 years ago because they are hidden by the deep oceans. The study of the deep oceans is very new even though they cover a major portion of the Earth's surface. Small regions of the chain appear on land at one or two

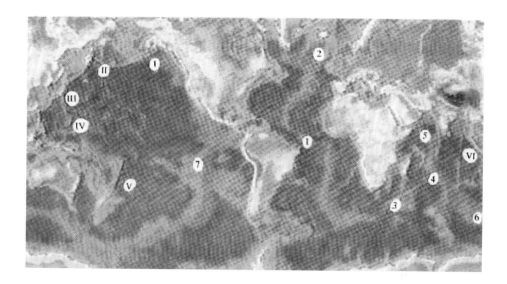

Fig. 11.5. The ridge and trench system of the Earth. The system is global but some parts have been named in the past. The trenches are largely in the Pacific while the ridges are elsewhere. **Ridges**: 1. Mid-Atlantic — north and south; 2. Reykjanes; 3. Southwest Indian Ocean; 4. Mid-Indian ocean; 5. Carlsberg; 6. Southeast Indian Ocean; 7. East Pacific Rise. **Trenches**: I Aleutian; II. Kuril; III. Japanese; IV. Mariana; V. Tonga; VI. Java. (NASA Landsat)

places where they appear as a gigantic crack in the surface. One place is the rift valley in East Africa which appears inactive. An active crack is in Iceland where the island is divided by the fissure. In fact, material emerging from the fissure is forming new land which increases the area of the island. It is now clear that the structure covers the whole Earth in a way shown in Fig. 11.5. The map also contains the location of trenches in the ocean. The light blue regions represent higher ocean floor. It is interesting to notice that Fig. 11.5 was obtained by radar telemetering the ocean surface, which has a profile determined partly by what is below. An interesting comparison is the distribution of earthquake foci, shown in Fig. 11.6. The similarity between the ridge-trench system and the distribution of near surface earthquakes is very striking. The dominance of the rim around the Pacific land (the so-called Pacific Rim) is shown clearly by the plot of deeper earthquakes. No earthquake source has been reported below a depth of 700 km.

The ridge system is the source of basalt from the upper mantle which forms new sea floor continuously. This is, indeed, the way that the surface material is formed. New hot material moves perpendicularly away from the

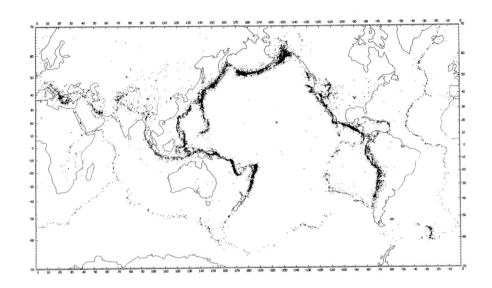

Fig. 11.6.   The distribution of earthquake epicentres in the depth range 0–100 km for the period 1961–7. (Seismological Society of America)

ridge at a rate of a few centimetres per year, cooling as it moves. A lower temperature means a higher density and so a sinking into the underlying crust material. The ridge is, in this way, uplifted from the general ocean floor. The system represents an efficient way of cooling the Earth's interior. The heat of the ridges has been found to have an extraordinary association, entirely unexpected and found only some thirty years ago. It contains a series of high temperature vents where very hot mineral rich fluid materials are forced out into the surrounding ocean under high pressure. These unique conditions, at a depth of 3 m or so, have given rise to a living eco-system involving a range of living things including giant sea worms, shrimps and other creatures living on microbial life there. Light from the Sun does not penetrate so far down and the pressure is as high as 100 atmospheres. There are other regions of greatly enhanced salinity covering a large area. The exploration of these conditions is only now beginning and present problems as severe as those of space exploration. The essential thing here is that the living systems receive no energy from the Sun and rely entirely on the energy emitted by the Earth itself. These creatures are still based on DNA but their evolutionary relation with surface creatures is entirely unknown. In particular, it is not known whether they pre-date surface life or not. It is interesting to review the possibility that living materials had to adapt to lower (surface) pressures and temperatures rather than the other way around. Or did the evolution of the two regimes occur independently?

11.4.   Plates and Plate Tectonics

Surface material is formed at the ridge system but the surface of the Earth remains constant. This can only happen if surface material is being returned to the inner Earth at the same rate that it is being formed. These regions are the subduction zones and are placed substantially away from the mid-ocean ridges. Between them is a layer of surface, some oceanic and some continental, which moves from the regions where the surface is formed to the locations where it is destroyed. These surface layers are called plates (they are more or less flat) and the movement is an example of plate tectonics. There are eight large plates and about 20 smaller ones, that move one relative to the other with speeds of some 5 cm per year. They are formed at the ocean ridge system and pass into the Earth at the subduction zones where one plate rides over another. These regions are the deep sea troughs. Plates can also slide past each other and one example is the San Andreas fault on the Californian coastal region (where the Pacific plate is sliding past the Atlantic plate — see Fig. 11.8). A schematic representation of the total plate system is shown in Fig. 11.8. This figure can be viewed remembering Figs. 11.6 and 11.7.

This recycling process, where hot rock appears at the ridges and cold sea floor passes back to the interior, is the main mechanism for the loss of heat by the Earth. It involves volcanism, some 90% of which occurs under water.

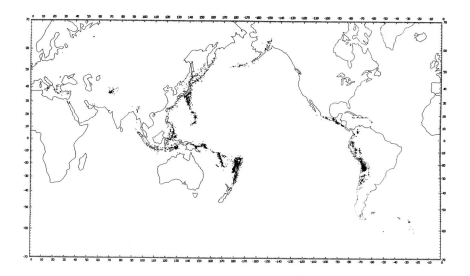

Fig. 11.7.   The distribution of deep focus earthquakes (100–700 km) 1961–76. (Seismological Society of America)

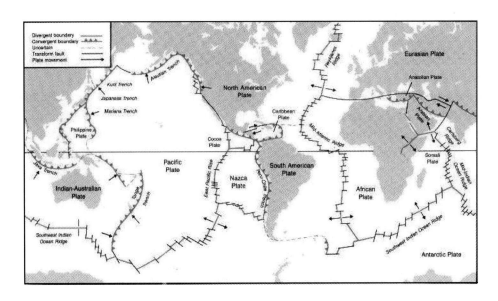

Fig. 11.8.   The worldwide plate tectonic system. The surface material is formed at the ridge regions and passes back to the interior at the trench regions.

The rest is found on continental masses and is called mid-plate volcanism. It is the source of new continental area. The activity is rarely continuous but more often the volcanic activity is spaced by intervals of quiet. The lava flows may be limited but geological history shows several examples of flash lava flows where vast quantities of material was ejected from inside over a very short time. The collision of plates gives rise to mountain chains which then suffer erosion by wind and rain. The collisions between the Indian and Asian plates some 60 million years ago have given rise to Mt. Everest, which is still very slowly gaining height.

11.5.   The Inner and Outer Cores

The densities of the two core components require a major constituent to be ferrous material, that is primarily iron with nickel and cobalt. The density of the out core is rather less and probably contains non-ferrous elements, the most likely being oxygen and/or sulphur. This surmise is based partly on the cosmic abundance but also upon the chemistry of elements in the proportions of the chondritic Earth. The outer core is unable to withstand shear (is unable to transmit S-waves) and so must be effectively a liquid as is understood in everyday terms. The surface spread of seismic waves is affected strongly if a central region cannot carry a particular type of wave, leaving a dead zone where seismic waves are not found. In practice, such a dead zone does not occur and this can be explained if the inner core is able to carry transverse waves. This means it is effectively solid. More precisely it can be said that the outer core has a low associated Prandtl number since the viscosity is low, whereas the inner core has a high Prandtl number, in common with the mantle. This strange situation is the result of the particular chemical composition, pressure and temperature in the central region. The melting curve for the material of the outer core falls below the temperature there whereas that for the inner core lies above it. This gives a constraint on the possible composition of these two regions. The inner core is likely to be essentially pure ferrous material.

The core rotates with the rest of the planet but there is some seismic evidence to suggest that the outer core rotates slightly faster than the mantle. This will have consequences to be discussed in Chapter 17 in the discussion of the westward drift of the non-polar magnetic field.

11.6.   A Dynamic Earth

The Earth is an active body supporting a range of physical processes. It is still cooling down from its formation. It contains radioactive elements but their natural abundance leads to the conclusion that the surface rock abundance will contain essentially the whole planetary budget. This will not affect the main volume of the planet.

The tectonic plate system can be interpreted properly only if hot magma is rising at the mid-ocean ridges while cold rock moves into the subduction zones. The temperature in the crust is above the melting temperature of the rock showing that the internal temperature is high. The compression within the Earth will ensure a central temperature of the order of 5000 K which is consistent with both a liquid outer core and a solid inner one. A temperature gradient within a body caused by compression alone is an adiabatic gradient, any very slow movement of fluid up or down bringing the fluid to the same temperature as the new surroundings. A flow of heat arises from a difference of temperature maintained by differences of heat: an adiabatic gradient does not involve such a difference. A heat flow must, therefore, involve heat beyond the adiabatic and so must be to some extent super-adiabatic. If a critical value of the Rayleigh number is passed, fluid will flow in order to move nearer the adiabatic condition. This is convection.

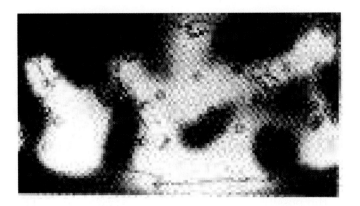

Fig. 11.9. A typical seismic tomograph of the interior of the Earth for the depth of 250 km. This represents the perturbation of the shear wave in relation to the Preliminary Reference Earth Model portrayed in Fig. 11.2.

The super-adiabatic temperature gradient is likely to be 1 or $2°C$ for the mantle. The Prandtl number perhaps of order 10^{40} showing the dominance of viscosity over heat conduction, and the critical Rayleigh number (perhaps 10^4 there) is surpassed (remember I.5). The result is a slow convection with a period of perhaps 200 million years. This, coincidentally, is the tine for renewal of the ocean floor. It is difficult to predict whether the convection covers the entire mantle or whether the observed transition zone at some 650 km depth marks a boundary between upper and lower mantle convection cells. Evidence from seismic tomography shows a cold (high velocity) subduction of materials down to about 600 km suggesting the latter possibility is perhaps the correct one. One such tomograph is shown in Fig. 11.9 (see Sec. 11.2.1). The dark regions represent an increase of the average shear (S) wave velocity by about 2% while the light regions show a decrease of S-wave velocity by about the same amount. The centre of the figure is over the Pacific Ocean: Australia and Asia is to the left while the Americas are to the right. The line at the bottom of the figure is the Antarctic coastline, under Mecartor projection.

On these ideas the interior becomes a region of convections at different scales all driven by the loss of heat by heat transfer. The rotation of the Earth will play only a minor part. A region accepts heat from below and its temperature is raised. This reduces the viscosity and allows a stronger convection. Heat is transported away as a consequence and the temperature falls. The viscosity increases and the passage of heat is reduced. The volume

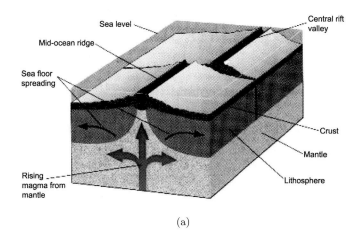

(a)

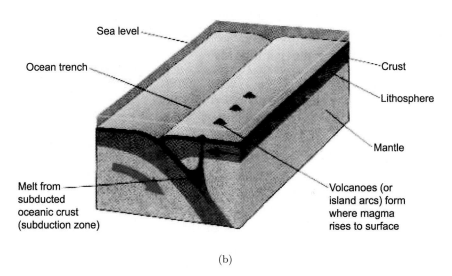

(b)

Fig. 11.10. The crustal motions showing (a) the creation of ocean floor (sea floor spreading) and (b) the elimination of ocean floor (subduction).

heats up again, reducing the viscosity and so on. In this way the interior is a closely controlled heat funnel. Strong convection can be assigned to the outer core region where the material is unable to withstand shear. Weaker convection is found in the mantle but on a different scale. The crust also shows convection with the upswelling which carries heat breaking surface at the mid-ocean ridges while the cold surface material moves downwards at the trenches. No ocean floor is older than about 200 million years. The crustal motions are shown in Fig. 11.10.

The association between continental volcanoes and the plate system can be appreciated from the figure. An essential feature is the movement of ocean water down the trench. Earthquakes result from the resulting slippage of material deep down and heating provides molten magma which rises to form continental volcanoes. This is the source of more land. The re-cycling of surface material with a period of about 200 million years is a characteristic of the Earth. Some continental regions have not been sub-ducted from the very earliest times and this gives the Archaen rock samples found today. The very earliest times have been obliterated by the constant activity.

One feature may transcend the divided convection regions and that is the thermal plume. This is a volume of mantle material that rises close to the core/mantle boundary and which rises through to the crust. Evidence for such a hot plume comes mainly from theory and there seems no direct seismic evidence for them. It is often conjectured that Iceland lies on a plume but seismic studies in that area show no evidence for it. The Hawaiian islands are sometimes believed to arise from a plume system which has moved relative to the crust over several hundred years.

11.7.   Comments on the Atmosphere

The Earth's atmosphere is the thin gaseous cover and has a mean sur-face pressure of 1.013×10^5 Pa: the meteorologist calls this 1013 millibars. The surface density is 1.217 kg/m^3. The total mass of the atmosphere is 5.27×10^{18} kg or about one-millionth of the total mass of the Earth. The main constituents are 78% N_2 and 20.9% O_2 (by volume). A list is given in Table 11.3. Water vapour makes the atmosphere wet. There is generally a 50% water cloud covering for the surface. The atmosphere acts as a green-

Table 11.3.   The normal composition of clean dry air for the Earth near sea level.

Gas	%	Gas	%
N_2	78.08	Kr	1.1×10^{-8}
O_2	20.95	H_2	5×10^{-9}
Ar	9.34×10^{-5}	N_2O	3×10^{-9}
CO_2	3.3×10^{-6}	CH_4	2×10^{-8}
Ne	1.82×10^{-7}	Xe	1×10^{-9}
He	5.2×10^{-8}	CO	1×10^{-9}

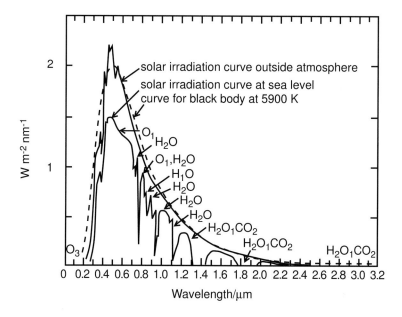

Fig. 11.11. The solar spectrum at the distance of the Earth and the absorption regions with the causes. The black body curve for the Sun's surface temperature is included for comparison.

house in raising the temperature of the surface, T_m, above that which would be found, T_a, were the atmosphere not present. The mean energy flux from the Sun at the mean distance of the Earth and normal to the solar beam is 1,370 W^{-2} giving $T_m = 288$ K. Correspondingly, $T_a \approx 256$ K. The mean temperature of the atmosphere is 245 K. Without the atmosphere the surface would be below the ice point while with it the temperature is above the ice point. In this way, the atmosphere allows life to exist on the planet. The composition of the atmosphere has changed due to natural circumstances over geological history but is also changing now partly through human intervention. It is worth remembering that pollution from living creatures affects the atmospheric composition quite separately from industrial installations. The solar spectrum of the incident radiation is given in Fig. 11.11. Shown in the figure are the absorptions by atmospheric constituents. Not all the energy that enters the upper atmosphere from the sun reaches the surface. The planet reflects incident radiation and the ratio of the reflected to incident radiation is called the albedo, A. For the Earth, $A = 0.30$ meaning that the flux entering the atmosphere is 913.3 W/m^2. Of this about 20% is absorbed by the atmosphere leaving some 730.6 W/m^2 that actually reaches the surface.

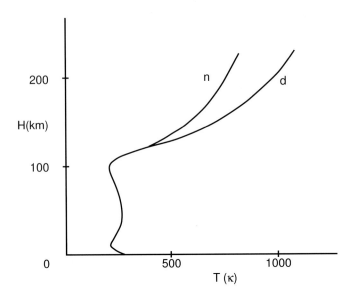

Fig. 11.12.   The temperature profile with height in the Earth's atmosphere. There is a distinction above about 100 km between night and day temperatures in the upper regions.

Fig. 11.13.   Moonrise over the Earth. Note how thin the atmosphere is. (NASA)

The atmosphere extends outwards with ever decreasing density until it is too tenuous to detect as a separate entity. The simplest model is an ideal gas in static equilibrium, the pressure decreasing exponentially with height. A measure of the decrease is the scale height, H, which is the increase in altitude necessary to reduce the initial value by the exponential factor $e = 2.74$, that is the pressure decreased by the factor 0.333. In the lower atmosphere $H \approx 8.5$ km, or about the height of Mt. Everest. The density becomes so low in the upper atmosphere that atoms are essentially free, the atmosphere not behaving like a continuum fluid. At a height of about 100 km the pressure is lower than that attainable in normal laboratory discharge tubes. (See Fig. 11.12.)

The atmosphere might at times appear to be a very powerful body, and indeed it can be, but its basic fragility and finiteness is very well represented in Fig. 11.13. This shows the Moon rising over an Earth of great beauty. The merging of the atmosphere into the surrounding space is clearly seen. It was photographs such as these that first made clear the vulnerability of the atmosphere to man-made pollutants and drew attention to the need to treasure our atmosphere.

*

Summary

1. The Earth is a unique planet in the Solar System for several reasons but especially because of its substantial quantity of surface water. The oceans cover a little more than 70% of the Earth's surface and have a mass of about 1.42×10^{21} kg.

2. The interior structure has been well explored using seismic waves. P-waves penetrate throughout the volume but S-waves are not able to propagate in a thick ring between about 1/4 and 1/2 of the radius of the Earth.

3. This gives the general internal structure. A central solid core covering the central 1/4 of the volume is enclosed by a liquid region out to about 1/2 the radius. This region is called the core.

4. The core is composed largely of iron/ferrous materials with perhaps a small amount of sulphur or oxygen.

5. The core is enclosed in a silicate mantle extending nearly to the surface.

6. The surface crust is some 8–60 km thick and contains the lightest rocks.

7. The crust has a worldwide plate tectonic system where continental masses are moved by the upwelling of mantle materials.

8. This is centred on a worldwide ridge and trench system called the mid-ocean ridge.

9. The mantle also has a structure. The majority of earthquake sources (focus) lie within a depth of 100 km. The number of (deep focus) earthquakes in the depth range 100–700 km is much smaller. An earthquake has not been found below 700 km.

10. This division is reflected in seismic wave velocities and is evidence for an upper and a lower mantle with a division at a depth of 640 km. This division is believed to be the effect of mineral change of phases under pressure rather than a chemical boundary.

11. The atmosphere is composed primarily of nitrogen (78%) and oxygen (21%) with a range of other elements at the level of a few parts per million (ppm).

12

The Planets: Mercury and Mars

Mercury and Mars are the two smaller members of the terrestrial planets. Mercury lies very near the Sun and has been observed in detail only by three automatic flybys, at high speed and at a distance in the 1970s. Mars, on the other hand, lies just inside the main asteroid belt and is much better known. It has been the object of orbital surveys and automatic landers and a growing body of data exists for it. Plans are developing for a manned mission at some point in the nearer future but the chances of visiting Mercury are remote. The two planets are reviewed separately now.

Mercury

Mercury is the innermost planet and always lies within 12° of the Sun. This is close when viewed from a distance. It makes direct observations of it from the neighbourhood of the Earth very difficult. Indeed, observation can be made only about 1/2 hour before sunrise and about the same time after sunset. This has left the planet very enigmatic and largely unknown in the past, and this situation still remains true in many ways today.[1]

12.1. Rotation and Temperature

The rotation period of 58.6 days was first established by the Mariner flyby and later confirmed by radar measurements from Earth. There is a link between the rotation period and the orbital period of 87.969 days. During two orbits of the Sun the planet has rotated 3 times. Why this particular commensurability has arisen is not clear. The link with Venus, the nearest

---

[1]Many illustrious observers of the past admitted they had not seen it through a telescope. Your author has.

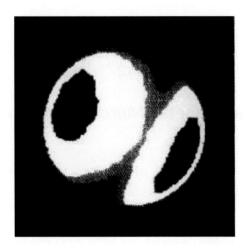

Fig. 12.1. The two hot poles (black regions) of Mercury observed from Earth in radio emission at 3.6 cm radiation. The day emission is on the right and the night emission on the left.

planet, is not significant. One orbit of the Sun by Venus takes 224.695 days which is a little greater than 4 Mercury rotations of 234.4 days.

The result of the particular rotation period for Mercury is that first one and then the other hemisphere is presented to the Sun at perihelion, shown in Fig. 12.1. The $0°$ meridian is chosen to be that sub-solar point on a perihelion at a time in 1950. The other sub-solar point is then the $180°$ meridian. The surface temperature at a hot pole at perihelion is about 740 K ($467°$C), found by microwave measurements made on the Earth. The temperature of the opposite hemisphere then can be as low as 90 K ($-180°$C). Shaded locations at the rotation poles, on the other hand, may well maintain a moderate temperature of around 250 K ($-23°$C, see later). The mean solar irradiance is 9,126.6 W/m^2 or 6.673 times the Earth value. With a visual goemetric albedo of 0.106 (about 1/3 the Earth's value) this gives a mean black body temperature of 442.5 K of the planet. There is a variation of temperature in the orbit arising from the ratio of the perihelial to aphelial distances of 4.6×10^7 km to 6.98×10^7 km respectively.

12.2. Surface Details

As we have seen, the proximity of the planet to the Sun has made the study of its surface very difficult from Earth. Even now it is not possible to use the Hubble telescope, for example, because of the risk of damage to the instrumentation from an accidental direct view of the Sun itself. The first

Fig. 12.2.   A mosaic picture of Mercury constructed from the observations by Mariner 10 of 1974/5. Features as small as 1.6 kilometres can be identified (U.S. Geological Survey). The bald patch signifies no information.

attempt to draw a map was made by Schiaperelli at the end of the 19th century but this was primitive and fanciful. He also attempted to draw a map of the Martian surface. The situation might have been expected to be transformed by the advent of space vehicles but the energy requirements to penetrate the regions nearer the Sun than the Venus are great and a sustained stay near Mercury has so far not proved possible (see Sec. 3.4 on transfer orbits). Nevertheless, Mariner 10 made three flybys of the planet in 1974 and 1975 when 45% of the surface was photographed providing the first reliable surface map. A mosaic constructed by the Astrogeology Team of the U.S. Geological Survey is shown in Fig. 12.2. The bare strip shows lack of information there. The ellipticity of the surface (the flattening) is taken to be zero, that is 0.000. The corresponding factor $J_2 = 6 \times 10^{-5}$ which is a factor 0.055 smaller than the Earth.

The first impression is that the surface is a copy of that of the Moon with craters having diameters from a few metres upwards. The surface divides easily into highlands and lowland plains. The highlands show very heavily

Fig. 12.3.   A region of the highlands showing a scarcity of very small craters. The largest craters have a diameter of about 180 km, the very smallest about 20 km.

cratered regions separated by inter-crater plains but there are relatively fewer smaller craters, with diameters less than 50 km, than on the lunar surface (see Fig. 12.3). There is, nevertheless, a common size-frequency relationship applying to all the terrestrial planets, including Mercury, which suggests a common impact history early on. There is evidence, however, that Mercury suffered a complete resurfacing early on, either by a covering of volcanic lava or from material ejected by impacts. The effect has been to bury the smaller craters accounting for the smaller proportion of small craters relative to the Moon. This might have happened some 4000 million years ago, after the initial period of intense bombardment common to all the terrestrial planets.

The most prominent surface feature is the Caloris basin which apparently resulted from a giant impact in the distant past. This is seen well from a photograph taken by Mariner 10 and shown in Fig. 12.4. It is disappointing that the basin is at one extreme of the picture but fortunate in another way. The anti-podal point to the basin shows a very scrambled terrain with heights up to about 1 km (see Fig. 12.5). It is accepted that the two regions are inked. The impact which formed the basin set up compression and shear seismic waves within the body of the planet (see II.10) which moved through the volume, and were focussed (by the spherical geometry) on the opposite surface distorting it to the presently observed chaotic form. The area of the Caloris region is large with a diameter of about 1340 km and a rim consisting

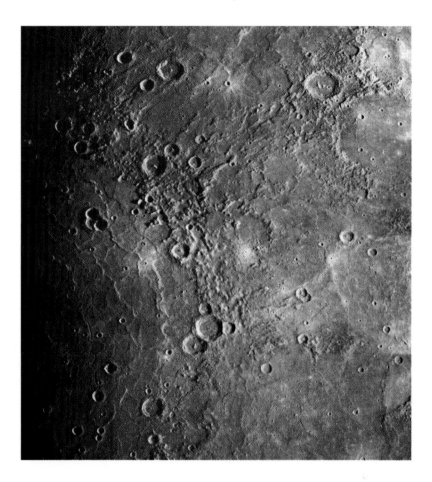

Fig. 12.4. The Caloris basin (left) with centre 30° north, 195° west. East is to the right. The Van Eyck formation is to the northeast and the Odin formation to the east of the basin. The smooth plains lie beyond.

of features up to 2 km high. The area is very nearly that of western Europe. It would seem that the collision very nearly split the planet into two pieces, very early in its history. With a longitude of 195° west it lies close to a hot pole and so is near the sub-solar position on every other perihelion. It represents a high temperature region of the surface — hence its name.

Smooth plains, or lowland plains, lie within and around the Caloris basin. They form the floors of other large craters and are also evident in the north polar region. They are characterised by a low density of small craters and seem to have formed some 3,800 million years ago. The origin of these plains is not yet agreed. One suggestion is volcanic lava flows but there is no immediate evidence of volcanism. Alternatively, they could be molten,

Fig. 12.5.   Some of the scrambled terrain on the hemisphere opposite to the Caloris basin.

or pseudo-molten, ejector from high energy impacts. This is one of many problems which still surround the planet. The surface shows geological faults, some appearing to be thrust faulting being the result of compression within the surface due to cooling.

12.3.   Internal Structure of Mercury

There is no direct evidence of this but the very high density and low mass make the general structure clear. The high density must indicate a very high proportion of ferrous materials and especially iron. This must account for some 75% of the radius. Outside is a silicate mantle of broadly Earth composition. This is consistent with an inertia factor of about 0.33. No more knowledge is likely to be forthcoming until seismic data eventually become available.

12.4.   The Mercury Atmosphere

The planet is too small and the temperature is too high to sustain a stable, dense atmosphere. Solar wind particles striking the surface can become trapped by the magnetic field (see III.19) to form an atmosphere of very low pressure which is continually being renewed. Estimates suggest a pressure of about 10^{-15} bar which is 0.01 picobar. The average temperature has been seen to be 440 K although this can be as high as 725 K on the sunward side. The main constituents are in this order by percent: oxygen, sodium, hydrogen, helium and potassium. These elements account for about 99.5% of the whole. The remaining elements are in trace quantities only and are Ar, CO_2, H_2O, N_2, Xe, Kr and Ne.

Mars

Mars was the Roman god of war and the planet named after him has fascinated mankind for generations. It has sometimes been supposed to be populated by intelligent creatures in a more advanced state than ourselves. At the end of the 19th century Schiaperelli attempted to draw telescopic maps of the surface which showed a complicated canal structure. Modern observations have revealed all this to be entirely wrong. Mars has a barren surface devoid of water now but with a complex geometry which has evolved over the life of the planet. A general view of the planet is shown in Fig. 12.6. The major valley structure is clearly visible across the equator, the largest such feature in the Solar System.

The figure is closely spherical with a flattening (ellipticity) $f = 0.00648$. The moment of inert factor is $\alpha = 0.366$ and $J_2 = 1.9605 \times 10^{-3}$ which is a

Fig. 12.6. The Valles Marineris of Mars. This is a gigantic valley rather more than 3,200 km long, over 600 km wide and 8 km deep. (NASA, Viking Project)

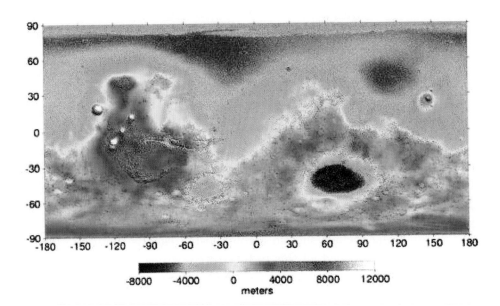

Fig. 12.7.   The topography of Mars. The range of topology (the overall height) is rather larger than for Earth. (MGS, NASA)

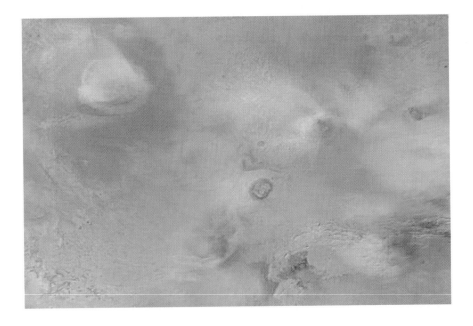

Fig. 12.8.   The Tharsis region showing clouds above the volcanoes. (MGS, NASA)

factor a little more than 30 times that for Mercury. Mars has two natural satellites, Phobos (fear) and Deimos (chaos). Its albedo is 0.150 which is less than half that of Earth. The black body temperature is $T_B = 210.1$ K, which is well below the freezing point of water.

12.5. The General Topology of Mars

The topography of the surface is shown in Fig. 12.7. The different shades of blue denote regions below the mean radius, with purple being about 8,000 m below. The yellow regions straddle the mean radius and the shades of red show regions above the mean radius. White regions are the highest with an elevation in excess of 12,000 m. The most striking property is the great difference in elevations between the northern and the southern hemispheres. The north lies below the mean radius whereas the south rises above it. The

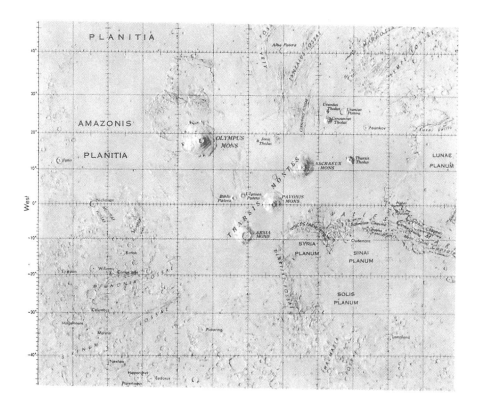

Fig. 12.9.   The Tharsis region including its relation to Valles Marineris. (NASA)

northern hemsiphere is generally smooth but the southern is rough with craters. There are one or two exceptions. One is the Valles Marineris (longitude 39–90°W, 15°S). This is an enormous gorge over 3,000 km long and 600 km wide. Its depth is about 8 km. It forms a gigantic gash in the surface. Another exception is the Hellas region 45–90°E, 30–50°S) which is a large deep depression. It is some 2,080 km across and, if an impact crater, is the largest in the Solar System. The objects straddling the equator in the region 120° W are four shield volcanoes of the Tharsis Province, of which Olympus Mons is the highest — indeed, at over 12,000 m it is the highest object in the Solar System and dwarfs the geologically corresponding Hawaiian complex on Earth. Its volume is more than 50 times that of the largest Earth volcano: the caldera is 70 km wide and it is surrounded by a cliff 10 km high. A general view of the Tharsis region is shown in Fig. 12.8. A surveyor's plot of the region is shown in Fig. 12.9.

12.6. Martian Geology

Mars has a complicated geology indicating a complex past. Pictures of restricted areas of the surface are well known. For instance, Fig. 12.10 is a natural colour view of Planitia Utopia taken by Viking 2 Lander. The reddish tinge of the soil is due to the presence of iron oxide (rust) on the

Fig. 12.10.   In the area of the Plane of Utopia. A part of one foot of the lander is seen at the lower right. There is a small trench in the centre foreground made to collect a sample of the surface soil. (NASA)

Fig. 12.11. A layered terrain on the floor of west Candor Chasma within the Valles Matineris. There are over 100 beds each about 10 metres thick.

surface. The impression is one of the abandonment of material after a flood but thus might be a subjective conclusion. The foot of the lander is seen in the right-hand bottom corner and part of the instrument for collected soil samples is seen just above it, on the right. There is more direct evidence, however, of Martian surface water of long ago. A layered terrain is shown in Fig. 12.11 with over 100 smooth, uniformly layered beds each some 10 m thick. Many other layered regions are also found elsewhere, especially in the equatorial region, and the indications from local crater densities suggests they are all some 3.5×10^9 years old. Similar structures are found on Earth resulting from sediments deposited by large bodies of water over long periods of time. It is suggested, therefore, that the Martian examples may have been deposited by ancient stretches of water on the planet. Some other evidence is shown in Fig. 12.12 which shows what seem to be dried up water channels. Craters are superimposed on the channels suggesting that the channels are ancient. Channels of this type are not uncommon on the surface.

There is one other piece of evidence supporting the presence of water. This results from a neutron scan of the surface. The neutrons result from cosmic ray impacts on the surface material. Silicate soils reflect the neutrons into space but the neutron scattering is inhibited by the presence of hydrogen atoms. Free hydrogen is not to be expected in the sub-surface regions of Mars and frozen water is the most likely source. A neutron map of Mars is shown in Fig. 12.13. The blue regions show where the neutron emission is most inhibited suggesting particularly that water is concentrated below

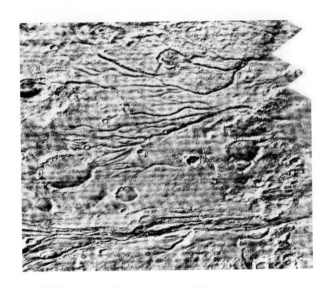

Fig. 12.12. Showing a system of channels in the Chryse region, centred around 55°N, 20°W. (NASA)

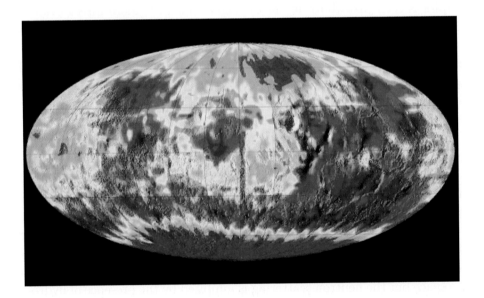

Fig. 12.13. Neutron scattering from Mars. The blue regions indicate where hydrogen can be found which is interpreted as indicating the presence of water. (Mars Odyssey, NASA)

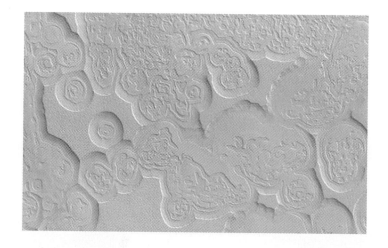

Fig. 12.14.   An unusual landscape covering the southern polar region. (MGS, NASA)

the Martian south pole. Such a concentration would be important for any manned stay on the planet.

Not all terrain features are understood. There are many examples where Earth parallels seem non-existent. One example is the microbial-looking landscape of the southern polar region shown in Fig. 12.14. The circular indentations are more than 100 m across while the flat topped mounds (mesas) are about 4 m high. One suggestion is that they result from a dry ice (frozen CO_2) composition but the mechanism for producing the observed pattern is not known. There might be a useful comparison with the giants Causeway in Ireland where a tight packed set of hexagon cells resulted from the rapid cooling of molten lava.

12.7.   Thermal Mars

Many features unique to Mars may arise from its low temperature. Its black body temperature of 210.1 K results from a solar irradiance of 589.32 W/m^2. The visual albedo is 0.150, which is under 1/2 the Earth value. A thermal image of the planet is shown in Fig. 12.15 taken soon after Martian midnight. Summer is in the northern hemisphere with winter in the southern. The temperature range is remarkable showing the full severity of a Martian winter. The temperature variations in the equatorial regions are due to the material composition. The cold blue areas are covered by a fine dust whereas the warmer regions have a covering of rocks and coarser sands.

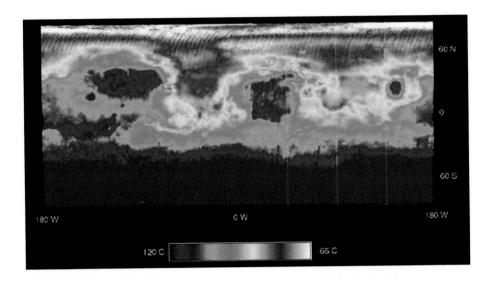

Fig. 12.15.   An early morning temperature scan of Mars on the night side. (MGS, NASA)

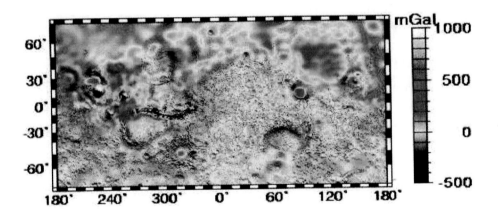

Fig. 12.16.   The free-air gravity of Mars measured by Mars Global Surveyor. (NASA)

12.8.   The Internal Structure of Mars

No reliable seismic data have been obtained so there can be no quantitative knowledge of the interior at the present time. One attempt to probe beneath the surface was made using a laser altimeter in the automatic Mars Orbiter. Small variations of the spacecraft orbital speed are linked to small variations in the planet's surface gravity. These variations can themselves be linked to particular latitude/longitude co-coordinates. Variations in the Mars gravity

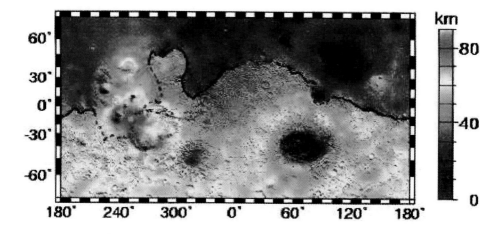

Fig. 12.17.   Regions of higher and lower mass in the surface region of Mars interpreted as thickness of crust. Red is thickest and blue thinnest. (MGS, NASA)

field are shown in Fig. 12.16. The yellow/green areas are regions of the mean gravity. Increased gravity is shown in red while decreased gravity is shown in blue. These regions are interpreted as having a thinner (decreased gravity) or thicker (increased gravity) crust. These variations are shown in Fig. 12.17. The figure shows a crust thickness in a range from essentially zero (purple) to 100 km (red/orange). It appears that the northern hemisphere generally has a thinner crust than the southern hemisphere. The Hellas basin is also a region of thin crust. The thickest crustal region is in the Tharsis region, supporting the very large volcanic structures.

In attempting to assess the internal conditions of Mars it is useful to make comparisons with Earth, where the interior is known in detail. The overriding feature is the mass difference — Mars is 0.11 times the Earth mass. This leads to a surface acceleration of gravity which is lower by the factor rather more than 1/3 with the internal pressure reduced by a factor of about 1/5. The central pressure of Mars corresponds to that in the deep mantle of Earth. Although the same general atomic/low pressure mineral structure can be expected as for Earth, the higher pressure variants will be absent on Mars. The compression temperature will be correspondingly less inside and, with weaker gravity, there will be no loss of mineral accumulations below the crust as for Earth. Earth is continually shedding minerals from a depth of about 20 km to maintain a thin crust whereas the Martian crust can grow to greater depths. Heat loss on Mars has probably always been via volcanism and a plate tectonic system seems not to have arisen. There is probably no internal convection system similar to that of Earth. In particular, the

Tharsis region, for example, is actually supported by the crust and not by an underlying convective upwelling.

The moment of inertia factor $\alpha = 0.366$ implies the presence of a core, presumably composed of ferrous materials. Different models will give different details but all will predict a core of general radius about $1/2$ the planetary radius. For the Earth this is correspondingly about 0.58 the radius giving core mass of about 32% of the whole. For Mars this ratio is nearer 28%. There will be very great interest when seismic data for Mars become available.

12.9. The Atmosphere of Mars

The surface pressure is very low, about 6.1 mb although this seems variable. The surface density is 0.02 kg/m^3 and the scale height is 11.1 km. The average temperature is 210 K but this varies by several percent in the diurnal range. The wind speeds at the Viking lander site are 2–7 m/s in the summer and 5–10 m/s later in the year. A dust storm is associated with speeds in the range 17–30 m/s.

CO_2 is the dominant component, accounting for 95.32% by volume. N_2 is next (2.7%), followed by Ar (1.6%) and then O_2 (0.13%). There is some CO. Apart from these elements there are trace constituents at the parts

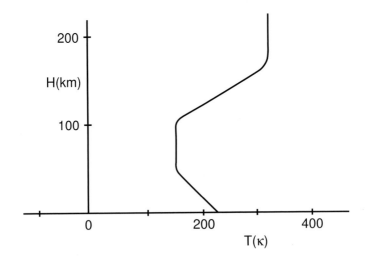

Fig. 12.18. The temperature profile of the Martian atmosphere with height from the surface. There is essentially no distinction between day and night temperatures.

per million (ppm) level. These include H_2O, NO, Ne, heavy water (HDO), Kr and Xe. The mean molecular weight is 43.34 gm/mole compared with 28.97 gm/mole for Earth.

The temperature is not uniform through the atmosphere but shows a profile indicated in Fig. 12.18.

12.10.   A Tentative History of Mars

Present knowledge of the planet allows a tentative early history to be constructed. It must be stressed that this is very tentative but can act as a framework for more detail later. Four distinct periods can be identified.

(i) The **Noachian Period**. This covers the early days of the Solar System when the surface had just solidified from the melt. It is presumed that Mars formed at the same time as the other terrestrial planets so this Period is between 4,500 millions years ago and about 3,000 million years ago. It shows now heavily cratered terrain and impact basins such as Hellas. This period is chronicled throughout the inner Solar System. The division between northern and southern hemispheres probably formed at this time. The northern crust was sufficiently thin to allow more rapid cooling than the southern and the outgassed atmosphere probably allowed water to condense in the north to form seas and lakes. That this terrain is still seen today suggests that the surface has not been substantially replaced since that time. It probably ended to give way to

(ii) The **Hesperian Period**. This is the period of extensive lava flows from cracks and fissures in the crust. The older volcanoes date from this time and the large shield volcanoes, such as the Tharsis group, began to form now. There was substantial crustal deformation and the Valles Marineris formed during this period. The surface shows a large number of volcanic structures that have been weathered down to fossil structures. The next period was

(iii) The **Amazonian Period**. This stretches from about 2,000 million years to a few million years ago and was a period of deposition of water sediments. The younger volcanoes flourished in this period and there was wide erosion. Then came

(iv) The **Modern Period**. This is Mars as it appears today. Volcanic activity has probably stopped entirely and the atmosphere has thinned.

Atmospheric gases have been absorbed into the surface regions as permafrost and ices. The planet will slowly cool. The main activities now are wind erosion and landslides.

The correctness of this history and more detail will come from future exploration of the surface and the interior.

*

Summary

1. Mercury and Mars are the two smallest of the terrestrial planets.

2. Mercury has a radius rather less than the largest of the satellites of the major planets but its density is several times greater, being comparable to that of the Earth.

3. The Mercury rotation period is 2/3 of its time period to orbit the Sun. This peculiar situation presents one hemisphere and then the other to the Sun.

4. The surface temperature at the hot pole is high, about 740 K.

5. The surface is divided into highlands and lowland planes. The most significant surface feature is the Caloris basin which resulted from a giant impact early in the planet's history. It has a diameter of 1349 km.

6. The highlands are quite heavily cratered though there are fewer small diameter craters than for the Moon.

7. The high density implies that the interior of Mercury is a mass of ferrous material core coated by a silicate mantle of about 1/4 the core size.

8. Mercury has very thin atmosphere which presumably is being replaced continuously.

9. While Mercury appears to be a rather inactive planet, Mars has obviously had an active history.

10. It has the largest volcano in the Solar System (Tharsis Mons) and the largest valley gorge (Valles Marineris) which is some 3,200 km long, 600 km wide and 8 km deep.

11. There are strong evidence for free liquid water on its surface in the distant past.

12. The internal structure involves a ferrous core of rather less than half the radius encased in a silicate mantle. The crustal thickness varies from virtually zero up to 100 km. There is no evidence of plate tectonics as on Earth. The existence of volcanoes shows that the mantle carries convection processes and that heat is transported from the lower regions upwards to be radiated into space.

13. The large volcanoes are supported by the strength of the crust alone and there is no contribution from mantle upwelling.

14. The atmosphere is very thin but has CO_2 as the major constituent.

15. A tentative history of Mars is offered. This is subject to considerable modification in the future.

13

Planet Venus

Fig. 13.1.   Venus taken in ultraviolet light. The pattern is due to the circulation at the top of the atmosphere and arises from differences of concentrations of SO_2 and S. (Pioneer Venus Orbiter, NASA)

Venus appears as a twin of the Earth though not quite an identical twin. Its radius is 0.949 that of the Earth and its mass is 0.815 that of the Earth. This makes its mean density 0.951 that of the Earth and its surface gravity 8.87 m/s^2 (rather than 9.78 m/s^2 for the Earth or 0.907 of the Earth value). It has zero flattening (that is $f = 0.00$ as opposed to the Earth with $f = 3.35 \times 10^{-3}$). This is associated with $J_2 = 4.458 \times 10^{-6}$, which is about 3 orders of magnitude less than Earth. Its geometric albedo is 0.65, which is 1.8 times that of the Earth. These properties would suggest a body not unlike the Earth except that it is observed to have a very dense atmosphere.

13.1. First Views of the Surface

The first picture of the surface was obtained by the Russian automatic Venera 13 craft which landed safely in March 1982. A second automatic lander, Venera 14, also returned photographs of the surface (see Fig. 13.2). One obvious comment is the large amount of light at the surface in relation to the cloud cover above. The illumination is similar to that on Earth on an overcast, rainy day. Again, the rock slabs have a sharp appearance and are probably slabs of lava that fractured after solidification. They appear relatively fresh with little erosion, suggesting a young surface. The Venera craft made elementary analyses of the composition of the rocks and found them very similar to Earth basalt material (see II.8). Both craft reported very hostile conditions and each operated for rather less than one hour. The surface temperature was found to be very high (later shown to be on average 737 K) with a very high surface atmospheric pressure (92 bar). Later, the orbiting Veneras 15 and 16 mapped the planets northern hemisphere but detailed mapping was made only during 1992–1994 by the orbiting American

Fig. 13.2. The surface of Venus photographed by the Russian Venera 14 at lat.7°S, long. 304°W. The object at the bottom is the foot of the lander. The horizon is at the top right and left corners.

Fig. 13.3.   Computer generated image of the surface of Venus revealed by radar. (Magellan Project, NASA)

Magellan automatic spacecraft using radar techniques. An image is shown in Fig. 13.3. Figure 13.3 can be compared with the visual view of Fig. 13.2. A relief map highlighting the elevations across the surface is shown in Fig. 13.4. The highest region is Ishtar Terra in the north containing the highest point on the planet named Maxwell Montes. This rises some 10 km above the mean surface level. The Mercator projection of the figure makes this feature appear rather larger than it actually is although it is about the size of Australia. A larger highland region across the equator is Aphrodite Terra. The highland regions are the equivalent on Venus of continents on Earth. The blue and deep blue regions are below the mean radius and correspond to sea and ocean regions on Earth. The light blue and dark green regions are rolling upland terrain close to the Venus "sea level". This feature accounts for about 2/3 the total surface area. The topological range is about 15 km which is only 3/4 that for the Earth. This means that Venus appears smoother than Earth. The surface contains no water.

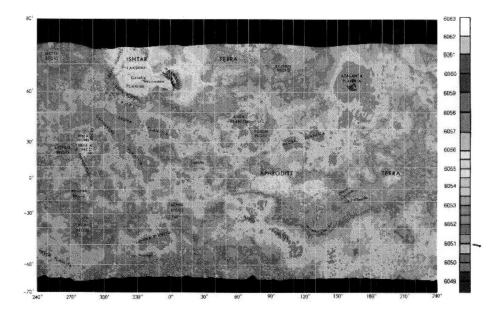

Fig. 13.4. A relief map of Venus showing the various elevations. (NASA)

13.2. Surface Details

The surface temperature is high, with a mean value of 737 K. This is above the melting point of a number of elements but below that of the minerals that might be expected on the surface. The atmospheric pressure is found to be at a depth of 1/3 km within the Earth. These factors would suggest an environment of strong erosion. The observations of the surface, however, do not provide evidence for this. Two features are immediately evident.

One is the lack of regions of heavy cratering such as which occur on Moon and Mars. The 1000 or so craters that are there appear recent, with little or no degradation. They have diameters in excess of 3 km but rarely greater than 35 km. This demonstrates a filtering effect of the atmosphere. Small objects from space do not reach the surface and very large objects are slowed sufficiently to strike the surface with low energy, too low to form a crater. There is evidence of clusters of objects reaching the surface suggesting that a primary body has broken up in the dense atmosphere into several smaller ones. The craters are spread randomly over a surface that cannot, from the evidence of crater counts, be even as old as 500 million years. Apparently the surface has been replaced relatively recently and perhaps frequently. Whether this is the latest of a succession of re-surfacings

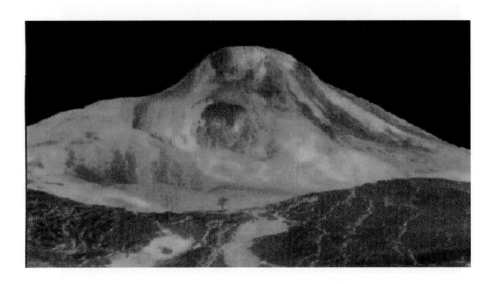

Fig. 13.5.   A simulation of the volcano Maat Mons. (JPL Magellan Project, NASA)

Fig. 13.6.   A Venus Rift Valley. (MP, JPL, NASA)

(the so-called equilibrium hypothesis) or whether the present surface is the result of a single recent event (the global catastrophe hypothesis) is not yet clear. What evidence there is does not point to an active planet on the scale of the Earth so the latter hypothesis might seem to be in favour but the matter remains open. The same problems arose in the early geological debates about the Earth and were resolved (in favour of the equilibrium hypothesis) by more information.

The other feature is the lack of rapid erosion. This can be understood because there is no water on the surface (one of the great sources of erosion on Earth) and the wind speeds there are very low (unlike on Mars). This is a little surprising because the surface is coated with clouds of sulphuric acid which suggests a highly corrosive rain.

The overriding feature of the surface is volcanism. Many volcanoes are high. Figure 13.5 is a simulation of the volcano Maat Mons which has a height of about 8 km. The colours again follow those shown by the Venera photographs. The barrenness of the surface is very clear. The volcanic landscape is accompanied by rift valleys which were formed by faulting of the crust. An example is shown in Fig. 13.6. The 3.5 km high volcano Gula Mons is seen on the right distance, some 800 km away. On the left in the distance is the volcano Sif Mons. The vertical scale in the picture has been enhanced.

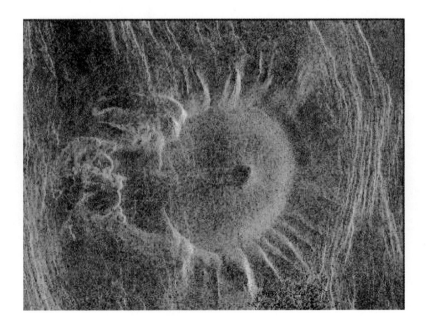

Fig. 13.7.   A volcano with ridges and valleys. (Magellan, NASA)

Another well recognised general feature is a volcano about 30 km across with ridges and valleys radiating down its sides. One example is shown in Fig. 13.7.

Although many surface features can be recognised from Earth there are some that cannot. Among these are large circular domes, a few metres

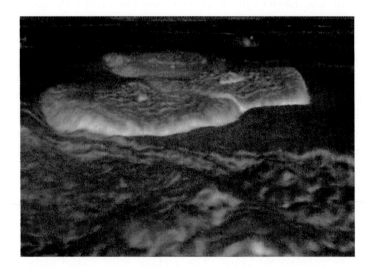

Fig. 13.8.   A simulation of a Venusian volcanic dome. (Magellan, NASA)

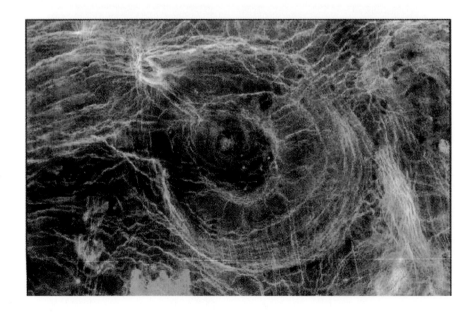

Fig. 13.9.   A fine structure on the Venusian surface (Magellan, NASA)

high but up to 25 km across. One such is shown in Fig. 13.8. It is believed that these features have a volcanic origin but the details remain unknown. They are found widely over the surface.

A feature which seems to have no Earth counterpart is shown in Fig. 13.9. It is presumed to have a volcanic origin and may well involve a slip plain structure. It has a span of about 200 km. Nearly 50 of these structures have been recognised so far. There is much to be learned of Venus but the hostile conditions of the surface will make landers very difficult to maintain.

13.3. The Venus Interior

The surface shows a range of volcanoes distributed widely over its surface and this gives the clue to the internal structure and how it differs from Earth. Like Earth, it will have a crust, a mantle and a core. The basic composition will be similar except for a lack of water in the crustal regions. Venus is, however, a slightly smaller version of Earth and the internal pressure will be lower, at any depth, by a factor of about 0.94. This small difference can have important consequences here as for Mars; it will lead to a thicker crust, beyond the strength of internal motions to break up as plates. With no plate motions, heat rising from the interior will not be able to escape through the tectonic boundaries and the equivalent of the mid-ocean ridge structure. There will be no subduction region volcanoes. The heat associated with the interior will be similar to Earth so a convection regime will certainly exist inside. The hot material moving upwards will discharge its heat through conduction in the crust but excess heat will escape through volcanic outlets of various kinds. A thick crust can lead to lava outflows through rifts and cracks and can be expected to use local weak spots in the crust. The regions of upwelling will be the centres of a volcano system of different types while the downwelling regions will show surface contractions as material is pulled down. This structure will lead to the type of features seen on Venus.

Like Mars, Venus shows no evidence of possessing an intrinsic magnetic field although its upper atmosphere will be ionised by solar ultraviolet radiation and so show a very small transient field. The lack of the intrinsic field may be the result of a core structure different from the Earth. With a moment of inertia factor $\alpha = 0.33$ a central ferrous core can be predicted but it may have a slightly different composition from Earth. In particular, an outer region may not have a lighter contamination and so may not be in the liquid phase. The slightly lower pressure in the central regions will also lead to different conditions there because the pressure effects on melting

points will be different. It is quite possible in consequence that the core will
not have the two layer system found on Earth with a liquid outer core and
a solid inner core. The problem of chemical composition and physical state
will only be finally solved when seismic data become available.

13.4.   Venus Atmosphere

The Venus atmosphere is very dense and hot. The black body temperature
at the outside is 231.7 K but the surface temperature is 737 K. This seems
to be constant throughout the year. The surface pressure is 92 bars and the
surface density 65.0 kg/m^3. The scale height is 15.9 km.

13.4.1.   Composition

The mean composition is very simple: 96.5% CO_2 and 3.5% N_2, each by
volume. There are also trace elements present in parts per million. These
include SO_2, S, Ar, H_2O, CO, He and Ne. The ratio of the hydrogen isotopes H and D provide water and heavy water and in the cosmic abundance
these two "isotopes" of water occur in the ratio D/H $= 1.56 \times 10^{-4}$. The
value of the ratio for Venus is about 1.6×10^{-4}, very similar to the value in
the Earth's oceans. The value for comets would seem to be about two orders
of magnitude higher making deuterium relatively more abundant there. This
suggests that the hydrogen on each planet did not originate in comets. Why
the ratio for comets should be higher is not known (remember II.7, 7.4 and
7.5).

There are as yet no indications of the history of the Venus surface and
atmosphere. The relatively recent resurfacing of the planet has obliterated
earlier information. The associated volcanic activity will have released considerable quantities of CO_2 into the atmosphere which, in the absence of
water, will have remained in the atmosphere as a greenhouse gas. There are
no indications of water channels on the surface so presumably this process
must have occurred long ago in the past if at all.

13.4.2.   Temperature profile

The average surface temperature of the atmosphere is 737 K but the thermodynamic temperature measured at the cloud tops on the day side is 231.7 K.
The equilibrium temperature in the absence of an atmosphere would be
227 K. The temperature profile is shown in Fig. 13.10. The temperature

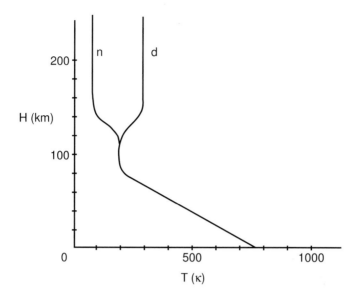

Fig. 13.10.   The temperature profile with height, H, in the atmosphere of Venus. d shows the daytime temperature while n shows the night time temperature.

falls with height essentially linearly up to a height of about 80 km. Then it stays constant at about 180 K until at a height of 120 km it bifurcates. By day it rises slowly up to 165 K when it stays constant at about 300 K up to 300 km or so when it decreases to the measured value 231.7 K. The pattern is different on the night side. The difference comes at the 180 km level when now the mean temperature falls to 100 K and then remains sensibly constant with height. The difference between day and night temperatures is also found in the case of the Earth.

13.4.3.   Structure

The motion of the dense atmosphere is driven by the solar irradiance to dissipate the energy received. Rising columns of atmospheric gas rise on the day side and move both in longitude east and west and both in latitude north and south. The latitude currents move to the poles where they have cooled and descend, passing back to the sub solar point to be heated again. The longitude currents move round to the anti-solar point where they descend and return to the starting point. These motions form a set of gas cells named after Hadley who first described this form of heat transfer.

These motions are associated with a distinct cloud structure. The surface is covered by a thin haze up to a height of about 44 km where a pre-cloud layer of thicker haze forms to a height of rather less than 50 km. A lower cloud layer then fills the narrow range from 50 to about 52 km encased in a middle cloud layer up to about 58 km. Beyond this are the upper clouds extending to perhaps 70 km. The composition of the clouds is not water but a mixture of sulphur (S) and sulphuric acid (H_2SO_4). This corrosive composition does not appear to act adversely on the surface.

*

Summary

1. Always supposed a twin planet of the Earth, Venus has turned out to be rather different.

2. Its very thick atmosphere hides a world of high temperature (average value 737 K) and pressure (92 bars, surface density about 65 kg/m^3).

3. The atmospheric composition is CO_2 (96.5%) and N_2 (3.5%) with trace amounts of other elements. The mean molecular weight is 43.45 g/mole.

4. The outer atmosphere has a rotation period of 4 days which is in contrast to the surface rotation period of 224.701 days. The atmosphere is driven by solar irradiance which is 2613 W/m^2. The surface temperature of 737 K is reduced to 231.7 K at the cloud tops.

5. There is an internal convection structure giving a surface haze up to about 44 km above which is a weak cloud layer leading to a substantial cloud structure in the upper regions. The clouds there are composed of sulphur and sulphuric acid.

6. This atmosphere obscures the surface to visible wavelengths with an impenetrable cloak, making it necessary to use radar frequencies to find the surface details.

7. The studies show a surface with high mountains, up to 8 km above the mean radius, and low lying areas, 1 or 2 km deep, below it. If there were water on the surface this would certainly provide continents and oceans.

8. The surface has, in fact, been photographed as visible on two occasions by Russian landers. In each case is was covered with broken rock pieces having apparently sharp edges. This indicates a lack of erosion.

9. The sharpness of the rock edges is understandable because of a lack of water on the planet but the lack of erosion is not in sympathy with the expectation of a general weak corrosive sulphuric acid drizzle at the surface.

10. The main feature of the surface is past volcanism. Many features can be understood in terms of earth experience but others, such as the volcanic domes, cannot.

11. The main source of heat loss by the interior is volcanism. There are some possible examples of plate tectonics but certainly not on a global scale, as for Earth.

12. Conditions inside must be very similar to those on Earth but with a consistently lower pressure. The material composition is most likely the same as for Earth. This will lead to a restricted range of mineral phase changes due to pressure and possibly a simple entirely liquid ferrous core unlike the Earth which has a solid central core. This can affect magnetic conditions inside (see III.19).

14

The Planets: Jupiter and Saturn

Fig. 14.1. An image of Jupiter as seen by the Cassini spacecraft moving to Saturn. The dark spot is the shadow on the surface of the satellite Io. (NASA)

Beyond 5 AU from the Sun lie the four major planets of the Solar System. Between them they carry the greater part of the mass of the planets and much of the angular momentum of the Solar System itself. Jupiter and Saturn have been observed continuously over more than 300 years, beginning with Galileo who pointed his new telescope at them. Since then they have been observed from Earth-based telescopes, from high flying aircraft and balloons, and more recently from space vehicles (particularly including the orbiting Hubble Space Telescope) both at a distance and *in situ*. One probe (the Galileo probe) actually penetrated the upper atmosphere of Jupiter. The four planets show certain similarities in pairs. Here we compare and contrast Jupiter and Saturn. Images of the planets are shown in Figs. 14.1 and 14.2.

Jupiter is the largest planet in the Solar System by far, being larger and more massive than all the others taken together. Saturn is the second largest, and is still massive, having a greater mass and volume than the

Fig. 14.2. A magnificent image of Saturn as seen by the Voyager 2 spacecraft in 1981 July at a distance of 21 million km. Three satellites are visible together with the shadow on the planet's surface (due to the Sun) of one of them. (NASA)

Table 14.1.   Some bulk properties of Uranus and Neptune.

Property	Jupiter	Saturn
Obliquity to orbit (deg)	3.13	26.73
J_2	1.4736×10^{-2}	1.6298×10^{-3}
Ellipticity, f	0.06487	0.09796
Gravity (at 1 bar) m/s^2	23.12	8.96
Mom. Inertia factor, α	0.254	0.210

remaining planets with Jupiter excluded. It has closely 1/3 the volume and 0.58 the mass of Jupiter. Its mean density is, therefore, about 1/2 that of Jupiter. The mean densities of Jupiter and Saturn are consistent with their compositions being very largely hydrogen. It was seen in I.4 that the Jupiter mass is, indeed, close to that of the largest hydrogen body without internal pressure ionisation and so near the largest hydrogen body that can be regarded as a planet.

Some basic bulk properties of the planets are shown in Table 14.1. There is a disparity between the obliquity angles (the angle made by the rotation axis to the plane of the orbit), in the equatorial gravity and the f_2 values. The moment of inertia factors show a strong concentration of mass towards the centre of each planet.

14.1.   Surface Features

The immediate images are of striated bodies with some spots. Jupiter appears grand with an active surface: Saturn appears elegant, smooth and quiet. The rings of Saturn are very obvious and have been a feature described since telescopes were first used in astronomy. Those for Jupiter are more tenuous and have only recently been discovered.

The surfaces of each planet show a series of brighter zones and darker, thinner, bands. They have the immediate appearance of clouds and this is confirmed by closer scrutiny. Jupiter has fifteen regions spanning its surface parallel to the equator. A wide Equatorial Zone gives way to three (wide) northern zones separated by three thinner northern belts capped by a North Polar Region. There are corresponding regions in the southern hemisphere. There is a symmetry but it is not complete. Saturn shows a comparable zone and band structure. The zones are high cloud layers while the belts are gaps where lower regions are visible. The zones and belts are in rotation

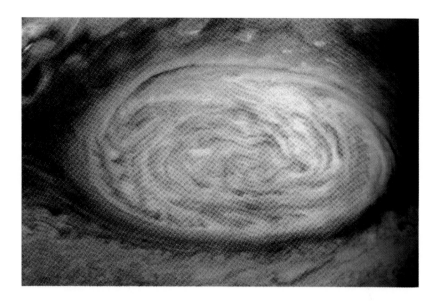

Fig. 14.3. The Great Red Spot of Jupiter recorded in the infra-red by the Galileo orbiter in 1996. It is colour coded, the white and red regions having an altitude of at least 30 km greater than the clouds marked in blue. (NASA)

about the rotation axis, some prograde and others retrograde with respect to the mean surface motion. An obvious feature of both planets is a series of "spots", resulting from cyclonic/anticyclonic atmospheric rotations. These are seen for Jupiter in Fig. 14.1 with the largest and most well known spot being the Great Red Spot (GRS) in the southern hemisphere. This was first reported early in the 17th century by the astronomer Galileo using his new telescope and has lasted uninterrupted since then and presumably was there before. The current Galileo space orbiter around Jupiter has obtained close-up pictures such as that in Fig. 14.3. It is known now to be a high pressure cloud formation rotating anti-clockwise with a period of several weeks. Winds at the boundaries reach speeds as high as 110 m/s. Apart from the GRS, smaller whitish spots appear and disappear from time to time lasting many tens of years. A number of these are seen in Fig. 14.3 and also arise from atmospheric circulation. Analogous regions appear also on Saturn though there is nothing there comparable to the GRS. The precise mechanism for feeding energy into these features is still unknown. That supplying the GRS has been operating for a very long time.

The circulation of the Jupiter and Saturn atmospheres involve winds of substantial speeds which vary with latitude. Some indication of the profiles are seen in Fig. 14.4. The motions are relative to the mean rotation of the

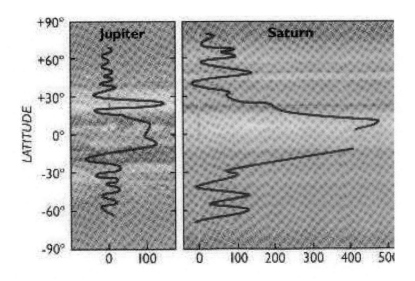

Fig. 14.4. The distribution of wind speeds on the surfaces of Jupiter and Saturn. The ordinate is the speed in units of metres per second. (NASA)

planet, positive being eastwards (the direction of rotation). In each case, the maximum speeds are found in the equatorial regions where wind speeds of 150 m/s are found for Jupiter but nearly 500 m/s for Saturn. Reverse (prograde) speeds are also found, the variations arising from the different layers of cloud that are observed. Regions about latitudes ±30° show winds of magnitudes of a few tens of metres per second against the direction of rotation. Computer simulations have allowed some of the observed features to be understood, surprisingly even on the basis of Earth atmosphere models. A central feature could well be the formation of small scale unstable eddies which feed their energy into the dominating zonal jet streams. There is the difference from Earth, however, that the primary driving force for the Earth's atmosphere is the solar irradiance but for Jupiter the Sun is five times further away and the internal heat must be more important as an energy source. One supporting observation for this conclusion is the greater wind speed for Saturn, where a three-fold increase of wind speed is associated with rather more than a three-fold decrease of the solar irradiance there.

The variation of wind speeds with depth is largely unknown. The Galileo probe gave some information for Jupiter where it seems that the surface speed increases down to the 5 bar level where it reaches 170 m/s. It then appears to remain constant down to the 20 bar level where the Galileo stopped transmitting data. Conditions in Saturn may be expected to be similar.

14.2. The Heat Budgets

The transfer of heat between each planet and the environment is very different from that for the terrestrial planets. The essential difference is that Jupiter and Saturn each have an internal source of heat comparable to the (rather weak) solar irradiance there. The conditions are listed in Table 14.2 where it is seen that the measured temperatures at the cloud tops are higher than can be accounted for by the incoming solar radiation alone. Jupiter radiates about 1.7 times that received from the Sun while Saturn radiates about 1.8 times. The possible sources of the heat will be considered later.

The distributions of temperature in the atmospheres are sketched in Fig. 14.5. The zero level is the 0.1 bar level which is the location of the lowest temperature for each planet. The "surface" is taken to be the 1 bar level where the temperature is 165 K and the mean density 0.16 kg/m^3 for Jupiter and 134 K and 0.19 kg/m^3 for Saturn. It is seen that the profiles are rather more compressed for Jupiter with the larger gravity.

The observed atmospheric layers can be identified with particular compositions. The main components are NH_3, NH_4SH and H_2O in each case and the depths are listed in Table 14.3. The layers are lower in the atmo-

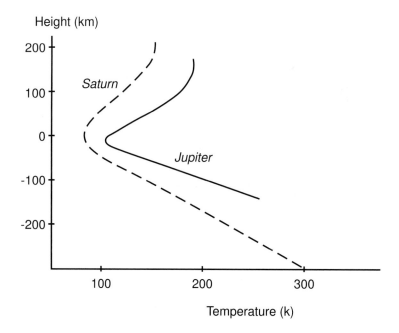

Fig. 14.5. The distribution of temperature with height for Jupiter and for Saturn. The zero level is the 0.1 bar pressure.

Table 14.2.   Various thermal properties of Jupiter and Saturn.

Property	Jupiter	Saturn
Solar irradiance (W/m^2)	50.50	14.90
Albedo	0.52	0.47
Mean absorbed irradiance (W/m^2)	10.0	3.0
Emitted radiation, (W/m^2)	13.5	4.6
(Internal/solar) source	1.35	1.53
Blackbody temp. (K)	98.0	76.9
Measured temp. (K)	112.0	81.1
Temp. 0.1 bar level (K)	112	84
Temp. 1 bar level (K)	165	134

Table 14.3.   The depths of the three main atmospheric layers for Jupiter and Saturn.

Materials	Altitude (km above/below 0.1 bar level)	
	Jupiter	Saturn
NH_3	0	−100
NH_4SH	−50	−150
H_2O	−100	−250

sphere of Saturn than of Jupiter and more widely spaced. These are again consequences of the lower Saturn mass.

14.3.   Visible Surface Compositions

The low densities of the planets suggest a largely hydrogen composition and this is found in the surface regions by observations. The details are shown in Table 14.4, being percentages by volume. The data for the Sun are added for comparison since this is the solar abundance which presumably was the initial abundance of material out of which the Solar system formed. The measured data refer to the surface layers where the light originates and can be analysed spectroscopically by the observer. This implies that the compositions of the interior regions are not included in the analyses. The solar abundance of the elements was listed in I.5 where it was seen to involve

Table 14.4.   The compositions of the atmospheres of Jupiter and Saturn by volume. That for the Sun is added for comparison. The values for Jupiter from Galileo probe — others by remote sensing.

Body	Jupiter	Saturn	Sun
H_2	89.8%	96.3%	84% (H)
He	10.2%	3.25%	16%
ppm	CH_4, NH_3, HD, C_2H_6, H_2O (?),	CH_4, NH_3, C_2H_6	H_2O, CH_4, NH_3, H_2S
Others believed there	Ne, Ar, Si, Mg, Fe	Ne, Ar, Si, Mg, Fe	–
Aerosols	Ammonia ice, water ice, ammonia hydrosulphide	Ammonia ice, water ice, ammonia hydrosulphide	–
Mean molecular weight (g/mole)	2.22	2.07	–
Scale height (km)	27 km	59.5	–

mainly H and He but only some 2% of the remaining heavier (rarer) elements. The solar composition will be of this general type and so also will those of the fluid planets. It will be realised that the total inventory of elements will not be represented on the surface unless the mixing is total. Any accumulations in the interior will show deficiencies in the surface and this test will be clear by comparison, assuming the overall compositions are the same. Referring to Table 14.2, it flows that both Jupiter and Saturn have more surface hydrogen than the Sun but less helium. These are relative statements and taken together imply that the Sun has more internal hydrogen and helium than either planet and that Saturn has a greater accumulation of internal helium than has Jupiter. The consequences of this deduction will be clear in relation to the internal constitutions of the planets.

The comparison of the minor constituents is difficult because in the Sun, unlike the planets, the outer layers are at very high temperatures. Molecules cannot exist there and atoms will be very highly ionised.

14.4.   General Comments on Internal Conditions

There is no direct evidence of conditions in the main bodies of the planets and even indirect evidence is difficult to come by. Being fluid planets, seismic studies cannot be applied. Helio-seismology, so useful for exploring the interior of the Sun, has not been able to be applied to the fluid planets so far. The source of the associated body motion is not clear so the amplitudes to be expected cannot be estimated although it is certain they will be very small. It is necessary to rely on theoretical analyses, using such basic quantities as density and calculated pressure to provide a series of models from which the most likely can be chosen.

In the most general terms, the composition can be supposed at first to be cosmic. This would mean that 98.4% of the mass would be H and He but 1.6% would be of heavier materials. These heavier materials form a central core composed of silicate and ferrous materials, much like the terrestrial planets but now under very high pressures. The central feature is the moment of inertia factor, α. For the case of Jupiter this might imply a core mass of about 3×10^{25} kg which is some 5 Earth masses (M_E). The mass of the Saturn core would be less, perhaps 1×10^{25}, or 1.5 M_E. These core masses based on the cosmic abundance of the elements appear, in fact, too low for models which match the observed gravity distributions. Apparently, neither Jupiter nor Saturn can have a simple cosmic composition. It is more likely that the core mass for Jupiter is closer to 10 ME (that is 6×10^{25} kg or about 3% of the total mass) and that for Saturn about 12 M_E (that is about 8×10^{25} kg or perhaps 10% of the total mass). The remainder in each case will be a mixture of H and He. These features must reflect the solar-type density for Jupiter and the very low density of Saturn.

The major constituent in both planets is hydrogen and it is necessary to know the properties of this element under a range of temperatures and pressures (the equation of state) appropriate to the interiors. Considerable effort has been expanded over recent years to obtain these data experimentally wherever possible but otherwise theoretically at the higher pressures and temperatures beyond the experimental range. This is a complicated quantum mechanical calculation. The composite equation of state for molecular hydrogen (H_2) is sketched in Fig. 14.6 in the form of a phase (p, T) diagram. This covers the range of pressures up to more than 10^{10} bar and temperatures up to about 10^5 K.

The diagram shows the normal solid, liquid and gas phases of molecular hydrogen but shows other more extraordinary things. For pressures beyond $p \approx 10^6$ bar (atmospheres) the simple form becomes metallic, either in solid or in liquid form for higher temperatures. This is a most important feature from the point of view of planetary science. Jupiter and Saturn contain a mixture of H and He and a novel feature of such a mixture is also shown in Fig. 14.6. Whereas the two elements will mix normally over the lower pressures and temperatures there are indications that a range of p-T exists for which He is rejected by H and is released in this form of droplets making

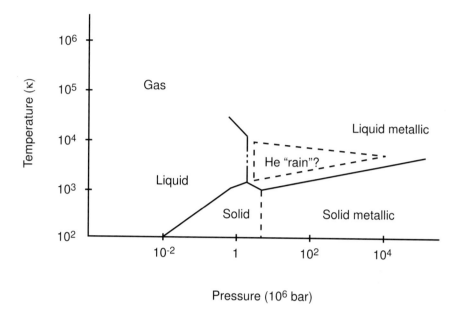

Fig. 14.6. The equation of state for molecular hydrogen. Also included is a region referring to a hydrogen/helium mixture where it is thought the helium will "rain out" from the
hydrogen in the form of droplets.

a "rain". The region where this phenomenon is thought to exist is shown
dotted in the p-T diagram. A corresponding diagram for He is not generally
established and is anyway not necessary for our present purposes.

14.5. Detailed Model Interiors

It is now possible to glean some quantitative information about the interiors.

14.5.1. Jupiter model interior

The interior will be dominated by the properties of hydrogen over the full
range of pressure. The central pressure on any model will be of the order
5×10^{11} Pa, or 5×10^7 bar. This will correspond to a temperature of about
2×10^4 K and a density about 2.3×10^4 kg/m^3. The central core of heavier
materials will have a mass $M_c \approx 6 \times 10^{25}$ kg and radius possibly of order
10^3 km. The structure, then, is relatively simple.

The surface composition suggests that He is not a major component of the
interior and that H_2 is the primary component. The hydrogen is in molecular
form down to a pressure of a little over 10^6 bar, which for hydrogen is at a

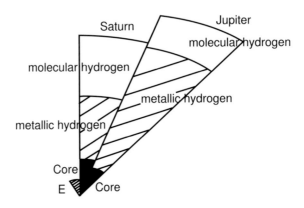

Fig. 14.7.   The interior structures of Jupiter and Saturn. The Earth ism marked for scale.

depth of about 0.1 R_J where $R_J = 7.15 \times 10^4$ km. The density will be of order 2×10^3 kg/m^3 and the temperature a few times 10^3 K. Below that depth the hydrogen is in a metallic liquid state where it remains to the boundary with the core. The core will be in a malleable, liquid state. Agreement with the details of the external gravitational field could require a thin zone outside the central core that might contain He.

14.5.2.   Saturn model interior

This is a less massive planet and will have rather milder internal conditions. It was seen earlier to have a larger internal proportion of He. The central pressure is now about 2×10^{12} Pa, that is about 2×10^7 bar. The central temperature will be of order 1.1×10^4 K and the central density about 1.9×10^4 kg/m^3. These values are obviously rather lower than for Jupiter. They do, however, place the interior with a lower region in the metallic hydrogen range. This is likely to begin at a depth of about 10^7 m or perhaps rather less than 1/5 the radius below the surface. At this point the H-He mixture will not be in thermodynamic pressure equilibrium and He will condense from the mixture to fall under gravity towards the centre. This will release gravitational potential energy and can be the origin of the large intrinsic heat flow that is observed. The Saturn interior is included in Fig. 14.7.

14.6.   Comment on Interior Heat Flow

The He rain mechanism may well be the source of the intrinsic heat for Saturn but the Jupiter heating remains unexplained since this planet is unlikely to have an excess of internal He. Another mechanism involving the release of gravitational energy is the cooling effects of contraction of the whole body. This is sometimes called the Kelvin-Helmholtz mechanism. Modern theories of the early history of the major planets view the formation process as one of contraction as heat is lost at the surface to space outside. The contraction of the radius will lead to a smaller gravitational energy and the loss must be by radiation. It is likely that Jupiter is still in the final stages of its formation as is still contracting. It has been estimated that then observed intrinsic heating of Jupiter could be accounted for by a whole body contraction by 1 mm per year. Presumably Saturn should show the same effect. The heat radiated by Saturn is larger than for Jupiter suggesting that the Saturn heat of contraction is augmented by the "He rain" effect. The heat flow in each case will be associated with a super-adiabatic gradient leading to convection. This will extend through to the upper reaches of the atmosphere and ultimately provide the energy for the observed surface stratification. In particular, the bands will result from hot currents rising from the interior while the bands will show the regions where the now colder material descends. The horizontal striations result from the effects of strong rotation.

14.7.   Intrinsic Magnetic Fields

Both Jupiter and Saturn are the seat of strong intrinsic magnetic fields. The Jupiter field was first discovered before the first spacecraft visited the planet by Earth-based observatories through the reception of decimetre and decametre radio waves. These were interpreted as synchronous radiation in a strong planetary intrinsic magnetic field. The various details are collected in Table 14.3. It is seen immediately that the Jovian field is very strong but the Saturnian field is more moderate, compared with that of the Earth.

The representation of the intrinsic magnetic field by a succession of magnetic poles has the dipole as the leading term. This is the main term and the field will be found to be that due to a simple dipole at a sufficient distance away from the source. For such cases the features of the field can be characterised by the magnetic dipole moment and the strength and direction of the dipole field component. The Jupiter field can be represented adequately this way since the dipole component is the largest. The Saturn field is more complex with the quadrupole and octopole terms being only slightly less important than the dipole component. It is important to bear this in mind when reading Table 14.5.

Table 14.5. The data for the intrinsic magnetic fields for Jupiter and Saturn. Corresponding data are given for Earth as a comparison.

Property	Jupiter	Saturn	Earth
Mag. dipole moment (Am^2)	1.54×10^{27}	4.6×10^{25}	7.9×10^{22}
Mag. dipole moment $(E = 1)$	1.95×10^4	5.82×10^2	1
Equatorial field (gauss)	4.28	0.22	0.305
Equatorial field $(E = 1)$	12.23	0.721	1
Polar field (gauss)	15	?	0.70
Angle between axes	$9.5°$	$0°$	$11.5°$
Offset vector	$148.57°$ S : $-8.0°$ W	0.04–0.05 R_{sat} N	$18.3°$N; $147.8°$E
Polarity	Parallel	Parallel	Anti-parallel

The dipole moment involves the product of the pole strength and the length of the dipole, represented now by the radius of the planet. The large Jupiter radius makes the dipole moment large in this case, and the radii alone would lead to a factor of 11 between the Earth and Jupiter. It is, in fact, 19,500 showing that the magnetic pole strength is much larger than for Earth times that of the Earth. It is seen from the table that the equatorial field differ by a factor 12 and the polar fields by a factor about 20. The presence of significant non-dipole components suggests the source is not very far below the surface. In fact it is likely to be located in the metallic hydrogen region of the mantle.

The case of Saturn is different. Now it is not clear that a dipole representation is suitable at all. Insisting that it is applicable gives a dipole lined up almost exactly with the rotation axis. This moves outside the constraint of Cowling's Theorem and shows the source of the field must lie closer to the surface than for Jupiter. It is interesting to realise that the sense of the magnetism is parallel, unlike the present state of the Earth which is anti-parallel. The established variability of the direction of the Earth's field suggests this is not, however, important.

*

Summary

1. Jupiter is the largest planet of the Solar System and Saturn the second largest. They are each a fluid planet composed primarily of hydrogen but there are differences between them.

2. The surface ellipticity of Saturn is half as much again as that for Jupiter and the figures are both flattened.

3. The inertia factor of each planet is below 0.26 showing a considerable concentration of mass towards the central regions.

4. Jupiter shows the most surface detail with horizontal bands of light and dark across the surface. These move with considerable speed across the surface; the central regions for Jupiter having a velocity of rather more than 100 m/s over and above the rotation speed of the planet. Saturn has higher wind speeds, up to 450 m/s.

5. The driving force is the internal thermal heat budget and not the solar irradiance.

6. Jupiter radiates about 1.35 times the heat it receives from the Sun. Saturn radiates a little more, about 1.53 times that it receives from the Sun. The Jupiter heat source could be a very slow contraction of the planet (by about 1 mm per annum would release enough gravitational energy). The Saturn source could be the internal separation of helium from a hydrogen/helium mixture.

7. The internal pressure of Jupiter is sufficient to cause hydrogen to enter its metallic phase. The surface composition suggests that there is not a large quantity of internal helium.

8. The internal pressure for Saturn is less than for Jupiter and it is not likely that hydrogen enters its metallic phase there. There is an excess of helium, however, and this separates from the hydrogen, releasing gravitational energy.

9. Jupiter possesses a very strong intrinsic magnetic field with a high proportion of quadrupole component than the Earth's field. It presence was first recognised fom Earth in the emission of deca- and deci-metre radiations.

10. Saturn is unique in having a magnetic axis aligned precisely along the rotation axis.

11. The interaction between the Jupiter field and the solar wind provides the largest magnetosphere in the solar System.

12. The ring system of Saturn has been known since the early 17th century but the gossamer system around Jupiter was discovered much more recently.

13. The Saturn rings are constituted of metre-sized water ice particles. That for Jupiter is composed on ions ejected from the Galilean satellites and especially Io.

14. Jupiter and Saturn are examples of many of the exoplanets observed orbiting other solar type stars.

15

The Planets: Uranus and Neptune

Fig. 15.1. Uranus as seen by Voyager 2. (NASA)

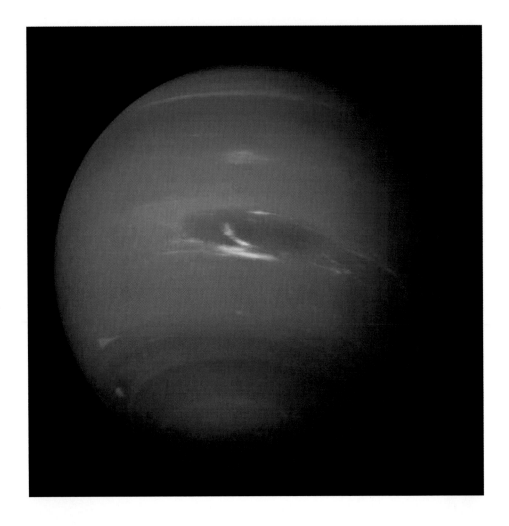

Fig. 15.2. Neptune as seen by Voyager 2. (NASA)

Uranus and Neptune are two very similar planets. They are giants in relation to the Earth but are much smaller than either Jupiter or Saturn. They show a range of properties which link them as a pair but distinguish them from others. Neither were known to the ancients requiring telescopes to see them. Uranus was discovered by direct observation (by Sir William Herschel, March 13, 1781) but Neptune was predicted mathematically (independently by John Couch Adams and Urbain Leverrier), on the basis of certain anomalies of the observed orbit of Uranus. Hearing of the prediction, Johann Gottfried Galle of the Potsdam Observatory turned his telescope to the predicted location in

Table 15.1.   Some bulk properties of Uranus and Neptune.

Property	Uranus	Neptune
Obliquity to orbit (deg)	97.77	28.32
J_2	3.34343×10^{-3}	3.411×10^{-3}
Ellipticity, f	0.02293	0.01708
Gravity (at 1 bar) m/s^2	8.69	11.00
Mom. Inertia factor, α	0.225	–

the sky and observed the new planet on September 23, 1846. Neither planet had been seen in any detail before the flybys by Voyager 2 in 1976.

Uranus is unusual in having a rotation axis which lies closely in the plane of its orbit. This means that each pole faces the Sun for roughly half an orbit. It also leads to some strange magnetic properties (see later). The basic data are given in Table 15.1.

15.1.   Surface Features

Unlike the surfaces of Jupiter and Saturn, the surfaces of Uranus and Neptune are not overburdened with details. Both appear blue because the red parts of the spectrum falling on the surfaces are absorbed by methane in the upper regions. Uranus is featureless in the Voyager photograph but surface features have subsequently been found by the Hubble Space Telescope. Bright clouds have been seen from time to time particularly in the equatorial region (which, remember, is a great circle perpendicular to the ecliptic). The Voyager photographs showed clouds above the Neptune surface. Such clouds have also been observed by the Hubble Space Telescope. Neptune also shows a series of dark spots, not unlike the Big Red Spot of Jupiter, although rather smaller and not as permanent, together with streamers. The largest of the spots seen by Voyager 2 and called the Great Dark Spot (GDB) was located at latitude 22°S circled the planet in 18.3 hours. The geometry is observed to change over a period of 4 or 5 days, the spot elongating and contracting in what appeared to be an 8-day cycle. A Small Dark Spot (SDS) at 55°S showed a rotation rate of 16 hours. The SDS was seen in close up by Voyager and appeared to be rotating clockwise. In this case the material will be descending and not ascending as for BDS. The problem of accounting for the spots is still unsolved.

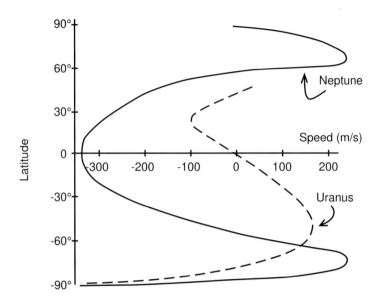

Fig. 15.3. The wind speeds on Uranus and Neptune. Data come from Voyager 2 and from the Hubble Space Telescope. (East is positive)

The surfaces of both planets show substantial wind speeds. As for Jupiter and Saturn, there are equatorial jets but here they are slower than the rotation speed of the planet. The (westward) speed of the jet for Uranus is about 100 m/s, as observed by the Hubble Space Telescope, but as high as 350 m/s for Neptune. Remember for comparison that the value for Saturn is about 500 m/s. The speeds become positive again at higher latitudes north and south, reaching 200 m/s on Uranus and 250 m/s on Neptune. The approximate data are shown schematically in Fig. 15.3.

15.2.   Heat Budgets

Data for the heat budgets for the two planets are collected in Table 15.2. It is seen that Neptune follows the pattern of Jupiter and Saturn in radiating more heat than it receives from the Sun but Uranus does not. Unlike Neptune, Uranus appears not to possess an internal source of heat. The surprising consequence is that, although the solar irradiances for the two planets are in the ratio 2.25:1, and the black body temperatures consequently in the ratio 1.3:1, the actual temperatures within the two planets are found to be very much the same. This can be seen in Fig. 15.4 which

Table 15.2.   Various thermal properties of Uranus and Neptune.

Property	Uranus	Neptune
Solar irradiance (W/m^2)	3.71	1.51
Albedo	0.51	0.41
Absorbed irradiance (W/m^2)	0.7	0.20
Emitted radn, (W/m^2)	$0.6 < R < 0.7$	0.7
(Internal/solar) source	≈ 0	2.5
Blackbody temp. (K)	58.2	44.6
Measured temp. (K)	112.0	81.1
Temp. 0.1 bar level (K)	53	55
Temp. 1 bar level (K)	76	72

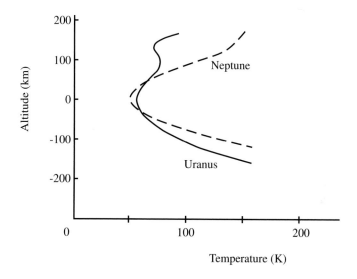

Fig. 15.4.   The variation of temperature with altitude for Uranus and Neptune. The two profiles are similar even though Neptune is much further from the Sun than Uranus.

shows the variation of temperature with altitude. The zero of altitude is taken to be the 0.1 bar level where the temperatures show minimum values.

Clouds are visible on the surfaces and the depth and composition were measured by Voyager 2. These were methane ice and floated at about the 0.1 bar level, roughly at the region of minimum temperature, which is the same altitude for both planets.

Table 15.3. The proportions (by volume) of constituents in the upper atmospheres of Uranus and Neptune. Data for the sun are included for comparison.

Material	Uranus	Neptune	Sun
Hydrogen	82.5%	80%	84%
Helium	15.2%	19.0%	16%
Methane	2.3%	1.5%	–
ppm	HD	HD, C_2H_6	H_2O, CH_4, NH_3, H_2S
Aerosols	Ammonia ice, water ice, ammonia hydrosulphide, methane	Ammonia ice, water ice, ammonia hydrosulphide, methane	
Mean mol weigh (g/mole)	2.64	2.6	–
Scale height (kg)	27.7	20	–

15.3. Visible Surface Compositions

The details are less well known than for Jupiter and Saturn. Table 15.3 list what details there are, the percentages being by volume. The proportions of hydrogen and helium are very similar to those for the Sun. The details of the minor constituents are preliminary. Uranus has very closely the same He abundance as the Sun.

15.4. Internal Structure and Conditions

Uranus and Neptune are very different from Jupiter and Saturn. Indeed, the masses of Uranus and Neptune are very similar to the predicted core masses of Jupiter and Saturn. The mass of Uranus is 14.536 Earth masses and 17.147 Earth masses for Neptune. Models suggest that some 10 Earth masses of Uranus are composed of elements heavier than hydrogen and helium while as much as 15 Earth masses of Neptune may be composed of such a mixture. The material here is largely water, ammonia and methane ices under high pressure and temperature. A central core of silicate/ferrous composition will have a mass of several earth masses. A schematic picture is shown in Fig. 15.5.

The central pressure and temperature are quite low by comparison with Jupiter and Saturn, the central pressure now probably being a few times 10^6 bar (around one million atmosphere) and the central temperature a

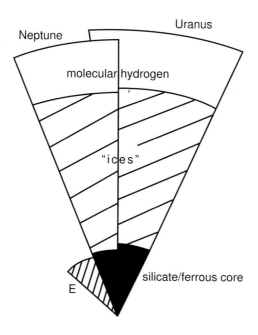

Fig. 15.5. A schematic view of the interiors of Uranus and Neptune. The Earth is shown for a comparison of size.

few thousand degrees Kelvin. There is no possibility under these conditions of hydrogen reaching a metallic state. The internal conditions are consequently simple. More detail must await more and better data in the future.

15.5. Comment on Interior Heat Flow

Neptune shows a heat flow similar to Jupiter and Saturn but Uranus shows no internal heat. Although this may seem to leave Uranus as the odd planet in fact it is the others that are unexpected. Radioactive heating will be negligible for all the planets and a simple contraction would be difficult to use as a heat source for Uranus and Neptune. One possible source of heat is the condensation of material at higher regions and the subsequent de-condensation lower down where the temperature and pressure are higher. The changes of latent heat in such processes will release gravitational energy due to falling of material in the magnetic field of the planet.

Table 15.4. The data for the intrinsic magnetic fields for Jupiter and Saturn. Corresponding data are given for Earth as a comparison.

Property	Uranus	Neptune	Earth
Mag. dipole moment (Am^2)	3.95×10^{24}	1.98×10^{24}	7.9×10^{22}
Mag. dipole moment $(E = 1)$	50.0	25.1	1
Equatorial field (gauss)	0.228	0.142	0.305
Equatorial field $(E = 1)$	0.754	0.459	1
Polar field (gauss)	1.0	0.8	0.70
Angle between axes	58.6	46.9°	11.5°
Offset vector	0.3 R_U along rot. Axis, S	Planet centre to dipole centre 0.55 R_N	0.0725 R_E 18.3°N/147.8°E
Polarity	Parallel	Parallel	Anti-parallel

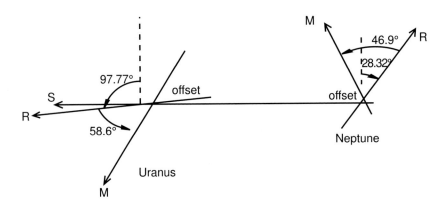

Fig. 15.6. The rotation (R) and dipole magnetic (M) axes of Uranus and Neptune. S gives the direction of the Sun. Each source is offset.

15.6. Intrinsic Magnetism

Both planets have intrinsic magnetism with details shown in Table 15.4. Neither field is as strong as the Earth's field and that for Uranus is nearly twice as strong as that for Neptune. Both fields make large angles with the planets rotation axes but that for Uranus is extraordinary because the rotation axis itself makes so large an angle with the orbit. The result is that alternate magnetic poles face the Sun during one orbit and this has consequences for the interaction with solar particle and magnetic emissions (the solar wind). The orientation of the axes is shown in Fig. 15.6.

The orientation of the rotation and magnetic axes for Neptune are unique in the Solar System. The larger magnetic dipole moments of Uranus and Neptune relative to Earth are due to the larger sizes of the major planets. The magnetic fields themselves are of comparable magnitudes.

*

Summary

1. Uranus and Neptune form a pair of similar planets, less massive than either Jupiter or Saturn. Neptune has a mass of about 1/5 that of Saturn and Uranus rather less than that.

2. Uranus is 14.5 times the mass of the Earth and Neptune is a little heavier at 17.1 times.

3. These masses are only slightly higher than what is believed to be the core masses of Jupiter and Saturn. They could be regarded as major planet cores with a slight covering of hydrogen. The internal composition will be similar to Jupiter but with the addition of ices, especially that of water.

4. For Uranus, the obliquity of the orbit (that is, the tilt of the equator to the orbit) is 97.86° meaning that the pole is directed towards the Sun. For Neptune, the obliquity is a "normal" 29.56°.

5. These rotation directions combine with the directions of the two magnetic dipole axes, −59° for Uranus and −47° for Neptune.

6. The interaction with the solar wind is complicated for Uranus because of the alignment of the dipole axis.

7. Neptune radiates more energy into space than it receives from the Sun, by a factor of about 3. Uranus, on the other hand, has not got an internal heat source.

8. Although there is no obvious banded structure on the surface of either planet, there are substantial wind speeds in the atmosphere, especially in the equatorial region. The origin for Neptune will be the internal source of heat but the source for Uranus is less obvious.

9. Each planet has a thin rings system orbiting about the equator.

10. More data are required before these planets can be understood properly.

16

Satellites of the Solar System

The Solar System is rich in satellites, especially associated with the major planets. One of those for the terrestrial planets, the Moon, has been visited by astronauts. We survey these various bodies here.

The Terrestrial Planets

There are three satellites, two of Mars and one for the Earth. Of these, that for the Earth, the Moon, is the only substantial body. Neither Mercury nor Venus has a satellite companion.

16.1. The Moon

16.1.1. Some historical background

The Moon has been a mystic in the sky since antiquity. Its motions were used as a clock for agriculture and for religion very early on and it has been the source of many myths and stories over the centuries. It is not easy to see clearly by eye but this changed with the introduction of the telescope by Galileo in 1609. The first map appeared in 1645, the work of Langrenus who was Astronomer to the Court of King Phillip II of Spain. It included some 300 features including 250 prominent craters. He named lunar features after earlier Spanish kings and nobles but all these names except one did not survive — the exception was his own named object. A second map was published two years later by Hevelius. In his map the mountains of the Moon are named after Earth mountains. In a third map in 1651, Riccioli introduced the naming system that is used today. The dark flat regions, regarded then to be seas, were given Latin names involving descriptions — Mare Imbrium (Sea of Rains), Mare Serenitatis (Sea of Serenity), Mare Procellarum (Sea of Storms) and so on. The craters he named after historical figures, scientist and men of letters. Today the names of astronomers and those involved in

Fig. 16.1. The cratered surface of the nearly full Moon, seen from Earth.

lunar studies are used particularly. This practice has been accepted by the ruling body of astronomy, the International Astronomical Union (IAU).

The Moon is the only body beyond the Earth that has been surveyed by human observers. The first landing was on July 20, 1969 by Apollo 11 in the Sea of Tranquillity (Mare Tranquillitatis). The last was with Apollo 17 in December 1972 in the Taurus Mountains and Littrow Valley. In the 1960–70s, the six Apollo lander missions[1] returned several kilogrammes of lunar

---

[1]The Apollo 13 landing was cancelled in flight due to the explosion of a concealed oxygen tank in the command module during the journey to the Moon. The astronauts were able to return to Earth safely.

Fig. 16.2.   The locations of the six Apollo landing sites. Apollo 13 didn't land. (NASA photo: 72-H-183)

material to Earth for study in the laboratories. Russian automatic probes also made landings about this time and returned a small quantity of material back to Earth. The lunar material composition is, therefore, fairly well known at several sites over the surface. The Apollo landing sites are shown in Fig. 16.2. Apollo 11 in Mare Tranquillitatis, 12 in Oceanus Procellarum, 14 in Fra Mauro, 15 in Hadley Rille and Apennine mountains, 16 in Decartes Plateau and 17 in the Taurus mountains and the Littrow valley.

16.1.2.   Bulk properties

The Moon has a mass of 7.349×10^{23} kg and mean density 3,350 kg/m^3. It has an ellipticity $f = 0.0012$ and a surface gravity of strength 0.166 that of Earth. It has an albedo of 0.12 or approximately one-third that of the Earth. This means more solar energy is absorbed by the Moon's surface than for the Earth. It has the same solar irradiance as Earth which gives a black-body temperature of 274.3 K. It also means that the lunar material will appear black, relative to that of the Earth. The moment of inertia factor is $\alpha = 0.394$ with $J_2 = 2.027 \times 10^{-4}$. The Moon is not quite a perfect sphere. The polar radius is about 2 km less than the mean equatorial one but this has a 2 km extension in the direction of the Earth. It has been

Fig. 16.3. The far side of the Moon. Only one maria, named Tsiokowsky, appears on the surface. (NASA)

found, from orbital variations of space craft, that the centre of mass of the moon is displaced by about 2 km in the direction of the Earth from the mean geometric centre. The Moon holds one face always towards the Earth, although there is a slight libration which allows a little more than 50% of the surface of the Moon to be observed from Earth. The rear side has been seen only by using space vehicles. A photograph of the far side is shown in Fig. 16.3 in Mercator projection.

16.1.3. Surface composition: The lava flows

The composition of the rocks on the Moon is very similar to those found on Earth. The lunar samples were, however, entirely lacking in water and even the iron in the minerals show no indications of oxide (rust). All Apollo rock samples are igneous rocks that formed from a once molten silicate material. There are no sedimentary rocks (formed by depositing minerals under water), nor are there any metamorphic rocks (formed when igneous or sedimentary rocks are buried deep underground and so experience high pressure and high temperature). The lunar rocks are of variable composition over the different sites visited by the Apollo missions.

The rocks from the Maria are volcanic lava called basalt. This material is well known on Earth, for instance in Hawaii and Iceland, both islands being of recent volcanic origin. This is the material composition of the upper part of the mantle. The lunar lava have ages between 3,700 and 3,300 million years. They form a very thin skin that covers about $1/5^{th}$ the lunar surface area. Their compositions are in many cases Earth-like. Pyroxene (Ca, Mg, Fe, silicate) is the most common mineral, making up about 50% of most specimens. Olivine, so common on Earth, is less common in lunar rocks and may, indeed, be absent from many flows. Plagioclase and Ilmenite are quite common but there are only small quantities of spinel. Although there are many similarities, lunar rocks differ from Earth rocks in four important respects. First, as has been said, lunar rocks do not contain water. Earth rocks contain usually 1% or so but lunar rocks have essentially none. Second, there is probably twice as much iron in lunar samples. This occurs as ferrous oxide, FeO, with some free iron. On the Earth's surface, iron occurs as Fe_2O_3 (rust) with some FeO, but there is no free iron. Third, lunar rocks contain more Ti than on Earth. In fact, Earth rocks are depleted in Ti in relation to other cosmic bodies, for example meteorites. With more Fe and Ti than Earth rocks, the lunar rock lava would have been much more fluid than Earth lavas. This would have allowed the lunar lava to cover rather greater areas when it was extruded. Fourth, there is a small amount of Na and K and other volatiles in lunar rocks.

These features have implications for the formation of the rocks. First, the temperature of formation must have been very high to boil off the volatiles, probably around 2300 K. This suggests a molten formation. Second, the rocks must have formed without oxygen. Estimates suggest an initial oxygen atmosphere with a partial pressure of less than 10^{-5} bar. The cooling of lavas in the absence of water and oxygen produced some new minerals not found on Earth. Three can be cited:

(i) pyroferroite — a mineral similar to pyroxene but with Fe;
(ii) armalcolite — a mineral like ilmentite being a Fe, Ti, Mg oxide. The name is an amalgam of the three Apollo 11 asrononauts who found it — ARMstrong, Aldrin and COLins.
(iii) tranquillityite — a mineral containing Fe, Ti and Si, together with the rare elements Zi, Yt and U.

16.1.4. Surface composition: The highland rocks

The highland areas account for about 4/5 of the surface area and are significantly older than the lava flow regions. Radioactive dating of Apollo highland rocks show ages between 4,000 and 4,200 million years. This is still

younger by 400 million years than the accepted age of the various members
of the Solar System. Supposing the age of the Moon is the same as that of
the other members this leaves an initial period for which no lunar record
has yet been found. There are several possibilities to account for the missing
first 4×10^9 years. The most direct assumption is that the lunar surface was
molten for much of this period and the highland rocks formed slowly. Other
possibilities involve a number of intermediate incidents for which there is no
evidence at the present time.

The highlands are light coloured and heavily cratered. The rocks are
all igneous and share with the lava flows a complete lack of water with
practically no volatile Na or K. The detailed chemistry and composition of
the highlands is, however, quite distinct from the lavas. For one thing, all
highland rocks are enriched in Ca and Al but depleted in Ti, Mg and Fe.
There is more plagioclase than in the mare; indeed, this appears to make up
at least 1/2 the highland rocks together with pyroxene, olivine and spinel.
The highland rocks, unlike the mare lava, appear to have cooled slowly,
presumably underground.

16.1.5. Surface composition: The lunar "soil"

The Moon is covered by a deep layer of rock broken into very small fragments. It has been liken to a layer of rubble with irregular particles of mean
dimension 10^{-1} m and is in no way comparable to soil found on Earth. There
are also glass spheres of the same size produced by melting. Also present are
small particles of meteorite that help form the soil — detected by the different chemical compositions shown by these objects. The layer covers any
bedrock that may be there. It has been formed by the continuous impact
of meteorites, solar wind and solar radiation uncontrolled by the presence
of an atmosphere. Over thousands of millions of years these processes have
produced a surface coating between 5 and 15 metres thick. The "soil" is
churned up by the incoming materials and is well mixed. This process has
been called "lunar gardening". Such cosmic gardening does not happen on
Earth.

16.1.6. The interior

The Apollo missions included seismicapparatus. That on Apollo 11 operated
as a single instrument and was not very effective but later missions formed
a group of instruments which worked remarkably well. The impact of meteorites can be recorded if the impacter has a mass in excess of about 10 kg

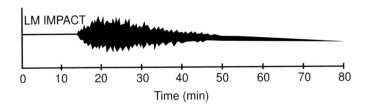

Fig. 16.4. A typical lunar seismic trace. (NASA Educational Publications)

and about 100 such events were recorded annually. It became clear immediately that the natural seismic activity within the Moon is substantially less than that on Earth both in frequency and magnitude. Natural events occur at an unknown time and an unknown place and, as on Earth, this uncertainty restricts the information that can be deduced. The situation is substantially improved by taking control of the source of the activity: astronauts also caused stages of the spacecraft to strike the surface so that the time and place of the surface impact, and the exact mass of the impacter, were known.

The moonquakes are nearly all of magnitude less than 2 on the Richter scale. A typical trace is shown in Fig. 16.4 taken by Apollo 15 and resulting from the impact of the discarded lunar module. The trace begins at the left. The interesting thing is that the disturbance builds up to a maximum and slowly decays. The full disturbance lasts for nearly two hours, which is some twenty times the length for the corresponding event on Earth. The Moon apparently rings like a bell showing little of the dissipation found on Earth. This is fully consistent with an outer layer of dry, fine shattered rock that allows the signal to reverberate for a very long time.

There are about 3,000 events per year compared with the Earth's tally of about 300 times this number and of greater magnitude. All the lunar activity has a source in the band between the depths of 600 to 800 km below the surface. For comparison, on Earth it was seen that very few are at depths as great as 600 km and none have been found below 700 km. More than this, the lunar activity seems to be associated with specific regions of the interior. Perhaps 40 centres have been actually identified, all except one lying on the Earth side of the Moon. They lie within two belts some 2,000 km across the interior with a width of about 300 km. These presumably mark regions where deformation is still taking place. Whereas all earthquakes can occur at any time, many (but not all) moonquakes have a periodic pattern linked to the orbital period of the Moon about the Earth. Both the Earth and the

Moon cause tides in each other as they orbit about their common centre of mass. The tides in the solid Earth due to the Moon are small but detectable: the tides in the Moon due to the Earth are approximately six times larger. The associated moonquake activity occurs periodically every 14 days.

The Moon, like other bodies, has a crust encasing a mantle with a central core. The value of the moment of inertia factor is $\alpha = 0.394$ with $J_2 = 2.027 \times 10^{-4}$. This value of α is sufficient to allow a small ferrous core, or slightly larger ferrous-sulphur core, perhaps with a radius as large as 600 km. The body of the Moon is shifted towards the Earth by about 1 part in 870 and this asymmetry is magnified in the inner structure. A schematic cross section is shown in Fig. 16.5, though somewhat simplified. The most obvious feature is the difference between the crustal thickness facing the Earth and facing away from it. On the Earth side the thickness is some 60 km but on the opposite side it is close to 100 km. With one exception (the crater Tsiokovsky, see Fig. 16.3) all the maria are on the Earth facing surface, presumably associated with the thinner crust. Presumably this difference dates back to the earliest times. The crust is rich in Al and Ca being composed of babbro and anorthosite. The mantle below is a denser material, probably composed of pyroxene and olivine and so rich in Mg.

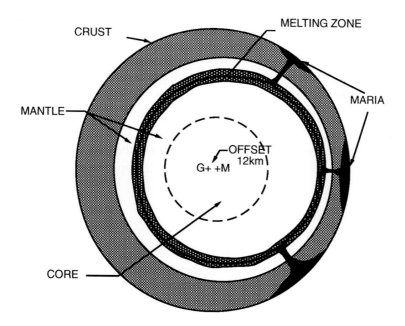

Fig. 16.5. The interior of the Moon deduced from lunar seismology. (After B. M. French, *The Moon Book*, Penguin, New York, 1977.)

16.1.7.   Lunar magnetism

Perhaps the most surprising discovery of the Apollo missions was the widespread occurrence of magnetism over the surface of the Moon. The magnetism has been shown to have the form of the stable remnant magnetism found in Earth rocks that have cooled encased in the Earth's intrinsic magnetic field. There is every indication that the Moon itself possessed a strong intrinsic field during the great period of cooling between 4.2 and 3.2 thousand million years ago. The measured fields now vary in strength between a few tens of gammas and a few hundreds. Generally speaking the highland regions show the largest fields now.

The origin of the remnant field is unknown. Various suggestions have been made for it but none are more likely than a present relic of an earlier feature of the Moon. The presence of the field early on would imply constraints on the lunar core but little is known about that. Indeed, the origins of the Earth's field still holds mysteries and the account of the lunar field must be left unfinished for the present.

16.1.8.   Transient lunar events

Sir William Herschel reported observing red glows on the Moon in 1783 and 1787. These were suggested to be small volcanic eruptions but this now seems unlikely. Since then many observers have reported a range of events involving brief colour events, hazes, glows and temporary obscurations of the surface locally. The events seem associated with particular locations such as the craters Aristarchus, Plato and Alphonsus. The edges of the maria have also been featured in such observations. The cause of these events is quite unknown at the present time. Perhaps a mission to one or more of the sites may allow the matter to be cleared up. These observations are still made from time to time which makes it interesting to report them now.

16.2.   The Satellites of Mars

Mars has two satellites, both very small bodies probably captured long ago from the asteroid belt. Each has a density below 2,000 kg/m^3 and they appear to be composed of silicate/ferrous material, probably very porously packed. Images of the satellites are shown in Figs. 16.6 and 16.7. The bodies have masses of 1.06×10^{16} kg (Phobos) and 2.4×10^{15} kg (Deimos) and are really large rock material. Deimos has an orbital period of 1.26 days and so will pass though all its phases each (Mars) day. Phobos with an orbital

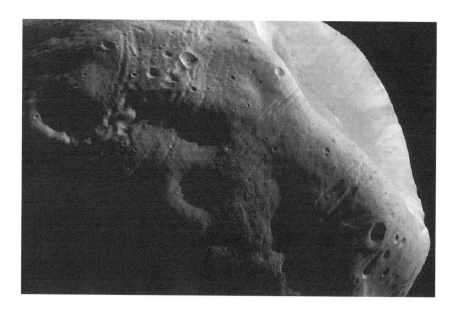

Fig. 16.6.   An image of Phobos showing an irregular body with a lightly cratered surface. (NASA)

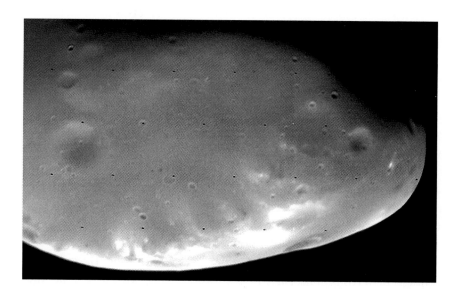

Fig. 16.7.   An image of part of Deimos.   The irregular surface carries some craters. (NASA)

period of 0.319 days will pass through the phases three times a day! Phobos is in an orbit that will eventually impact with the surface of Mars.

The Major Planets

The major planets between them are associated with a large array of natural satellites of many sizes: Jupiter — 61; Saturn — 30; Uranus — 21; and Neptune — 8. This will not, in fact, be the final tally because very small satellite bodies are still being discovered with the recently improved optical detection systems. The largest satellites of Jupiter, Saturn and Neptune are comparable in size to Mercury and Moon, confined to the equatorial plane of the planet. No more of these will be discovered. The interest in them is their general composition. All except one (Io) are a mixture of water ice and silicate materials, and especially water ice, which is a different type of hard "rock" for the long term construction of the surfaces. The one exception, Io, is a silicate/ferrous body with a hot interior which provides a plastic and ever changing surface.

The physical data for the satellites were presented in I.6 where it was seen that they fall into the groups of large, medium and small objects. It is convenient to review these bodies using these terms now.

16.3. The Larger Satellites

There are six members of this class, four for Jupiter (Io, Europa, Ganymede and Callisto), Titan for Saturn and Triton for Neptune. Their comparison with Mercury and Moon is shown in Fig. 6.4. It is seen that Ganymede is the largest of all, in fact being rather larger than Mercury. The mean densities vary throughout the group. This is shown in Table 16.1, Mercury being very high. The equatorial radii are more or less nearly the same.

The range of densities is taken to reflect the different compositions. Mercury is the outstanding member where the composition is silicates and particularly ferrous materials. The silicate density in the region of the upper Earth mantle has a density of about 3,330 kg/m^3 and, if this is broadly typical of the silicates in the inner Solar System, then there is a small percentage of ferrous materials in the Moon. The proportion in Mercury is very high to provide so high a mean density for the planet. The largest mean density for the Galilean satellites is that for Io which is slightly higher than that for Moon. This may be due to a higher proportion of free ferrous

Table 16.1. The equatorial radii and mean densities of the principal satellites and for Mercury.

Body	Eq. radius 10^3 km	Mean density 10^3 kg/m^3	Body	Eq. radius 10^3 km	Mean density 10^3 kg/m^3
Io	1.816	3.55	Titan	2.56	1.90
Europa	1.563	3.04	Triton	1.90	2.00
Ganymede	2.638	1.93	Moon	1.74	3.35
Callsto	2.410	1.81	Mercury	2.44	5.427

Table 16.2. The calculated proportions of rock and water ice in the principal satellites using approximate arguments. Data for Pluto has been added for comparison. The columns concerned with a core presume all the rock material is concentrated towards the centre. (Rock density = 3500 kg/m^3; ice water density 920 kg/m^3.) The mean surface temperatures are also included.

Body	Proportion of rock (%)	Mass of rock ($\times 10^{19}$ kg)	Core radius, R_c ($\times 10^2$ km)	Core radius R_c/R_s	Proportion of ice (%)	Mean surface temp. (K)
Europa	80.6	3928	14.0	0.88	19.4	103
Ganymede	38.0	5662	15.0	0.59	62.0	113
Callisto	33.0	3511	13.0	0.55	64.0	118
Titan	37.5	5095	15.0	0.59	62.5	93
Triton	43.0	925	10.3	0.76	57.0	58
Pluto	13.5	210.6	6.5	0.44	86.5	42.7

materials or to a heavier silicate. Either way, the difference is small and it is best to suppose that the mean density for Io is typical of the non-water ice material for all the Galilean bodies.

The densities for the other three members suggest a dilution of the silicates with a material of much lower density. Remembering that the first and (chemically) second most abundant elements in the cosmos are hydrogen and oxygen this lighter material is most likely water. Certainly water forms the surfaces of these satellites. If this is the case (and all the evidence to date suggests that it is) the proportion of silicates/ferrous (mean density 3500 kg/m^3) and that of water ice (mean density 920 kg/m^3) can be calculated. The result of approximate calculations is shown in Table 16.2. These data, although approximate, show that, with the exception of Europa, the bodies are very largely composed of ice. Pluto has been included for a

Fig. 16.8. The surface of Io showing its many volcanoes. (Galileo, NASA)

comparison and it is seen that that body is nearly 87% water ice. If it is an Edgeworth–Kuiper object and is not untypical then these objects as well are very largely balls of ice.

16.3.1. Io

Io is one of the most remarkable bodies in the Solar System. It is the one body of the Galilean satellites without water ice and a recent picture of the surface is shown in Fig. 16.8. It has a bright yellow colour which results from a thinly covered sulphur surface and sulphur rocks. Different shades of sulphur refer to different temperatures over the surface giving a rather crude indicator of the temperature distribution. The surface contains several hundred volcanoes, some recent but some (for instance the Prometheus[2] plume rising 70 kilometres above the surface) are still visible as they were when the first pictures were obtained by the initial Voyager flybys of 1979. A plume invariably rises above the volcanic caldera such as the 140-kilometre

[2]In Greek mythology, the God Prometheus gave fire to mortals.

bluish plume that rises above the volcano Pillan Patera. There are no impact craters on the surface showing the surface is renewed too frequently for these to be visible. Generally, the surface looks like an area full of boils and calluses (which in a way it is) although some people prefer to say it looks like a pizza.

The equilibrium thermal temperature of the surface is about 118 K but the volcanic regions are much hotter, perhaps as high as 1,000 K or higher. As many as 80 hot spots have been located on the surface, most showing a degree of permanence of location. The source of the internal heat driving the volcanic activity is gravitational friction. Io is under the strong gravitational effects of the elliptically shaped Jupiter together with interactions with the other Galilean moons. It was seen in I.3 that the result will be a stretching/compressing of the internal material of Io. The associated friction within the slightly inelastic materials produces heat which drives the volcanic system. This mechanism has been shown to provide energy of the correct magnitude and, indeed, was suggested theoretically in 1974 a little before the surface of Io was observed (in 1975) to be the result of substantial volcanic activity. The surface of Io is described as showing plumes and volcanoes and it is important to be clear about the distinction between these two phenomena because they are not the same.

The nature of the volcanism on Io was debated soon after it was discovered because sulphur was undoubtedly on the surface (and could be identified spectroscopically) but it is a weak material to support elevated surface material. Again, although sulphur is very abundant throughout the Solar System it is not so abundant that a substantial satellite can be supposed to be composed of it. More than 250 caldera with a diameter in excess of 20 km have been observed and will stand several kilometres above the mean surface level. The highest regions stand 10 km or 20 km above it. One example is Hæmus Mons near the south pole. This region is an area some 20,000 square km in extent which rises 10 km above the surface. The underlying rock must have sufficient strength to sustain such a mass. Sulphur is not strong enough but, of course, silicates are. It must be that the interior of Io is composed of silicates and ferrous mineral, like the terrestrial planets, but with a component of sulphur that can be recycled onto the surface. This structure will also account for the mean density of the satellite. The volcanoes are, therefore, silicate entities like those on Earth issuing molten larva, with temperatures around 1,000 K. There are many of them because the internal temperature is high and substantial quantities of heat must be radiated into space. The photograph of a typical caldera

Fig. 16.9. A close view of the caldera of a volcano. The flows of molten sulphur are clearly evident.

is shown in Fig. 16.9. The different colours of sulphur relate to different temperatures.

Separately from volcanoes the Io surface shows a system of long lived plumes that rise mushroom shaped high above the surface. These are the equivalent on Io of geysers on Earth. Instead of water they use sulphur. Sulphur and sulphur dioxide can act as a geyser fluid if heated under pressure below the surface. Nozzle velocities of 1,000 m/s can be achieved this way and these fit well with those observed on the satellite. Hot silicate magma will be the heating material. The effect will be a thin layer of sulphur deposited over the surface. The different colours over the surface would seem to be the result of suphur at different temperatures. The Io surface is unique in being a silicate surface covered with a thin sulphur "snow".

The emission of silicate gas lavas and sulphur compounds in geysers would suggest the presence of a substantial atmosphere but this is not observed. All indications are of a very tenuous covering of the surface. There are indications that it is a little more dense above volcanic regions and so appears rather patchy, especially at mid-day. Data from the Pioneer 10 spacecraft (of 1973) showed a substantial concentration of electrons in the vicinity of the surface suggesting the presence of a restricted ionosphere.

This will affect magnetic fields that will be part of the Jovian magneto-sphere. The environment of Io contains many suphur ions, among others, and Io's orbit is within the magnetosphere of Jupiter. The ions form a doughnut shaped ring around the equatorial region of the planet. The result is a complicated plasma-ionic link between the satellite and the planet pro-ducing complicated electric currents and discharges. Io and its associated links with Jupiter have been the object of sctudy during the Galileo space probe observations of Jupiter and its moons.

16.3.2. Europa

This is rather the opposite of Io. Here the surface is covered with water ice, as is shown by the strong absorption of radiation in the near infra-red part of the spectrum by the surface. That the surface is not simple has been realised for the last 40 years. It was realised using Earth-based telescopes that the leading hemisphere is covered by a water-ice frost. By contrast, the training hemisphere appears redder. These effects have been found to result from the bombardment of the rear surface by sulphur ions in orbit which arise from Io. The detailed close up pictures that resulted from the Pioneer and Voyager flybys of the 1970s (one of which is shown in Fig. 16.10) show a smooth surface almost entirely devoid of craters. In fact no more that half a dozen

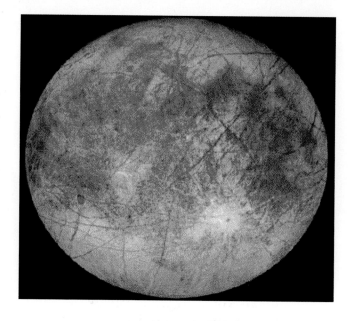

Fig. 16.10.   A full view of Europa. (NASA)

craters are seen on the surface. This implies a very young surface which itself implies heating in the relatively recent past. The crater in the lower right of the surface is called Pwyll, a very recent crater with a diameter of about 50 km. The long dark lines are fractures of the crust, some extending over 3,000 km in great circles.

It has been realised for some time that the gravitational heating of Io should also occur within Europa but at a lower intensity. The proportion of ice that is known to be in the surface region will be at a higher temperature than that lower down with the result that it will undergo tidal movements due to Jupiter and the other Galilean satellites. The ice covered surface is observed to have a form reminiscent of the ocean pack ice on Earth and this is believed to be the underlying cause of the observed surface detail on Europa.

The silicate/ferrous material will, presumably, contain a radioactive impurity which will produce heat. It is widely believed that this heating,

Fig. 16.11.   The face of Ganymede, the largest satellite in the Solar System.

together with the tidal heating, will provide a liquid water layer below the surface. There is some conjecture that such a water layer, should it exist, could have become the evolved home of elementary living materials. There seems no firm evidence for this but it is not impossible.

16.3.3. Ganymede and Callisto

Ganymede is the largest of the satellites of the Solar System and its surface shows every sign of being very old. Its has the appearance of being smooth and in some sense transparent, almost as if one is looking inside it, but with a spread of impact craters as is seen in Fig. 16.11. In fact, the surface is far from smooth as is clear from Fig. 16.12 where a cratered terrain is seen with smaller craters within larger ones.

The surface falls generally into two parts, one dark and the other lighter in colour. The darker regions are presumed to have suffered evolutionary changes originating in the interior while the smooth parts are presumed to be regions resurfaced with water turned to ice. This is the ice-equivalent of silicate lava and the process of water "volcanoes" is called cryovolcanism. The surface has many complex markings and the origin of these is not yet understood. One example is shown in Fig. 16.13. The appearance is of a complex flow of material. It contains craters and so must be the relic of an earlier age. The surface bears witness to considerable internal activity which is consistent with it, like Io and Europa, being subject to internal heat dissipation due to the gravitational effect of Jupiter and the resonance with Io and Europa.

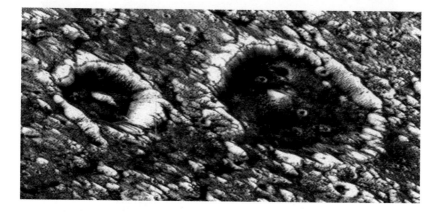

Fig. 16.12.   Craters on the surface of Ganymede. (NASA)

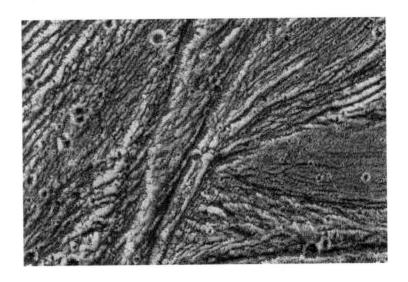

Fig. 16.13.   Complex surface structure of Ganymede. (NASA)

Fig. 16.14.   The full face of Callisto. (NASA)

Callisto is the last of the Galilean family and the furthest from Jupiter: a photograph of its icy face is shown in Fig. 16.14. Its surface is covered by craters of every size, some tens of kilometres in diameter. The largest is named Valhalla and is the central region which is some 600 km in diameter. The surprise is that the surface shows considerable smoothness at the very crater size. It is as if a mechanism is at work breaking up the surface; perhaps reminiscent of the "gardening" met on the Moon. There is, however, every indication that the interior has seen very little activity over the lifetime of the satellite.

16.3.4.   Titan

This satellite of Saturn has long been regarded as an intriguing object because it is the only satellite in the Solar System with a substantial atmosphere. It has a surface pressure of about 1.5×10^5 Pa or about one and a half Earth atmospheres. It is opaque to optical wavelengths but is partially transparent at infra-red frequencies. This means the surface had not been seen directly until recently when the first very hazy images were obtained using 1.6 micron radiation. Earth-based observations had shown previously that the atmosphere contains particles with diameters in the range 0.2 to 1.0 micron. Such particles cannot remain suspended in a stable atmosphere for any length of time which suggests a circulation within the atmosphere and presumably accompanying "weather".

The composition of the atmosphere was first studied by Voyager 1 and is unusual. Nitrogen is the major component (82–99%), with methane (6–10%) and probably argon (less than 5%). The Voyager 1 spectrometer also found other organic compounds in the parts per million category including ethane (C_2H_6), acetylene (C_2H_2), ethylene (C_2H_4), propane (C_3H_8), methylacetylene (C_3H_4) and diacetylene (C_4H_2). There is also a proportion (about 2,000 parts per million) of hydrogen. Oxygen compounds have since been found (CO and CO_2) and nitrogen compounds such as hydrogen cyanide, cyanogen, cyanoacetylene, acetonitrile and dicyanoacetylene. The processes that fire this chemical laboratory can only be guessed at the present time. One interesting fact involves naturally occurring hydrogen. This contains a small proportion of deuterium and the distinction between methane molecules with H and D can be distinguished. It appears that deuterium found on Titan has between 4 and 8 times the abundance of that found in the atmospheres of Jupiter and Saturn. The deuterium levels in Oort cloud comets are some 4 times that found on Titan and twice that found in the oceans on Earth. The measured temperature of the surface is 94 K

$(-179^\circ C)$. Water has a vanishingly low vapour pressure at this temperature. Although water ice is likely to be an important component of the subsurface material, it will not be expected to participate in the chemical processes of the atmosphere. It might be that the composition of the atmosphere of Titan has some resemblance to that of the early Earth although the temperature of the early Earth will have been some three times greater than that for Titan now.

So interesting an object was given its own part in the joint NASA/ESA Cassini space mission to Saturn launched from Earth in 1997. ESA devised and built a free lander vehicle named Huygens weighing 319 kg and designed to study the atmosphere and surface of Titan in several ways. It

Fig. 16.15. First impressions of the local surface of Titan. The "rocks" are composed of frozen water ice. (ESA/NASA/ISA)

was carried by the Cassini spacecraft and launched free from the orbit about Saturn. Huygens would have contact with Earth via the orbiting Cassini probe. This most daring and ambitious project has, in fact, achieved complete success. The Huygens craft landed on the Titan surface without incident on 2005 January 14. On entering the atmosphere it used successively three parachutes to reach the surface after a $2\frac{1}{2}$-hour descent. The various instruments worked perfectly providing a wealth of information about the atmosphere that will take many years to evaluate. Preliminary analysis shows that these data support entirely the earlier Voyager conclusions. There had been much speculation of whether the surface would be solid or liquid, or a mixture. In fact it turned out to be like malleable clay or wet sand at the landing site. The probe landed with a speed of about 4.5 m/s (about 9 miles per hour) and seems to have penetrated some 15 cm into the surface material. The batteries on the probe were expected to last only a few minutes but, in the event, lasted for about 90 mins.

The pictures wired back, via the Cassini orbiter, show a remarkable surface. The first impression was of a misty, rocky, terrain with an orange tinge as seen in Fig. 16.15. This is similar to the first impressions of the Martian surface although the two are very different. The clay-like surface would be consistent with it having been raining just before the probe landed but the rain would have been of methane (melting point 90.5 K, boiling point 111.5 K) and not of water.

Two other pictures of the surface show details of the landscape. The first was taken at a height of 8 km during the descent and is shown in Fig. 16.16. The other, Fig. 16.17, was taken much closer to the surface. Figure 16.16 shows a general undulating surface with a peak towards the right-hand side formed by geological processes familiar on Earth. Figure 16.17 shows a typical drainage system caused by precipitation, again very familiar on Earth.

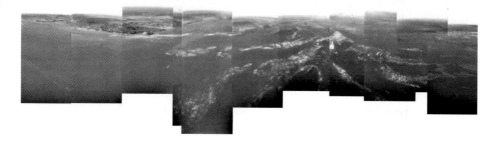

Fig. 16.16. The surface taken as a height of 8 km showing a rocky, undulating surface. (ESA/NASA/ISA)

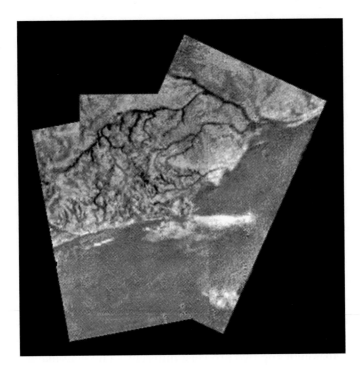

Fig. 16.17.   The surface of Titan showing a "river" system. (ESA/NASA/ISA)

The basin into which the drainage occurs is apparently not a liquid sea but rather a plain that becomes saturated with liquid draining off the elevated land, and which subsequently evaporates. The marks in Fig. 16.17 are clouds near the ground, presumably composed of methane as the only organic liquid of those available that has a triple point in the temperature range. These first pictures give a vision of a landscape with abrasion, erosion and precipitation very similar to Earth. A microphone aboard Huygens recorded the sound of gentle winds at the surface.

Descriptions of this type suggest a planetary body like the Earth and in some ways that is true. There is, however, an enormous difference in temperature between the two bodies and this is crucial. It has been suggested that Titan may lead to information about the formation of life on Earth. Terrestrial life seemed to have formed under hot conditions and probably under water. The Titan surface cannot match that but this does not necessarily prohibit elementary life having formed there. It will take many years to analyse all the data provided by the Cassini/Huygens mission and we can look forward to unexpected results in the future.

Table 16.3. The proportions of rock and the mass of rock in the secondary satellites.

Body	% Rock	Mass of rock ($\times 10^{19}$ kg)	Body	% Rock	Mass of rock ($\times 10^{19}$ kg)
Mimus	10.5	0.39	Miranda	45.7	1.64
Enceladus	7.4	0.55	Ariel	42.3	28.3
Tethys	3.6	2.25	Umbriel	41.5	31.5
Dione	19.0	19.95	Titania	43.3	52.0
Rhea	10.7	24.4	Oberon	42.3	34.7
Iapetus	11.5	22.2			

16.3.5. Triton

Triton is the coldest satellite body of the Solar System with a surface temperature of 58 K. The bed rock is water ice but the surface contains other molecules and especially methane, nitrogen, CO and CO_2 as well. There is a tenuous atmosphere of nitrogen with a mean pressure of about 15 microbars. The Triton surface is largely devoid of craters suggesting that it is young. This implies internal activity in the satellite which resurfaces the body by some form of cryovolcanism. Plumes have been observed issuing from the surface, presumably composed of liquid nitrogen, sending material 8 km or more above the surface. This is clearly a very active body inside whose properties have yet to be investigated. It is believed to have been captured by Neptune at some time in the past but its origin is unknown.

16.4. The Smaller Satellites

The physical data are given in Table 6.4. The low mean densities suggest they are mainly icy bodies and the entries in Table 16.3 confirms this. The quantities of rock are very variable and there seems no logic to the distribution in terms of orbital ordering around the parent planet.

16.5. Internal Conditions: Internal Differentiation

It is of interest to attempt an assessment of the possibility of internal differentiation of the interior. By this is meant the separation of the materials according to density, the ferrous material having sunk towards the centre forcing the less dense materials (water ice and silicates) upwards. The criteria chosen is the ratio of the gravitational energy of the material per atom to

Table 16.4. The ratio of the self-gravitational energy to an average interaction energy per atom in the icy satellites.

Body	e_d	Body	e_d	Body	e_d
Europa	1.5	Dione	0.33	Mimas	0.02
Ganymede	11.9	Rhea	0.34	Enceladus	0.03
Callisto	9.1	Iapetus	0.33	Tethys	0.05
Titan	11.7	Ariel	0.43	Miranda	0.06
Triton	6.8	Umbriel	0.46		
Pluto	1.15	Oberon	0.49		

an average binding energy linking the atoms together, written as e_d. Movement can occur if the gravitational energy is sufficiently great. The values of this ratio for different satellites is shown in Table 16.4. Differentiation is possible if $e_d > 1$. It is seen that criterion is met for the principal satellites of the first column and Pluto. These bodies could be differentiated. The satellites listed in the fifth column are very unlikely to become differentiated unless they were formed in that condition. The satellites in the third column form an intermediate case where differentiation is not impossible.

An important factor is the time for differentiation to occur. Internal heat, such as from radioactive elements, will allow the differentiation to occur more quickly. Calculations have allowed the estimate that differentiation of a body of radius 10^6 m will take of the order of few times 10^9 years to complete. This is of the order or the age of the solar System so differentiation may be well advanced for many satellites. Heating from gravitational compression will certainly allow more rapid differentiation. It might be expected that the satellites listed in the first column of Table 16.4 will be fully differentiated, excluding Pluto which is a rather different problem.

*

Summary

1. The satellites cover a very wide range of body. Two are silicate/ferrous bodies while the remainder are a mixture of silicates/ferrous materials and water ice.

2. They vary in mass from the Moon and the Galilean satellites on the one hand down to very small, irregular bodies largely of water ice.

3. The Moon is the only body beyond the Earth that has been visited by astronauts in the Apollo programme between 1969 and 1972.

4. The rocks brought back by the Apollo missions have given us a detailed inventory of the history of the Moon. Rocks can be found back to about 4,000 million years old but the first 4–500 million years of the Moon's history is not known. Indications are that the surface was molten during much of this period.

5. In general terms the Moon has less volatile material than Earth but is devoid of water.

6. The seismic activity is much less than on Earth, both in terms of frequency of occurrence and magnitude.

7. The lunar interior shows a core which is probably liquid, a pyroxene/olivine mantle and a crust of gabro and anorthosite. The crust has a depth of about 60 km on the Earth side but nearer 100 km on the far side.

8. The two satellites of Mars are very different. They are very probably captured asteroids and one of them (Phobos) has an orbit which will take it eventually into the planet.

9. Io is a silicate/ferrous body whose interior is kept fluid by the non-elastic behaviour of its material under the gravitational attraction of Jupiter and the other Galilean satellites. It has the only active volcanoes beyond the Earth.

10. Titan, of Saturn, is of special interest because it holds a substantial atmosphere containing a range of molecules including hydrocarbons. It is due to be visitied during the Cassini mission in the year 2004.

11. The larger satellites co-rotate with the parent planet but the very smallest have a retrograde motion, moving in the opposite direction.

12. It would seem that many of the smaller satellites are bodies captured into orbit by the parent body after their formation.

Part III

MAGNETISM WITHIN THE SOLAR SYSTEM

The intrinsic magnetism of the condensed planet and its atmosphere together with the interactions with the solar wind.

17

Intrinsic Magnetism of the Earth

We explore now what is known of magnetism of the condensed Earth. The magnetic field was known to antiquity and was used very early on as a means of navigation. Only more recently has it been possible to view it on a truly global scale and to begin to relate the Earth's field to those of other planets.

17.1. The Magnetic Poles

The study of the Earth's field has defined the parameters of the subject. The field is seen most easily by suspending a magnetic needle (originally made from lodestone, a magnetised rock, but it can equally be a magnetised needle) by a string about its centre of mass. The needle does not hang arbitrarily but assumes a particular orientation which is characteristic of its location. In general, it will hang with its long axis pointing along a line close to, but not coincident with, that linking the north and south geographic poles. The hypothetical geographical points define the direction of the spin axis of the Earth. The positive direction of the axis is from the south pole to the north pole and the direction of the spin is anti-clockwise as view from above the north pole. The spin can be represented by an arrow with the line joining the south geographical pole to the north geographical pole with the arrow coming out of the north pole.

It turns out that the magnetism of the Earth can also be represented by a hypothetical arrow. It was Sir William Gilbert who, in his treatise de Magnetisme published in 1600, expressed the view that the Earth acts like a great uniformly magnetised sphere. This is also equivalent to the field of a simple bar magnet in which two magnetic centres of opposite polarity, called poles, are joined together to form a small bar. One pole is called north and the other south, the north pole of one magnet attracting the south pole of another magnet but repelling its (like) north pole. The field of the two linked

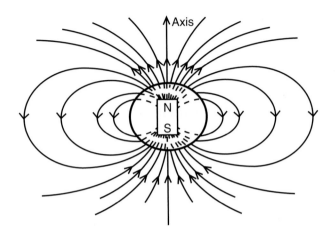

Fig. 17.1.   The magnetic field of a bar magnet has the same form as that of a uniformly magnetised sphere.

poles is called the field of the dipole, or simple the dipole field. Gilbert's statement can alternatively be said that the Earth behaves magnetically as if it contains a simple magnetic dipole. More complicated magnetic fields can be constructed by adding further dipoles. Thus, two bar dipoles linked together at their centres, with the axis one dipole at right angles to that of the other, form a quadrupole, with four poles, and so on. It is true that the magnetic field of the Earth most resembles the field of a dipole and the match is very good but, as will be seen later, is not perfect. The approximation of a dipole field is attractive as a method of description because it has a simple axis and so is convenient to describe apart from providing a good approximation to what is observed. The form of the dipole field is sketched in Fig. 17.1. The direction of the dipole axis for the Earth makes an angle of about 11.5° with the geographic axis. Observations over three centuries have shown that this angle changes slowly: the magnetic field is not static on time scales of that type.

A magnetised needle does not generally hang horizontally even when suspended precisely about its centre of mass but one pole points downwards, which pole depending on the hemisphere involved. It was this observation at a few locations over the Earth that led Gilbert to make his important generalisation about the magnetism of the Earth. In the northern hemisphere the north pole of the magnet dips downwards while in the southern hemisphere it is the south magnetic pole that is attracted downwards. While the angles of dip, as they are called, are generally small they have extreme values at two special points on the Earth. At a point near the north geographic pole the magnet hangs vertically with the north pole downwards, pointing into the ground and attracted by the magnetic pole of the Earth there. At a point near the south geographic pole the same situation is found but with the south magnetic pole hanging downwards. The Earth is attracting them with its magnetism and these two special points on the surface of the Earth are called the north and south magnetic poles. The magnetic axis through the Earth is the imaginary line joining these two poles. It is found that the magnetic axis is offset from the centre of the Earth.

This behaviour implies something special. Because the north magnetic pole of the suspended magnet is attracted by the north pole of the Earth and vice versa in the south, it follows that the north geographic pole is actually associated with a south magnetic pole while the south geographic pole is actually associated with a north magnetic pole. The directions of both the geographic poles and the geomagnetic poles are defined to act from south to north so it follows that the magnetic and geographic axes of the Earth are oppositely directed: the magnetic field is said to be anti-parallel. Recent accurate measurements of the location of the north magnetic pole show it to be a complicated region rather than a simple point. It extends, apparently, over a region of a few kilometres and is observed to be moving northwards at a rate of about 40 m each day. It is also found to execute a small daily elliptical path. The south magnetic pole has not yet been studied in detail but it is presumed that it will show the same features. This must imply that the magnetic axis is not a fixed line within the Earth nor, indeed, a simple phenomenon.

There is another special feature about the magnetic field. There is a line of points encircling the Earth in the equatorial region where the suspended magnet actually hangs exactly horizontally. Although it is a closed line in the vicinity of the lower latitudes it is not exactly a circle and it does not coincide with the geographical equator. This closed magnetic encircling line is called the magnetic equator. It separates the north magnetic hemisphere from the south magnetic hemisphere in exactly the same way that the geographical equator separates the geographical north and south hemispheres.

17.2. The Magnetic Elements

The strength and orientation of the magnetic field at any point is described by the magnetic elements, shown in Fig. 17.2. Three reference axes are defined — X pointing to the geographical north, Y to the east and Z pointing vertically *downwards*. The suspended magnet makes an angle with the

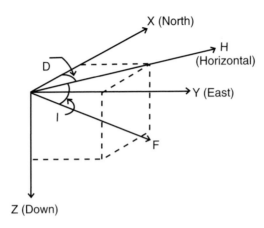

Fig. 17.2. The magnetic elements in relation to the geographical axes, north, east and vertically downwards.

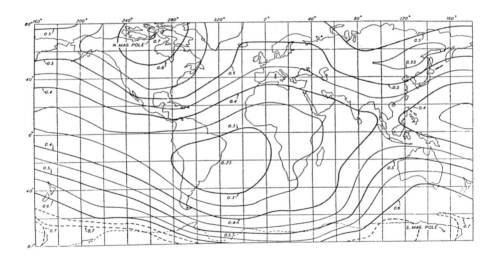

Fig. 17.3. The isodynamic chart of lines of equal total magnetic field intensity for the epoch 1922. The units are 10^2 μT (from British Admiralty Charts).

horizontal called the magnetic dip angle, I, or sometimes the magnetic inclination. The horizontal component of the magnetic field is denoted by H and makes an angle D with the north direction X, called the angle of magnetic declination. The strength of the magnetic field at some location is denoted by F and is summarised by the five quantities (F, H, D, I, Z) which are called the magnetic elements.

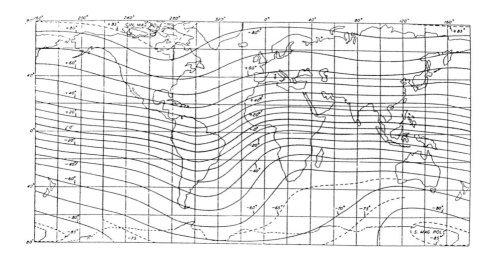

Fig. 17.4. The isoclinic chart of lines of equal magnetic inclination for the epoch 1922, in degrees (from British Admiralty Charts).

Edmond Halley first showed in the 17th century the great utility of representing a magnetic element as a plot on a map of the Earth with lines of equal intensity joined together. These are the magnetic charts. This is now a standard representation and some magnetic charts have a characteristic name. Plots of the total magnetic field, F, or the horizontal component H are called isodynamic charts. Plots of equal values of D are called isogonic charts while equal values of I are called isoclinic charts. An example of an isodynamic chart is shown in Fig. 17.3 for the total magnetic field for the year 1922.5. Lines of equal value are joined on this Mercator's projection. The isoclinic lines of equal inclination for the same year are shown in Fig. 17.4. If the field were a true dipole the lines of force would be straight lines making an angle of $11.5°$ with the lines of geographical latitude. An interesting plot is the isoclinic polar chart for inclination shown in Fig. 17.5, after Airy. The symmetry is very clear as is the departure of the field from it. The northern polar region is particularly distorted as would be expected for a complicated polar region.

17.3. Separation into a Dipole and Non-dipole Fields

The analysis of the details of the magnetic field has occupied many people over the last two centuries drafting maps and collecting data. The original approach was made by Gauss who showed mathematically that the observed

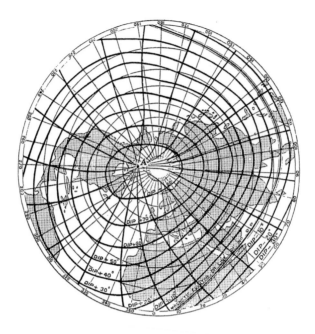

Fig. 17.5.   Lines of equal inclination and of declination for the northern hemisphere for the epoch 1835. The units are degrees (after Airy).

intrinsic field is of internal origin. He showed, more generally, that any field can be decomposed into an internal and an external component with the caveat that there be no electric currents in the surface dividing the regions if his method is to be effective. This is satisfied to a high approximation for the Earth. The data available to Gauss didn't allow him to recognise any magnetism of external origin but it is known now that there is such a contribution and that it has important consequences — it will be considered in the next Section.

The mathematical analysis of the internal field replaces the actual magnetic field data by a series of component field data due to dipoles, quadrupoles, octopoles and so on. The strengths of these various contributions are adjusted to represent the observed field as closely as possible. Usually only two or three terms are sufficient for this purpose. (This is the so-called Legendre polynomial expansion but details are not appropriate here.) The magnetic force from each component falls off with distance, and the decrease is greater the higher the component in the group. Thus the octapole contribution falls off faster with distance away from the source of the magnetism than the quadrupole contribution and the quadrupole contribution itself falls off more quickly than that of the dipole. Explicitly, the dipole force falls off as the third power of the distance, the quadrapole as the fourth power of the distance and soon. The measured magnetic details depend upon how far the observer is from the source. At a sufficient distance away any magnetic source the field appears to be dipolar. For the Earth (measured at the surface) the dipolar contribution accounts for some 85% of the total field, the quadrupole accounting for most of the rest.

It is important to be able to separate the dipole contribution from the remainder. This is not an easy thing to do with accuracy because the details of the whole field must be available. The dipole field obtained this way is approximate and so is called the best-fit dipole field. The field that remains forms the non-dipole field. Data from magnetic observatories over the whole earth are used these days for this purpose and the values of the fields are agreed upon internationally and published as the International Geomagnetic Reference Field (or the IGRF). It is updated every few years.

(a) The best-fitting dipole field, denoted by F_1, is currently displayed about the magnetic axis, which intersects the surface of the Earth at two points — north, 79°N, 71°W and south 79°S, 109°E. These are called respectively the north and south geomagnetic poles and are hypothetical points. The axis does not pass through the centre of the Earth and the magnetic axis is said to be off-set, in this case by about 550 km. The term geomagnetic is used to distinguish this hypothetical construct from the measured magnetic field which is real. There are, then, two distinct pairs of poles — the real magnetic poles including all the measured contributions and the imaginary geomagnetic poles of the best-fit dipole. Again, the measured magnetic equator has zero dip but the geomagnetic dipole equator is the hypothetical equator line for the best-fitting dipole.

A dipole has an associated magnetic dipole moment and the magnitude of this moment for the Earth is $m = 7.94 \times 10^{22}$ Am^2. The mean strength of the field is measured in microteslas: the strength of the equatorial field is 0.305 gauss while that for the pole is 0.70 gauss. Although a rather artificial quantity, the magnetic dipole moment is useful as a general measure of the strength of the total field. It is particularly useful for making the comparison between the field of different planets.

(b) The non-dipole component is the field that is left when the best-fitting dipole field has been removed from the total measured field. There is no analytical way of finding it. The non-dipole field for the year 1980 is shown in Fig. 17.6 where the quantity $F - F_1$ is plotted using the spherical geometry rather than the Mercator's projection. The units are 10^{-4} tesla

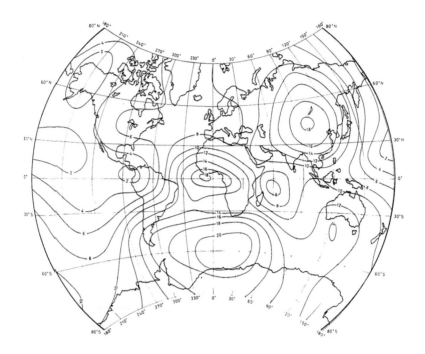

Fig. 17.6.   Showing the non-dipole magnetic field $[\mathsf{F} - \mathsf{F_1}]$ for the epoch 1980. The units are $10^2\ \mu\mathrm{T} = 10$ gauss (from Nevanlinna, Pesonen and Blomster, 1983, Earth's Magnetic Field Charts (IGRF 1980)).

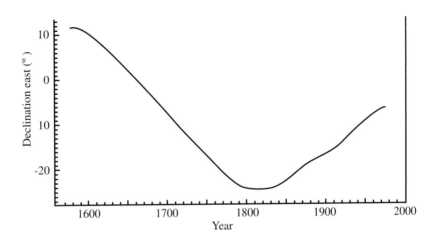

Fig. 17.7.   The magnetic declination measured at Kew over the period 1560 to 1980 (after Malin and Bullard). The change is not periodic.

(T). Units traditionally used in these studies are the gauss $= 10^{-5}$ T and the gamma $= 10^{-4}$ gauss $= 10^{-9}$ T.

17.4. Short Time Variations: The Secular Variation

The fact that the magnetic field changes with time was known to very early observers. Magnetic measurements have been made at several magnetic observatories, including Kew in England and Potsdam in Germany, since the 16th century. One set of data are shown in Fig. 17.7 showing the values of the magnetic declination measured at Kew between 1550 and 1998. The declination has fallen and then risen again during this period although the change is not periodic.

The variation of the full non-dipolar field is shown in Fig. 17.8 for the epoch 1922.5 and the epoch 1942.5. The two curves are similar but the pattern for 1922.5 has moved westwards by 1942.5 by a few degrees of latitude. This movement of the non-dipole field is called the secular variation. It is clear that the magnetic profile does not move as a fixed pattern nor does it move equally at all latitudes. In fact, it appears to move more quickly in the Atlantic and Americas regions than in the eastern regions even though all show a westward drift.

Another variation with time can be inferred from accumulated data of the best-fit dipole field. It is found that the calculated dipole moment has decreased over the 120 years between 1840 and 1960 and is still falling today. These data are plotted in Fig. 17.9. They show that the dipole field strength itself is falling in a corresponding way. While the dipole field strength is apparently decreasing this seems not to be the case for the non-dipole field.

17.5. Long Time Variations: Magnetic Field Reversals

Support for the idea that the dipole field does indeed vary with time was a most surprising event in the 1950s. Then the behaviour of the magnetic field over geologically long intervals of time became open for study for the first time. The study became possible due to recognition of a new feature of rocks.

Rocks laid down in the past are found to carry with them a signature of the conditions of the local magnetic environment at the time they were laid down. This applies equally to igneous or to sedimentary rocks.[1] This comes about because magnetized ferrous mineral grains in the materials

---

[1]This is also true of fired pottery.

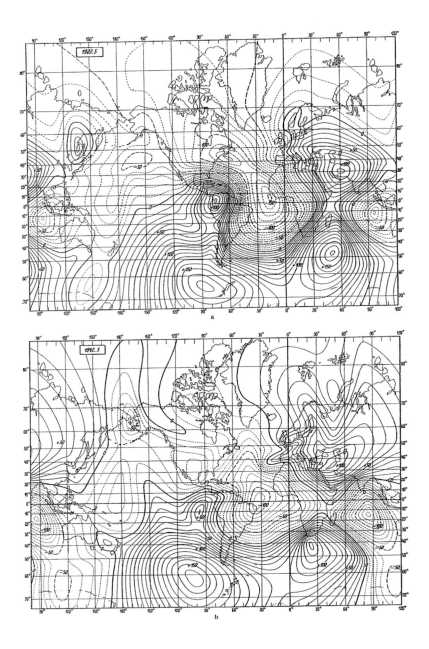

Fig. 17.8.   The secular variation in the Earth's magnetic field for the epochs 1922.5 and
1942.5 for the vertical component Z. The changes are in 10^{-9} T (gammas) (from British
Admiralty Charts).

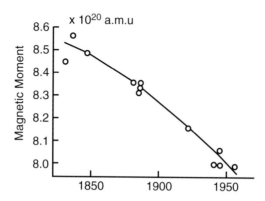

Fig. 17.9. The magnetic dipole moment for the best fit dipole field of the Earth has decreased over period 1840 to 1960.

are able to line themselves up in an average sense along the direction of the local magnetic field permeating the material. The directions of these mineral magnets become frozen at the time of hardening and the rock specimen shows a preferential magnetic direction. The material can be volcanic magma cooling, or sedimentary materials which harden under pressure or even clay used to make cooking pots. The magnetization is very weak but can be detected by magnetometers of sufficient refinement. The magnetization is called remanent magnetization. The technique, now well developed, measures the remanent magnetization accurately by measuring the energy required to remove it.

One early worry of the method was the stability of the magnetization held by the rock. Did it really come from the past or was it an artificial property of later origin? Happily, the magnetic stability of a range of rocks was firmly established and the technique of measuring the intensity of the remanent magnetization has become routine, though this must not detract from its delicacy. The results of these studies were quite unexpected. It was found that the polarity of the Earth's field has changed direction completely not once but many times over geological history. The reversals have not been periodic. In some epochs the reversals have been frequent but in other epochs they are rare or non-existent. A plot of the polarity of the field over the last 180 million years is shown in Fig. 17.10. Here the black entries refer to the present anti-parallel alignment while the white entries refer to the reverse parallel arrangement. No intermediate alignments have been found. The age is shown on the left-hand side of the pattern — the numbers on the right are for a different purpose that will not concern us now. It seems the dipole field component reverses its polarity completely when it changes.

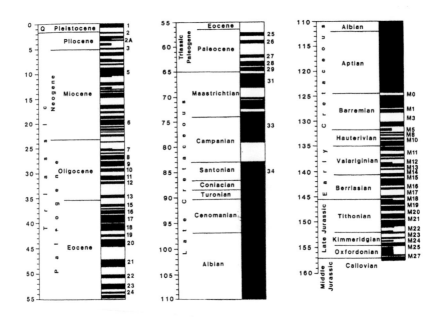

Fig. 17.10.   The pattern of reversals of the Earth's magnetic field over the last 180 million years. The black regions are normal polarity while the white regions are reversed polarity. There are no intermediate regions. The age is given by the numbers on the left of the plot.

It is seen that the different polarities have occurred with about equal frequency except for a period between 83 and 125 million years ago when the present polarity remained unchanged. There has not been an equivalent period when the parallel configuration was so firmly established. The evidence of Fig. 17.9 would suggest that a change of polarity will occur at some time in the nearer future. This variability of the field must have some deep theoretical implication but the matter is still obscure.

17.6.   The Geomagnetic Poles have Moved: Continental Drift

Still more startling results have emerged from the study of rock remanent magnetisation. These studies have shown that the geomagnetic poles have "wandered" over the continents in the geological past. Put in another way, the dipole axis of the Earth appears to have moved relative to the continents. To establish this movement it is necessary to locate the magnetic poles from remanent magnetization. In particular, the orientation of the measured field must be established. This is done by accepting that the conservation of angular momentum requires the rotation axis (that is the geographical axis)

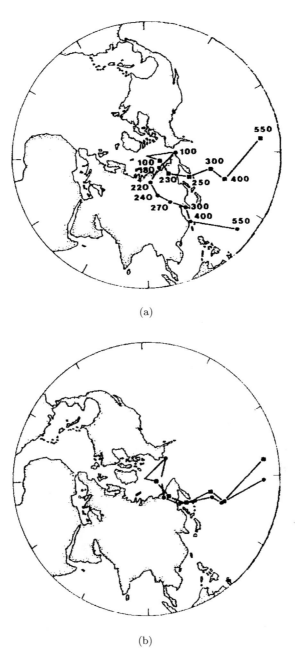

Fig. 17.11.   Pole wander as deduced from data of remanent magnetisation. (a) The plots of positions of the geomagnetic north pole for the remanent magnetic data from Europe (squares)and North America (circles). (b) The pole wander curves for Europe (squares) and North America (circles) over the last 550 million years when account is taken for the opening of the Atlantic Ocean. The fit is extremely good.

of the Earth to stay always in the same direction in space and so can act as datum for magnetic measurements to be made against that fixed axis. Some of the results are collected in Figs. 17.11.

Figure 17.11a shows the inferred north geomagnetic pole positions for different ages from North America (circles) and Europe (squares). The ages are in steps back to about 550 million years. It is clear that both sets of measurements suggest the location of the pole somewhere in the Pacific Ocean 550 million years ago but they do not agree on the exact location. Reconciliation between these data is made essentially complete in Fig. 17.11b when account is taken of the relative movement of North America over the

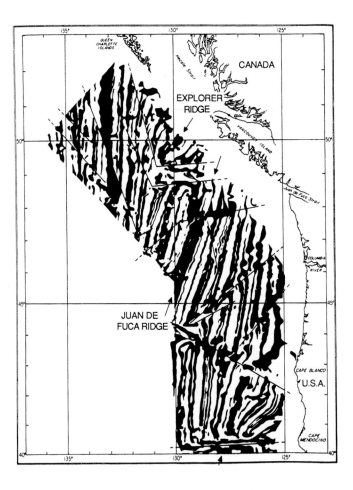

Fig. 17.12.   A magnetic profile for an active ocean ridge off the Pacific coast of Canada and the USA. Black refers to the current anti-parallel polarity and white to the reverse parallel polarity (Raff and Mason, 1961 Geol. Soc. Am. Bull. 72).

last 50 million years away from Europe, with the opening of the Atlantic Ocean. It is clear from such measurements that the surface of the Earth has been an area of great activity over the geological past.

17.7. Creation of Ocean Floor

The time scale of surface movements has been determined from the ejection of lava from the mid-ocean ridges. The lava is magnetized when it cools, so retaining remanent magnetization, and the pattern of polarity change is embedded in the lava sheet which is observed to spread away from the ridge. The first example of this action to be recognized was that involving explorer ridge and Juan de Fucca Ridge off the Pacific coast of Canada. The profiles are reproduced in Fig. 17.12. Black regions have the current polarity while white shows the opposite polarity. The details of the figure are rather complicated and were actually developed to study the dynamics of the ridge system and the details of the subduction of the ocean floor. The discovery of the magnetization allowed a time scale to be associated with the features. Comparison between the magnetic profile of the ridge system and the general profile of rocks collected in the data such as those of Fig. 17.10 provided a dynamic description of the ridge movements. It is found generally that the speed of the outflow is one or two centimetres per year. From our point of view the important aspect is the way that polarity allows the outflow of lava from the ridge to be measured which allows a link to be established between surface phenomena and those within the crust.

*

Summary

1. The Earth has a permanent intrinsic magnetic dipole field as a good approximation.

2. The details of the field are expressed in terms of five magnetic elements. Charts are drawn showing the profiles of points of equal values of each element over the surface of the Earth.

3. The separation of the total measured field into a dipole and a non-dipole component can be done only within the accuracy of the measured field data, there being no independent dipole field. The separation leads to the best-fitting dipole field.

4. The best fitting field changes with the total field and with the increasing accuracy of the measurement of the field. The values of the field are agreed internationally and published as the International Geomagnetic Reference Field (IGRF). The values are confirmed every few years.

5. The (hypothetical) dipole field obtained this way is called the geomagnetic field which must be distinguished from the measured magnetic field which does not have an exact dipole form.

6. There are, then, the details of the geomagnetic field, including the geomagnetic poles and the geomagnetic equator, which are separate from the corresponding details of the magnetic field.

7. Both the dipole and non-dipole components show variations over time though the time scales are different. The variation of the non-dipole field is called the secular variation and shows a westward drift of a few degrees of longitude per year. This gives it a time scale of decades to show substantial change.

8. The dipole field also changes with time but the change involves longer time intervals. Rock remanent magnetisation bears witness to the complete reversal of the magnetic field over geological time with a characteristic time of several millions of years.

9. Remanent magnetisation has led to the establishment of the movement of the continents over the surface of the Earth (called continental drift) and to their rearrangement. It has also allowed a time scale to be recognised for the production of new oceanic materials through the ejection of material through the mid-ocean ridge system.

18

The Earth's External Magnetism

The intrinsic magnetic field extends outside the surface of the Earth. If the Earth were isolated from other bodies this field would extend out into space with a simple form dominated by the dipole of Fig. 17.1. The agnetism outside the Earth is, however, more complicated than this due to the presence of the Sun, the Moon and the atmosphere, perhaps in that order of importance. The result is a complicated pattern of magnetic fields within the atmosphere and beyond dominated by the solar emission of radiation and of atomic particles.

18.1. The Effects of the Solar Emissions

The solar radiation heats the surface of the Earth and the atmosphere giving them a temperature a little above the freezing point of water. Without the atmosphere the mean temperature of the surface would be 256 K whereas with the atmosphere it is about 290 K. The full range of the electromagnetic spectrum enters the atmosphere including the ultra-violet components which ionise the atoms of the upper atmosphere. These effects change with depth into the atmosphere. At the outer reaches the number of atoms is low so there is little interaction with solar photons. This means that little ionisation can occur. The level increases with depth at first because the number of atoms increases and there are still plenty of photons. The number of photons of appropriate energy decreases with depth as atoms are ionised and ultimately very few photons are left. The result of these interactions is a shell of ionisation in the upper atmosphere centred a little below 100 km above the surface. This ionised layer is called the ionosphere. It is a characteristic of any atmosphere subject to bombardment by photons of sufficient energy. Being ionised the ionosphere is a good conductor of electricity and here lies another very rich grouping of magnetic phenomena to be explored further later. One very important consequence is its use in radio communications across the Earth. Being electrically conducting, it opposes penetration by electromagnetic radiation, except for certain frequencies. Looked at from the outside, incident electromagnetic energy is largely reflected at the surface although some is transmitted through the plasma. It is this mechanism that allows signals to be bent (reflected) around the surface of the spherical planet and so allows messages to be sent between the continents on Earth. This would not be possible on Moon or Mercury where there is no atmosphere.

The Sun also emits a stream of atomic particles called the solar wind. This is composed of ionised atoms according to the solar abundance of the elements which means it is primarily hydrogen but accompanied by trace quantities of heavier atoms. This ionised stream of particles, which carries solar magnetic field with it, acts as an electrically conducting continuum fluid as it moves from the Sun. It emerges as a spiral from the rotating Sun and approaches the Earth at an angle of about 45°. This is shown in Fig. 18.1. The speed of the wind at the Earth varies considerably. On a quiet day it may have a low of 300 km/s whereas on an active day it may be as high as 900 km/s or more. The solar wind temperature (actually the temperature of the protons) again varies between 100,000 K up to 800,000 K. This high temperature is not of great significance because the material density

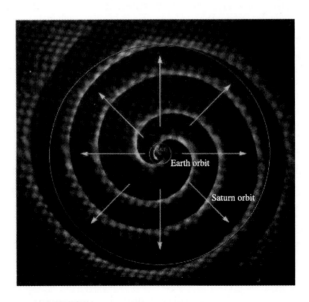

Fig. 18.1.   The spiralling out of the solar wind due to the solar rotation (the "hose pipe" effect).

is so low: anywhere between 6 and 40×10^6 particles per m^3. The moving wind has, of course, got a pressure which varies with a maximum of the order 11nP ($\approx 1 \times 10^{-15}$ bar). These various variations are a manifestation of what is called space weather, arising from the variability of solar conditions.

The solar wind is a region of high electrical conductivity and, like the ionosphere, rejects other magnetic regions. In particular, the planetary magnetic field opposes the solar wind that strikes it. As a result, the solar wind speed is reduced and ultimately stopped and deflected by the Earth's field lines on the Sun-side. This distorts the Earth's magnetic field lines until an equilibrium is reached on the Sun-side of the Earth. The point where the solar wind is stationary is called the stagnation point, and lies along the Sun-Earth line. The wind passes round the Earth, as a stream would flow round a sphere, passing downstream and carrying the high latitude portions of the magnetic field with it to form a magneto-tail. The profile of the solar wind/Earth interaction is shown schematically in Fig. 18.2. It is seen that the lower latitude magnetic field is little affected by the solar wind but that the high latitude profile is substantially distorted.

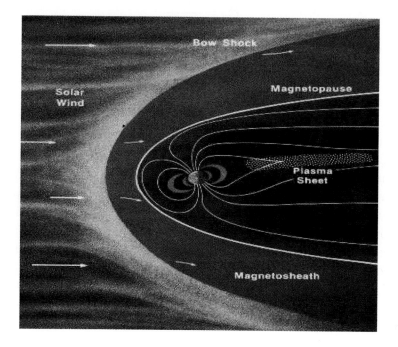

Fig. 18.2.   Showing the interaction between the terrestrial magnetic field and the solar wind.

The solar wind stands away from the surface of the Earth by an amount that can be estimated quite easily. The solar wind is atomic particles carrying a weak magnetic field with them from the Sun. The total plasma (positive and negative particles) carried momentum and the ram pressure is defined as that pressure necessary to bring the plasma to rest. The ions in the solar wind which will be the most resistant to deflection are the ions, and the ram pressure of the stream given by the solar wind pressure P_{sw}

$$P_{sw} = nmV^2$$

where n is the density of protons, m the proton mass and V the solar wind speed. The opposing magnetic pressure P_m associated with the Earth's field of strength B is given by

$$P_m = \frac{B^2}{2\mu_0}.$$

Here μ_0 is the permeability of the free space, $4\pi \times 10^{-7}$ Hm^{-1}. If the dipole moment of the Earth is D then at distance r in the magnetic equator the field is D/r^3 giving

$$P_m = \frac{D^2}{2\mu_0 r^6}.$$

The condition for the distance of the bow shock is the equality $P_m = P_{sw}$ so that

$$r = \left(\frac{D^2}{2\mu_0 nmV^2} \right)^{1/6}.$$

For a numerical estimate, the dipole moment of the Earth is 7.9×10^{15} Tm^3 and taking $n = 5 \times 10^6$ m^3 with $V = 4 \times 10^5$ m/s we find $r = 5.1 \times 10^7$ m or about 8 R_E or $r \approx 4.8 \times 10^7$ m. The tilt of the magnetic axis of the Earth relative to the equatorial plane of the Sun means that the interaction with the solar wind is not symmetric. In fact it comes closest to the Earth's surface in the region of the South Atlantic Ocean. The solar wind raps round the Earth, compressing its field on the day-side and stretching it out on the night-side to form a tail.

The relevant wave speed in the solar wind plasma is the Alfvén speed defined as $V_A = \frac{B_{sw}}{\sqrt{4\pi\rho}}$ with B_{sw} being the solar wind magnetic field strength. The Alfvén speed plays the rôle here of the sound speed in the ordinary flow of gas. The magnetic conditions outside the solar wind are different, with a lower wave speed so the solar wind suffers from a discontinuous change encasing the Earth — this is called the bow shock. Particles pass through the bow shock into the magnetopause which has a boundary encasing the Earth's magnetic field. The lower latitude lines of force are simple extended downstream but the high latitude lines are different. The magnetic poles play a special role because the magnetic lines of force are open and are extended downstream into the magneto-tail. The tail extends downstream some 100 Earth radii, beyond the orbit of the Moon and so affecting the lunar surface It is a region of great importance in the electromagnetism of the ionosphere.

18.2.   The Interplanetary Magnetic Field

The solar wind is an example of a plasma system in which the particles are dominant and the magnetic field is secondary. The charged particles carry magnetic lines of forces with them from the solar surface into the regions of the planets outside. The magnetic lines of force act analogously to thin elastic cotton threads carrying many beads: the movement of the beads on the same piece of string twists the string although the same beads remain on it. The retention of a given set of charged particles on the same magnetic line of force is an example of the preservation of a field line. This is sometimes explained in terms of the magnetic lines of force being frozen into the fluid.

The ejection[1] by the Sun of charged atomic particles in the solar wind carries magnetic lines of force with it into the Solar System. A given line of force, with its origin in the Sun, is extended to the Earth and beyond and the wide range of particles linked by the line carry a weak magnetic field with them away from the Sun. This complex of magnetic lines of force permeates the whole Solar System with a weak chaotic magnetic field called the Interplanetary Magnetic Field. This is seen to arise in this rather strange way and to fill the Solar System with a solar influence that can arise no other way. Remember the Sun does not possess a general intrinsic magnetic field extending beyond its surface but the field of the solar wind is second best.

The solar wind permeates the whole solar system and beyond, extending far beyond the confines of the System. It has, in fact, been tracked well beyond the orbit of Pluto by the Voyager space probe now leaving the Solar System. The region covered by the interplanetary field forms the regime of the Sun. Eventually the solar magnetic field will merge with the interstellar

---

[1]The precise way in which the Sun ejects the solar wind is not yet known.

magnetic field which lies between the stars. The boundary between these
two weak magnetic fields mark the outer limit of the solar influence. The
barrier between the solar and interstellar fields is called the heliopause. Its
distance from the Sun is not known precisely but it is likely to be some
100 AU away.

18.3. The Polar Aurorae

The magnetic lines of force of the Earth blown downstream on the night-
side and those compressed on the day-side are separated by a region which
has the form of a cusp, rather like the profile of a parting in your hair.
This allows incoming charged particles to gain access to the magnetotail
and through this to the lower atmosphere in the regions of the magnetic
poles. It is this access that allows charged particles, electrons especially, to
move to the auroral zones in the polar regions and provides the link between
the observed magnetically disturbed days on Earth and events on the solar
surface. One of the most beautiful manifestations of the arrival of a magnetic
disturbance on Earth is the polar aurora. It consists of constantly moving
sheets and curtains of colour, and particularly red and green. That there
is a link between the auroral displays and disturbed conditions on the Sun
had been surmised although the nature of the link was unknown until the
magnetotail had been discovered. Certainly electric currents in the upper
atmosphere were regarded as the origin.

   The electric current induced to flow in the polar ionosphere ionises oxygen
atoms in the low pressure of the upper atmosphere. A green emission with
the wavelength 5577 Å is seen at heights between 110 and 250 km. A ruby
emission with the pair of lines 63 and 6364 appears at lower gas pressures,
with heights in excess of 800 km. The situation is closely analogous to a
vacuum tube and the atoms glow in the optical wavelengths. These are the
northern lights (aurora borealis) and the southern lights (aurora Australis),
both emissions of great beauty. The properties of these natural displays have
only become clear with the use of space vehicles. For example, it was not
known whether the north and south displays were entirely independent or
whether they in fact occurred simultaneously until Earth satellites were able
to photograph the two poles simultaneously. That they do occur together
is seen in Fig. 18.3. The aurora has been photographed from the surface of
the Earth but space vehicles allow it also to be photographed from above,
beyond the Earth's atmosphere. One such example is shown in Fig. 18.4.
Comparable displays are found on planets with an intrinsic magnetic field.

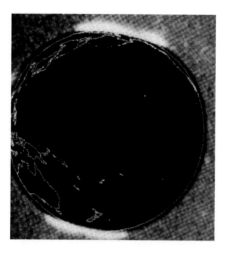

Fig. 18.3. The simultaneous occurrence of aurorae at the north and south magnetic poles of the Earth. (NASA)

Fig. 18.4. The aurora Australis seen from above from a spacecraft. (NASA)

They have special interest there of allowing the magnetic axes to be located with some accuracy.

18.4. Magnetic Storms and Transient Disturbances

It has been known for a century or more that the intrinsic magnetic field has an underlying weaker field of apparently random transient components.

The magnitude of this secondary field can be as small as a few gammas ($1\gamma = 10^{-9}$ T) but at other times of strong disturbance in may be as large as 10% of the main intrinsic field. The secondary field is variable over a wide range of time scales. The longest is 11 years, the period of the solar cycle, but the shortest can be hours or even minutes. The effects vary from one surface location to another and also with the solar and lunar times. One component depends on whether the location is on the sun-side or not. This is the daily diurnal variation and denoted by S. It is generally cyclic. There is another component with a period of about 28 days and follows lunar time. This is the lunar variation, L. There is an annual component and a component which varies with the sunspot number.[2]

The level of magnetic activity at the surface of the Earth is plotted vertically against time as horizontal axis on a magnetogram. If the magnetogram shows a smooth variation with time, the day is said to be quiet. Other days show erratic variations and are said to be disturbed, denoted by D. A very strong or protracted disturbance is called a magnetic storm. Understanding these effects at all was a very difficult task until the advent of automatic space vehicles. Now worldwide measurements can be made virtually instantaneously — during an interval of 100 mins or so if one orbiting space vehicle is involved or shorter times if more than one satellite is used. There was a certain confused mystique about the early work but this has now been largely overcome. The principles of physics are becoming clearer but the subject remains very complicated and there is much yet to learn.

18.5.   The Special Effect of the Moon

The semi-diurnal tides in the Earth's oceans due to the Moon are well known. Tides of much lower magnitude are also induced in the solid Earth and corresponding tides are also induced in the atmosphere. The Sun induces analogous tides but of smaller magnitude. The total effect of these tidal disturbances depends on the relative locations of the Sun and Moon, whether they act in consort or are opposed. The atmospheric tides are particularly significant in the region of the ionosphere because the electrically conducting ionospheric material is moved relative to the lines of force of the intrinsic magnetic field. The result is the induction of an electromotive force (emf) which induces electric currents — these are associated with their own magnetic fields which supplement those of the intrinsic Earth

---

[2]The number of sun spots on the solar surface varies essentially periodically through the solar cycle.

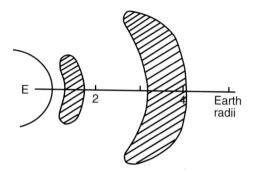

Fig. 18.5. The location of the van Allen radiation belts around the Earth, E.

field. These weak magnetic fields have periods corresponding to the lunar and solar time.

18.6. van Allen Radiation Belts

One of the first discoveries after the introduction of Earth orbital vehicles in 1957 was the identification of an entrapped region of plasma in the lower latitudes, contained by the intrinsic field. This was achieved in 1958 by van Allen and his students using the results from the Explorer 1 and 2 vehicles. They discovered two doughnut structures around the lower latitudes containing high energy particles, mainly electrons, and now called the van Allen radiation belts. Their positions are shown in Fig. 18.5. The figure shows a cross-section of the belts, each forming a doughnut shaped region with the earth's magnetic axis as the axis of symmetry for the belts. They circle the earth such that the central plane is close to the magnetic equatorial plane. This means they do not line up with the geographical equator. The inner belt is composed primarily of protons with energies in excess of 10 Me and electrons with energies in excess of 500,000 ev (0.5 eV). The lifetimes of these particles are several decades. The belt extends each side of an axis with a radius of about 12,000 km from the centre of the Earth but is fuzzy with a thickness of some 6,000 km. This means that the inner boundary is only about 500 km above the surface. The outer belt has a radius of some 5 Earth radii and contains lower energy particles, again protons and electrons but with energies less than 1.5 MeV. The offset of the Earth's intrinsic field means that the region over the central and northern Pacific ocean is further from the Earth's surface and the region over the Southern Atlantic Ocean is nearer than the average distance. This occurrence is called the

South Atlantic Anomaly. These belts restrict the possible orbits of space vehicles. The high particle intensities provide high dosages to instruments and passengers alike and must be avoided.

The origin of these charged particles is interesting and perhaps unexpected. A major source is neutrons produced by the collision of cosmic rays and solar particles with atmospheric atoms. While most of the neutrons formed this way, being uncharged, escape the Earth's magnetic field, a small fraction decay within the field to protons and electrons.[3] The charged particles are trapped and populate the belts. The number of such particles produced at any time is small but their life time in the belts allows their number to build up and become large. The behaviour of the charged particles in the belts is complex. They move along magnetic lines of force in a helical motion bouncing from one point in the northern hemisphere to the mirror point in the southern hemisphere and vise versa. Electrons show a longitudinal drift eastwards while protons show a longitudinal drift westwards around the Earth. The electrons emit radiation of high frequency which can be detected as whistles — the motion and emission is called a whistler by radio hams. This behaviour is common to all planetary atmospheres and will certainly be replicated in those of exo-planets. No electrical-magnetic effects from them have been detected so far.

*

Summary

1. The intrinsic magnetic field extends outwards from the Earth and allows the Earth to be connected directly to the Sun through the solar particle/radiation stream.

2. The Earth acts as a blunt body in the solar wind causing a magnetic tail to develop downstream, called the magnetotail.

3. The Earth's magnetic lines of force are compressed on the sun side and stretched downstream on the night side. The magnetotail extends beyond the orbit of the Moon.

4. The dividing line between the upstream and downstream field configurations provides a cusp which allows direct access to the upper atmosphere of the Earth.

[3] A neutron is a radioactive particle and decays to a proton and electron with a half life of about 20 minutes.

5. The introduction of solar charged particles into the upper atmosphere causes electrical disturbances which can appear as an aurora at the north and south polar regions.

6. These displays of largely mobile red and green light can, for severe disturbances, even be visible as low as the equatorial regions, though this is rare.

7. The external magnetic field holds a double doughnut shaped zone of charged particles about the equatorial plane called the van Allen radiation belts after their discoverer.

8. They represent an intense region of radiation between one Earth radius and about 5 Earth radii which must be avoided by electronic equipment and astronauts alike.

9. This is one reason why launching sites are best located away from the equatorial regions.

19

The Magnetism of the Other Planets

Intrinsic magnetism has been found in other planets of the Solar System at first from measurements made in Earth observatories but later, and in more detail, from observations from automatic space probes. Less detail is known about the magnetism of the other planets than for the Earth but a number of interesting conclusions have been drawn from the data available so far.

The first detection of a magnetic field in another planet was for Jupiter using Earth based detection devices. As early as the 1960s deca- and decimetre radiations were detected from the planet and these were interpreted as synchrotron radiation from electrons orbiting the planet in the presence of a magnetic field. The orientation and strength of the radiation allowed estimates of the strength and orientation of the Jupiter magnetic axis to be deduced. Early efforts to detect a magnetic field for Venus using the aurora displays failed to find any evidence: at the time this was accepted as observational inaccuracy rather than, as is now known, the essential absence of a planetary field there.

19.1. The Intrinsic Magnetic Fields

The details of the magnetic fields are collected in Table 19.1. It is seen that two planets (Venus and Mars) have not got a measurable magnetic field but all the rest have. It must be stressed that the field is not the total intrinsic field but the hypothetical dipole best-fit component which defines the magnetic axis. The angles between the magnetic and rotation axes are listed in column 2. The axes for the Earth and for Jupiter were the first to be identified and measured. They are comparable and led to the thought that all angles would be similar. The dynamo theory showed that the angle

Table 19.1. Data for the intrinsic magnetic dipole fields of the planets of the Solar System. The second column gives the angle between the rotation and magnetic axes.

Planet	Angle bet. axes (degrees)	Equatorial field (gauss)	Dipole moment (Tm^3)	Dipole moment ($E = 1$)	Polarity	Magnetospheric distance (RP)
Mercury	14	0.003	5.53×10^{12}	7×10^{-4}	Anti-parallel	1.5
Venus	–	$< 3 \times 10^{-5}$	$< 3.2 \times 10^{12}$	$< 4 \times 10^{-4}$	–	–
Earth	10.8	0.305	7.9×10^{15}	1.0	Anti-parallel	10
Mars	–	$< 3 \times 10^{-4}$	$< 1.58 \times 10^{12}$	$< 2 \times 10^{-4}$	–	–
Jupiter	9.6	4.28	1.58×10^{20}	2×10^{4}	Parallel	80
Saturn	≈ 0	0.22	4.74×10^{18}	6×10^{2}	Parallel	20
Uranus	58.6	0.23	3.95×10^{17}	50	Parallel	20
Neptune	47	0.14	1.97×10^{17}	25	Parallel	25

must not be zero (to give complete symmetry between the rotation and magnetic characteristics) but is likely to be small. The next angle to be determined was that for Saturn and to everyone's surprise turns out to be essentially zero, contrary to the predictions of the dynamo theory. Special circumstances may apply but they have not yet been found. Following this, the large angles for Uranus and Neptune were discovered which added further confusion. It was believed that the presence of a magnetic field was linked in part to the rotation characteristics but the two outer planets showed this not to be the whole story.

The other interesting result is the polarity of the fields. The polarity of the major planets now is parallel to the rotation axis. The polarity of the two terrestrial planets that show a measurable field is anti-parallel to the rotation axis. It is known from the case of the Earth that that the polarity changes from time to time and that the polarity at a particular epoch is in a sense accidental. It is not known whether this is also true of other planets although it is suspected that it is. This is a topic where more information is necessary but it is not likely to come easily. There are no rocks on the major planets to appeal to.

Also listed in the Table in the last column is the distance of the stagnation point upstream of the surface, that is, the point where the ram pressure of the solar wind stream is neutralized by the intrinsic planetary field. These values will be considered in a later Section.

19.2. The Magnetospheres

It has been seen that the Sun is the source of a solar wind of charged atoms which is ejected into the planetary environment of the Solar System, carrying magnetic field lines with it. This plasma-field combination passes through the Solar System ultimately meeting the interstellar magnetic field well beyond the edge of the System. The passage of the solar wind which causes magnetosphere around the planets and that around the Earth was studied in III.18. The major planets are accompanied by substantial magnetospheres and those for Jupiter and Saturn are especially large. This large sacle is partly due to the large radii of the bodies but also partly because their magnetic fields are strong. Only the magnetosphere for Jupiter has been studied in detail so far (early 2003) but the Saturn magnetosphere is now being studied as the Cassini mission develops after its arrival at the planet in 2004.

(a) Jupiter

Jupiter has a magnetic dipole field greater than that of the Earth by a factor 20,000 but the strength of the solar wind at the distance of Jupiter is only 4% of that at the distance of the Earth. It is not surprising, therefore, the magnetosphere of Jupiter is very substantially larger than that for the Earth because the magnetic field will dominate the solar plasma. The surfaces of the Galilean satellites are also targets for the solar wind, emitting electrically charged particles as a result of the interaction. The volcanically active

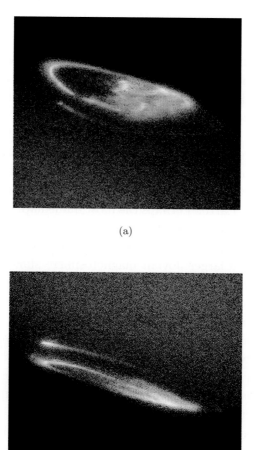

(a)

(b)

Fig. 19.1. Aurorae on Jupiter. (a) is the northern lights and (b) the southern lights. (Univ. Michigan, NASA)

surface of Io is especially rich in emitting sulphur and sodium ions. Many of these emitted charges are also drawn in to enhance the Jovian magnetosphere, making it one of the most complex of the Solar System as well as the largest. Indeed, the diameter of its circular cross-section at Jupiter is larger than the diameter of the Sun. One effect of its great size is that the magnetotail extends a great distance downstream — to some 6.5×10^8 km which is beyond the orbit of Saturn.

The first recognition of Jupiter as a magnetic planet was made in 1956 when non-thermal variable decametric radiation was detected, with a frequency of 22.2 MHz. A correlation was soon established between this radiation and the longitude of Io in its orbit about the planet. Further investigations led to the discovery of a constant decimetric radiation with frequencies between 300 and 3000 MHz. This was interpreted as synchrotron radiation from relativistic trapped particles trapped by an intrinsic Jovian magnetic field, and entirely analogous to the van Allen belts encircling the Earth. The symmetry of the radiation shows further that the magnetic axis is inclined at an angle of about 9.5° to the rotation axis. Twenty years later measurements from the Pioneer 10 and 11 automatic space vehicles confirmed entirely these Earth bound observations. It showed further that the density of particles in the radiation belts is orders of magnitude greater than for the Earth. Perhaps the most immediate manifestation of the Jovian magnetic field is the observation of polar aurorae, shown in Fig. 19.1. The northern and southern emissions are found to occur simultaneously. In general terms the larger Jovian magnetosphere shows a close similarity of structure to that of the Earth. A representation of the magnetosphere is shown schematically in Fig. 19.2. The magnetotail encasing the plasma sheet spreads downstream perhaps as far as 10 AU.

The Galilean satellites are enclosed in the inner radiation belts (analogous to the van Allen belts on Earth) which means that their surfaces are bombarded continually by charged particles. Io is itself a major source of charged particles for the plasma torus. The interaction between Jupiter and Io is very complicated. Electric currents flow in Io, linking it with Jupiter where the current flows in the ionosphere. Other currents flow in the doughnut of Io's orbit. The radiation belts co-rotate with the planet giving them a rotation period of 9 hours 59 minutes. The associated Coriolis force is strong and draws the equatorial plasma out along the equatorial plane to form a plasma disc along the magnetic equatorial plane. It turns out from arguments of dynamics that the co-rotation of the plasma and the planet cannot be maintained throughout the magnetotail and the downwind structure of the magnetosphere remains unclear at the present time. Many problems remain in

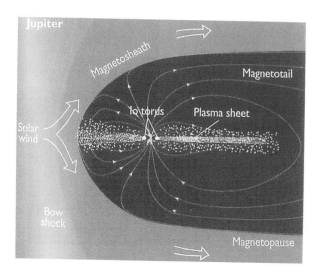

Fig. 19.2. A schematic view of the magnetosphere of Jupiter (after Beatty, Petersen and Chaikin, *New Solar System*, CUP).

understanding the system and probably can be elucidated only by further study of the system in situ.

(b) Saturn

There had been no direct indications from Earth of magnetic conditions on Saturn before the flyby of Pioneer 11 in 1979. That mission confirmed that Saturn has an intrinsic magnetism and that the interaction with the solar wind is strong. Subsequently, aurorae have been observed in ultraviolet light by the Hubble Space Telescope, as is seen in Fig. 19.3. The rings structures of the aurorae are plainly visible, and appear simultaneously. These observations confirm the surprisingly exact coincidence between the rotation and magnetic axes, a coincidence forbidden by the general dynamo theory of planetary fields. Theory and observation need to be reconciled. Unfortunately, these measurements do not lead to details of the structure of the magnetosphere. Happily, some preliminary data were returned by the Pioneer 11 spacecraft during its flyby.

A bow shock was detected at 24 Saturn radii upstream (that is about 1.5×10^6 km). The magnetotail spreads downstream as for the Earth and Jupiter but the composition of the Saturnian plasma sheet is different. This is brought about by the ice particles in the Saturn ring system replacing the charged particles familiar in Earth and Jupiter, by an agglomeration of

Fig. 19.3.   The simultaneous aurorae at Saturn's poles, in ultraviolet light. (HST)

particles of which the major component is neutral particles. As for Jupiter, the satellites act as a source of charged and uncharged particles for the magnetosphere. The details are still to be found and it hoped that the Cassini mission commencing in 2004 will provide more details.

(c) Uranus and Neptune

The large angles between the rotation and magnetic axes for each of these planets and the orientation of the rotation axes themselves provide asymmetric geometries of the intrinsic magnetic fields. In this respect these planets differ from Jupiter and Saturn. Added to this, the weakness of the solar wind at their orbital distances allows their fields to have a much stronger effect. The magnetospheres for both Uranus and Neptune were observed during the flyby of Voyager 2 in 1987–9.

The case of Uranus is the more complicated because the magnetic field makes an angle of 59° with the rotation axis which itself makes an angle of 98° with the plane of the orbit. This means the angle between the magnetic axis and the solar wind changes between the angles 41° and 149°. The orientation changes throughout the 84 years of the orbital period of Uranus, leading to a substantial change in the configuration of the field during this period. The plasma tail is in the plane of the magnetic equator near to the planet but relaxes to lie along the direction of the solar wind some 250,000 km downstream. The whole tail structure rotates about the axis formed by the line joining Uranus to the Sun. Detailed analyses

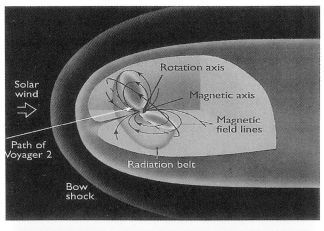

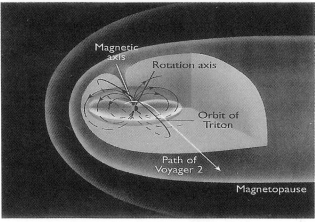

Fig. 19.4. The magnetosphere structure for Neptune as indicated by Voyager 2 in August 24, 1989 (after Beatty, Petersen and Chaikin, *New Solar System*, CUP).

will be possible when a detailed survey of the environment of the planet becomes possible, hopefully in the not too far distant future. Less can be said about Neptune. This planet also has a magnetosphere but with a peculiar property due to the large angle between the magnetic and rotation axes. Again the details are lacking in the absence of data but it does seem that that the structure of the magnetosphere shows significant changes over the rotation period of the planet which is 16 hours. An indication of the magnetospheric structures is given in Fig. 19.4. The path of Voyager is shown entering the bow shock and leaving the magnetosphere covering a period of 38 hours. The changes during this period are quite large.

19.3.   Other Examples for Planetary Bodies

The are a variety of possible combinations of magnetospheres involving intrinsic magnetic fields and atmospheres. We consider these now.

(a) Intrinsic magnetism but no atmosphere: Mercury

In this case the magnetic field is compressed on the day side and extended on the night side, as for the Earth. With no atmosphere, and consequently no ionosphere, the tail is largely devoid of charged particles. Mercury gives an example of this case. The solar wind field compresses the intrinsic magnetic field of Mercury on the day side but the field is sufficiently strong to hold the solar wind away from the surface. A bow shock and a magnetotail form as for the Earth. The distance of the bow shock is determined by the strength of the intrinsic planetary field. The magnetotail has a restricted form because there is no atmosphere to act as a source of additional charged particles. The fields are shown schematically in Fig. 19.5.

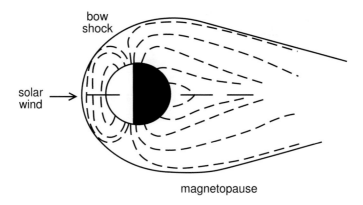

Fig. 19.5.   The magnetosphere for a mercury-like body.

(b) No intrinsic magnetism but an atmosphere: Venus

The more energetic photons of the solar radiation induce an ionosphere with its own weak magnetism. With no detectable intrinsic magnetic field to make a contribution only the ionosphere can oppose the solar wind plasma albeit perhaps only weakly, preventing it from actually reaching the surface. The magnetopause is entirely symmetrical about the direction of the solar wind. The planet acts as a blunt body in the wind flow, familiar in fluid mechanics.

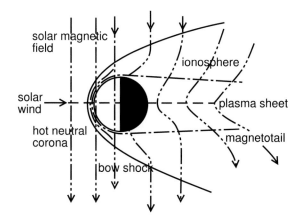

Fig. 19.6.   The magnetopause for a Venus type planet.

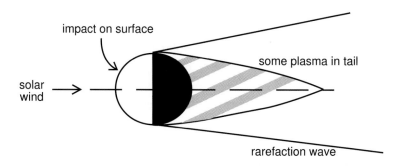

Fig. 19.7.   The encounter between the solar wind and a moon like object.

Venus is typical of this case and the magnetosphere for this planet is shown in Fig. 19.6. There is a plasma sheet downstream.

(c) No intrinsic magnetism and no atmosphere: The Moon

The solar wind now penetrates to the surface. This is the case for the Moon, shown in Fig. 19.7. Now the surface is bombarded by charged particles of the solar wind directly, causing radiation damage and fragmenting ("gardening") the surface to produce a surface of micro-particles. This can become quite substantial over the lifetime of the body. A heavily fractured surface is an important feature of the Moon. There is a shadow zone behind the main body leading to a region with few, if any, solar wind particles.

19.4.   Motion through the Interstellar Medium

The Solar System moves as a unit through the interstellar medium, which is permeated by a weak interstellar magnetic field. This results from a super-position of the stellar winds of all the stars in the region. The solar magnetic field acts as an obstacle to the interstellar field leading to a bow shock and magnetopause for the whole System as for a planet, and called now the helio-sphere. There is a bow wave thought to occur at a distance of some 230 AU ahead of the Sun. A tail lies behind extending until the magnetic pressure is the same as that of the interstellar field. This point may be more than 150 AU ($\approx 2.3 \times 10^{10}$ km or 4 times the Sun/Pluto distance) downstream. The uncertainty arises from a lack of detailed knowledge of the interstellar field and its distribution in space. The magnetic field of the solar wind meets the interstellar field at a region called the heliopause. Charged particles can enter the Solar System from interstellar space at this end of the heliopause. The interstellar field varies locally with time providing "weather" just as the solar wind has its own weather system. This leads to changing conditions within the interplanetary regions of the Solar System, affecting the dynamics of the solar wind to a greater or lesser extent. The huge magneto structure for the Sun contains within itself the magneto structures of the various plan-etary and other bodies. It is this composite complex structure that moves through the interstellar magnetic and charged particle field.

The Explorer and Voyager spacecraft have moved through the Solar System making close encounters with the major planets and have now reached the outer regions of the Solar System. This is close to the esti-mated distance of the heliopause. There is some communication with the craft though the communication time is about 2 days. The craft have been operating for some 20 years, working on electrical systems fuelled by the ra-dioactive decay of plutonium atoms (providing heat energy). These sources are finite and at some time in the future the craft will first emit a signal too small to be detected from Earth and then finally will stop emitting al-together. They will, in the fullness of time, meet another stellar system and may even be found by intelligent aliens. It would be interesting to know what the aliens will make of them.

19.5.   Companions to Other Stars

There is no direct evidence of the magnetism of companion bodies to other stars but it is known that the stars themselves are associated with magnetic fields of the same general type as the Sun. From the studies of the Solar System it would seem safe to assume that the companions possess intrinsic magnetic fields of some kind and that the dominant component will most likely be the dipole. It can also be safely supposed that the star emits a stellar wind — strong stellar winds are known to be emitted from some stars and a weak version is likely to be emitted by the remainder. The orientations of the rotation and magnetic axes are unknown so little can be said about the details of the magnetic interactions with the stellar winds. For those systems with very small semi-major axes, or periastron distances, direct magnetic interactions must be important. The companion is only 10 stellar radii from its stellar centre at its nearest and the very large stellar mass emissions can extend to a third of this distance. The bow shock for the companion could be a distance 10^9 m upstream giving the novel situation of a bow shock near the surface of the source of the stellar wind. There are no parallels to this within the Solar System.

*

Summary

1. With the exception of Venus and probably Mars, all the planets of the Solar System are associated with an intrinsic magnetic field. The Moon has not got an intrinsic field now.

2. This means the intrinsic fields interact with the solar wind in ways similar to the Earth with magetospheres and magnetotails.

3. The structure for Jupiter is by far the largest with a magnetotail that stretches downstream beyond Saturn. The case of Uranus is different because of the orientation of its magnetic field.

4. Bodies with no intrinsic magnetic fields but with atmospheres (and so ionospheres) and those without interact with the solar wind in different ways which are considered.

5. The Sun itself also shows a magetotail structure as it ploughs through the interstellar magnetic field. The tail ultimately meets the interstellar field as the downstream side at a distant point called the heliopause.

6. The magetosphere structure of the Sun ultimately marks the boundary of the Solar System in space.

Part IV

STARS AS A CONTINUING SOURCE OF ENERGY

With Special Reference to
Solar Type Stars and the Sun

The planetary environment is determined by the physical state of the central star. The state of animate matter on the planet must depend on it.

20

Evolution of Stars

A star is the central member of a planetary system and its behaviour is, therefore, of importance in the understanding of the behaviour of such a system. Here we survey briefly the ways in which we obtain information about stars and the general properties that are observed. Of importance for planetary systems is the lifetime of a star, the characteristics of its radiation and the form it takes when its energy sources are used up.

20.1. Observations and Measurements

Deducing the properties of stars and planets is not a simple task. Everyday experience is extended into new realms and new phenomena must be surveyed carefully to establish links with that that is already understood. It is a fundamental assumption that nature is broadly the same everywhere so that local experience here is valid elsewhere. The difficulties met with in everyday life still apply there. For instance, a moving object can be either a small body moving slowly but close by or equally well it may be a large object moving quickly and far away. One very obvious problem is that of deciding who is moving — the relative motions of the Earth and Sun fooled observers for generations in the past. The resolution of problems must be made by means of comparisons and interpretations of separate information. We are familiar with nearby objects changing their perspective as they move passed us and the movement will generally be fast. More distant objects change only slowly as we move and very distant ones change hardly at all over quite large times. These observations are called parallax and are one of the most important methods for discovering the form of the world around us. We use this everyday. An example is the view from the window of a fast train. Rushing through a station the passenger sees very little of detail — it is usually impossible to read the station name. The features of the middle distance change much more slowly and the passenger can look at the changing features in detail. The far distance seems hardly to move. Although the various objects appear to move in fact, of course, they do not: houses, hedges, trees and hills stay still. It is the observer in the train who moves, as is very obvious in this example. Again, the distances can be gauged by the relative sizes of the objects in view. We know that hedges are smaller than houses and houses smaller than hills. We subconsciously structure our view to suit what we know would be found were the various components measured accurately.

The same principles apply to astronomical observations except, not knowing so much about the objects that are observed there and the relations between them, the observer cannot be as pragmatic. As in much of everyday life, size is important but can be confusing. For instance, the Sun and Moon occupy about the same area in the sky and it might be guessed, with no other information, that they at the same distance from the Earth. A solar eclipse, where the Moon covers the face of the Sun, shows without ambiguity that the Sun is actually further away. How much further away is not easy to say. The distance of the nearest objects can be found by trigonometric triangulation. Beyond the distances where angles can be measured with appropriate accuracy, other distance measures must be used. For instance, it may be assumed that a group of objects being observed are all essentially the same in size and brightness. Then differences of size or of brightness allow some spatial order to be applied to the members of the group. These problems are especially relevant in astronomy.

Measurements are subject to uncertainties of several kinds. Observations from the surface of the Earth are disturbed by the turbulence of the Earth's atmosphere. This can be corrected now by the processes of speckle interferometry. There are, of course, a range of frequencies that are absorbed by the atmosphere so working in a window near the edge can cause problems.

The instruments themselves provide inaccuracies. It is often necessary to wind a screw system: there is always a certain play in any system, however carefully constructed, and this introduces errors on reading a scale and reproducing an initial setting. Again, reading a dial involves a pointer and an etched scale both of which have a finite thickness however thin the markings may be. There are ways of making the associated uncertainties small but they can never be eliminated entirely.

These considerations lead us to recognise a precise relationship between theory and experiment or observation. Any measurement must involve errors, which may be systematic or random. An experimental result cannot, therefore, be claimed to be exact. Any exact theoretical statement cannot apply directly to an experimental situation. The representation of the real world by theoretical concepts must involve a generalisation of experience which may not be unique. Any statement which might be interpreted as a "law" is at root tentative although the more generally it is found to apply the greater the confidence that the generalisation is valid. Certainly the description of the world, using theory, presumes a simplicity which is itself a remarkable generalisation. It is quite remarkable that science has been as effective as it has been in describing the physical world and allowing us to achieve a growing insight into it. One of the essential requirements of observational data (or experimental for that matter) is an assessment of the errors that are believed to be involved. This should included the errors expected in that observation together with an assessment of systematic errors that might be expected to be present.

What, then, can be done with data so acquired? The aim is to find relationships between the variables that have been isolated as a way of storing data in a simple and economical way. As an example, in I.1 it was seen how Kepler discovered three laws describing the motion of a planet around the Sun. This allowed a host of numerical data to be replaced by a single formula, or paradigm, to allow all such cases to be described at once. This simplification then allowed the physical force controlling the motion (that is gravity) to be isolated and described. Finally it was realised that the simple motion described by Kepler was an approximation when several planets are present together. It might be said that Kepler would not have been able to deduce his laws if the planetary data had been more accurate or of wider validity. The orbit of any planet is now not exactly a closed ellipse, the deviation being the greater the larger the masses involved and the nearer the bodies are together. These refinements were added later: the initial discovery must be sufficiently crude to hide such features. Again referring to Kepler's laws, they may well not have been discovered had the path of Mercury been the central evidence or had the orbits of Neptune and Pluto been the dominant examples. With these thoughts in mind we turn to a resumé of the observations of the stars in the firmament.

20.2. Galaxies and Stars

The Universe is a vast assemblage of many components; some visible but others apparently not. The main visible components are galaxies and quantities of gas, mainly hydrogen. Each of the visible galaxies contains on average about 10^{10} stars and the majority are of about the solar mass ($\approx 2 \times 10^{30}$ kg). There are probably of the order 10^{11} or 10^{12} galaxies in the visible Universe. This gives a total number of 10^{21-22} stars with an estimated total mass rather in excess of 2×10^{51} kg. According to the special theory of relativity, the self-energy E associated with a mass m is $E = mc^2$ where $c = 2.99 \times 10^8$ m/s is the speed of light in a vacuum. This implies the enormous self-energy for the stars in excess of $E \approx 2 \times 10^{68}$ J.

There is the strongest observational evidence that an invisible dark matter shadows the normal matter that we can see. The evidence comes from the comparison between the kinetic and potential energies in situations where the visible matter is unaccountably stable although it should not be. The kinetic energy arises from motion of the matter while the potential energy arises from its gravitational attraction. As an example, galaxies are in rotation and the observed strong stability that they show must involve a larger gravitational component than can be accounted for by the visible matter alone. The deficit is very substantial and is supplied by dark matter, so called because it cannot be seen or otherwise detected except through gravity. To make sense of observations it must be there. Again, it is observed that galaxies occur in stable clusters (sometimes containing many thousands of members) and their stability must involve many times more gravitational energy than can be accounted for by the visible galaxies alone. The deficit is again to be supplied by dark matter. These, and other, arguments, give the strongest evidence that dark matter exists and that it is several times more abundant than visible matter. There is as yet no firm indication of the nature of dark matter even though it appears to interact with ordinary matter through the normal gravitational field. Experiments are underway to test the hypothesis that it consists of very massive sub-atomic particles. Excluding visible matter from detection is a major problem and many experiments are conducted deep underground for this reason. No candidate particles have been detected so far.

20.3.   The Life Expectancy of a Star

Individual stars have a finite lifetime and the processes of the birth and the death of stars is now believed to be generally understood.[1] The characteristic thing about a star is its radiation of energy. The birth involves the

---

[1]One can never be sure — science continues by discovering new and unexpected things.

aggregation of gas (very largely hydrogen but heavier elements are important as well) into a body of sufficient mass to achieve a high enough central temperature for the thermonuclear energy source to begin. The end of the life of the star comes when the energy source is exhausted. In the most general terms, the life of most stars is between 10^{10} or 10^{11} years. The age of the present phase of the Universe that we see has a time scale ("age of the Universe") of about 1.37×10^{10} years which is a comparable age. Our Sun was formed about 5×10^9 years ago when the Universe had had its present form for some 9×10^9 years. Not all stars are the same. The rate of consumption of energy is dependent on the mass with very heavy stars consuming energy the fastest and their lifetime being correspondingly short. It can be as short as 10^6 years if the mass is large enough.

The star is a self-correcting system, as will be seen in IV.21. It will be seen that the equilibrium radius is that which allows the energy produced at the centre to pass through the body of the star and be emitted at the surface. Some regions will carry energy by radiation while others, nearer the surface, will carry energy by convection but throughout the star is balanced to radiate the energy it produces. The cessation of energy production in the central region affects the entire structure of the star. When the energy stops it ceases to be a star but the form it takes then depends on the mass.

The less massive stars, like the Sun, first expand (becoming red giants) with a radius of perhaps 2 to 3 AU, which is rather beyond the orbit of Mars. The central region, which has a strong component of helium, then collapses to form a body of rather less mass than solar mass but a radius of broadly Earth size. The density is very high, as much as 10^{12} kg/m^3, and the surface temperature is low, perhaps a few thousand degrees K. This is called the white dwarf stage. The temperature slowly falls through radiation and the body ultimately becomes a cold sphere. The outer region is gaseous and is ejected into space with high speeds. The expanding gases viewed from afar can be very beautiful and forms a planetary nebulae. Planetary bodies initially with orbital semi-major axes within 2 to 3 AU of the star are absorbed into the star and disappear. This will most likely be the fate of the terrestrial planets, including our Earth, when the Sun eventually has used up all its central energy supply.

High mass stars show a very different evolution. The central pressure is high enough to provide a temperature sufficient to allow nuclear "burning" to continue beyond the helium stage. Successive rings of burning appear around the centre, making an onion structure shown in Fig. 20.6. It is seen that there is a difference between the death of less massive and more massive stars.

Whereas the low massive star cools to become a dead body, the high massive star explodes in a supernova explosion. The heavier chemical elements form during the explosion due to an abundance of neutrons then. We shall now look at these processes in a little more detail.

20.4.   The Hertztsprung–Russell Diagram

A star radiates energy and the Stefan-Boltzmann radiation law states that, in the ideal case to be considered in a moment, the quantity of energy radiated through unit area of the star in unit time (that is the flux of energy through the surface) is proportional to the fourth power of the surface temperature. The constant of proportionality is known as the Stefan-Boltzmann constant. This would be the actual surface temperature were the star to radiate as a black body[2] but, although it does so to a very high approximation, it does not do this precisely. The surface temperature determined this way from the Stefan-Boltzmann law is, therefore, not the actual temperature although it is usually very close to it: the calculated temperature is called the effective temperature of the star, T_e. The surface temperature also determines the state of the atoms in the surface and so to the particular distribution of frequencies that are radiated into space. The particular form of spectrum is called the spectral class.

It was shown by Hertzsprung of Denmark and independently by Russell of America that the plot of the luminosity L of a star against the type of spectrum it shows (its spectral class) allows various properties of the evolution of stars to be portrayed conveniently on a single diagram. This plot has become known as the Hertzsprung–Russell diagram (the H-R diagram). A plot for the nearest and brightest stars is shown in Fig. 20.1. The data for the Sun are included. The form of the H-R diagram used now is somewhat different from that introduced by Hertzsprung and Russell. The luminosity that we measure, called the apparent luminosity, is related directly to the actual luminosity of the star, and called the absolute luminosity of the star. The absolute luminosity is easily found from the observed luminosity if the distance of the star is known.[3] This can be used as the ordinate in the diagram. Again, the spectral class can be replaced by the effective temperature, T_e but these are only cosmetic changes for convenience. A special feature

---

[2]A black body radiates the full range of wavelengths according to a precise spectrum of the energy for each wavelength range called the black-body (or Planck) spectrum.
[3]Remember the intensity of radiation falls off as the inverse square of the distance of separation.

Evolution of Stars

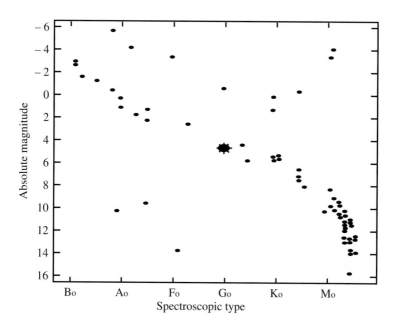

Fig. 20.1. The H-R diagram for the nearest and brightest stars. The position of the Sun is marked with an asterisk. The luminosity is represented by the absolute magnitude.

appears if T_e is used. The spectral class as traditionally arrayed, with hot stars on the left and cooler stars on the right, means that when T_e is used the temperature scale is the *wrong way round*. Increasing temperature goes from right to left and not from left to right as would normally be expected.

It is clear that the position of the stars is not random but follows some pattern. Most stars in the H-R diagram fall on a diagonal across the diagram from the lower right-hand corner (low temperature and low luminosity) to the upper left-hand corner (high temperature and high luminosity) as seen in Fig. 20.2. This diagonal band is called the main sequence. As a mass of gas condense to form a star it enters the main sequence at a point depending on its mass because the mass determines the initial luminosity and surface temperature. The line containing these starting points is consequently called the zero-age main sequence, or ZAMS. As the star evolves it tends to move towards a somewhat higher luminosity but somewhat lower surface temperature so the stars already formed lie in a small band forming the main sequence. The main sequence condition involves stars which are chemically homogeneous and produce energy by converting hydrogen

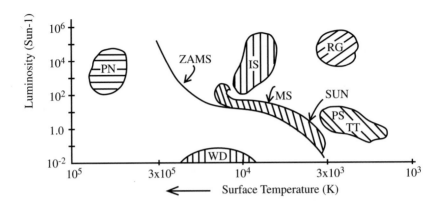

Fig. 20.2.   The main categories of H-R components. ZAMS is the line for zero age main sequence stars and MS is for the general main sequnce; WD is white dwarfs; PS-TT is protostars and T Tauri stars; RG is red giants; IS is instability stars; and PN is the nuclei of planetary nebulae. These categories are explained in the text. The position of the sun is shown.

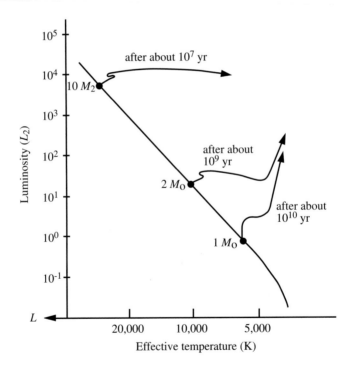

Fig. 20.3.   The movement of stars of solar mass, twice solar mass and ten times solar mass on the H-R diagram away from the initial ZAMS positions, according to theory. It is seen that the larger the mass the faster the evolution.

into helium through the thermonuclear processes. They are considered in IV.21.

The full H-R diagram has been developed over the last 50 years and its basic structure is shown schematically in Fig. 20.2. It is seen that the different groupings of stars lie in particular regions of the diagram and are not distributed at random. The fresh zero age main sequence stars lie along the diagonal ZAMS line as we have seen already. The initial composition there is believed to be closely 74% hydrogen and 26% helium. Heavier elements will be present but in small proportions. Energy production converts some hydrogen to helium and the star moves towards higher temperatures and higher luminosity on the diagram. This gives the thicker main sequence line. The thickness corresponds to a time of evolution of about 10^{10} years. The movement of low and high mass stars from ZAMS is shown schematically in Fig. 20.3.

The representation of the state of stars on the H-R diagram is a snapshot of the body of stars including all levels of evolution from the very beginning to those at the end. It is to be understood, therefore, in terms of evolution. It can be realized at once that the likelihood of observing a particular state

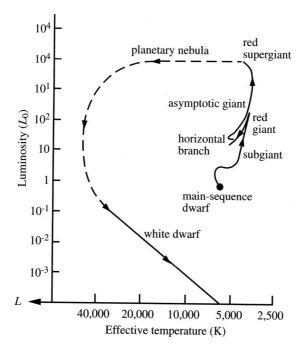

Fig. 20.4. The evolution of a star of about solar mass on the H-R diagram. Its end point is a white dwarf star.

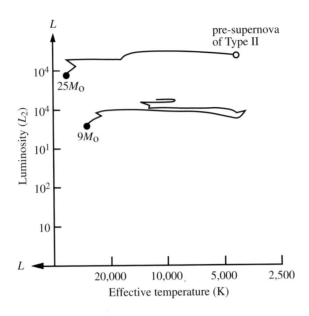

Fig. 20.5.   The evolution of massive stars on the H-R diagram.

will depend on how many stars are at that stage of development and also on how long such a stage actually lasts. Expressed the other way round, the gaps in the diagram indicate that stars in these states pass through very quickly. The movement through the different states is not a uniform process. The differences are already apparent in Fig. 20.3 where comparable distances from the ZAMS line have taken 10^{10} years for a solar mass star but only one-thousandth this time for a 10 solar mass star. The differences between the evolution pattern for low and high mass stars are well displayed in Figs. 20.4 and 20.5.

A star of solar mass begins on the ZAMS line and there is called a dwarf because it has a smallest radius then than it will ever have as a hydrogen star. When the hydrogen fuel begins to become exhausted the star moves off the main sequence (the post main sequence phase), the interior separating into a concentrated central helium core (the spent fuel) enclosed in an extended hydrogen atmosphere. There is a thin shell of hydrogen burning at the core surface. The radius to the old radius of the star (called the photosphere) increases and, because the same quantity of radiation is leaving an expanding surface, the surface temperature falls. The star appears more red because the energy flux decreases. At its most extended state the radius will be some 50 times the core radius — the red giant stage has been reached on the H-R diagram. Further rearrangement increases the radius further to make the star become a supergiant. Now the initial radius of perhaps 7×10^8 m expands to become about 2×10^{11} m, or about the orbital radius of Mars. The terrestrial planets will have been absorbed. The outer, convective, mantle is ejected leaving a small object of about 0.8 solar masses but with a planetary type radius, encased in an expanding gas cloud. The planetary nebula is one of the most beautiful objects in the sky. The disappearance of the cloud leaves the central star as a white dwarf composed very largely of helium which steadily cools into a cold body with ever decreasing luminosity. It sits at the very bottom of the H-R diagram. The time scales for some of the phases are too short to enable us to actually observe these processes taking place now.

The decay of a low mass star is seen to involve several regions across the H-R diagram. The decay of a high mass star is less spectacular on the H-R diagram but is much more spectacular in reality. The decay modes of a 9 solar mass star and a 25 solar mass star are shown schematically in Fig. 20.5. The paths lie along essentially horizontal branches of constant luminosity but decreasing surface temperature as the interior thermonuclear processes develop to form a series of ever more complicated shells about

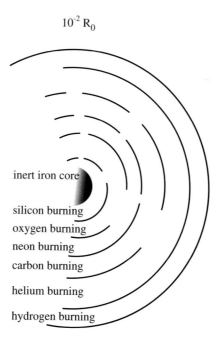

$10^{-2} R_0$

inert iron core
silicon burning
oxygen burning
neon burning
carbon burning
helium burning
hydrogen burning

Fig. 20.6.   Shell burning in a fully developed high mass star involving the chemical elements from hydrogen up to iron. R_o is the initial radius of the star.

the centre. The shell structure of an advanced high mass star is shown in Fig. 20.6. The high temperature allows the helium decay product of hydrogen burning to ignite the helium ash. The region where hydrogen burning can take place has moved out to a shell which encloses the central region where helium burning has occurred and then successively carbon, neon, oxygen and silicon burning become possible leaving an iron core as the final ash that can burn no further. This layered region occupies about 10^{-2} of the total stellar volume. Further from the centre the temperature is too low for hydrogen burning. The end of the thermonuclear burning leaves the star unstable — it explodes as a supernova. The explosion releases large numbers of neutrons which take part in a range of nuclear fusions allowing the formation of the higher members of the chemical periodic table. Many elements are destroyed in the process as they are formed but many remain with the net balance of a range of elements heavier than iron that are ejected into the interstellar medium. This is the origin of the abundance of the chemical elements.

The supernova explosion ejects large quantities of gas into interstellar space leaving behind a very tiny though massive object as the supernova remnant. With a mass of several solar masses the object has a radius of only a few kilometres. The internal pressure is very high allowing a composition of neutrons as the only possibility. This is the neutron star as a very dense object. Neutron stars can be associated with very strong magnetic fields and rapid rotation. These features give rise to pulsars which are rotating neutron stars where there is a strong electromagnetic radiation emitted along the magnetic axis. The rotation of the star gives a "lighthouse effect" of a rotating beam of radiation along the rotation axis. The times of rotation are very small, in the millisecond region. These matters are outside our study.

*

Summary

We have derived a number of very important results, fundamental to the description of two bodies in orbit under the mutual action of gravitation.

1. It is accepted that the details of our knowledge of the stars is constrained by the observational techniques that are available at a particular time. As observational techniques improve and new approaches are developed, our knowledge correspondingly increases.

2. The stars and the collections of stars in galaxies have a mechanical equilibrium involving the balance between the forces of motion and gravitation.

3. The observed stability of galaxies and collections of galaxies can be accounted for only if there exists a non-luminous matter, called dark matter. The luminous component represents a small proportion of the whole.

4. A star has a life between formation and the end of its nuclear fuel which is shorter the more massive the star. For a star of solar mass the lifetime is probably about 10^{10} years.

5. The evolution of a star is conveniently represented on the Hertzsprung–Russell diagram where the luminosity of the star is plotted against the class of spectrum displayed by its surface. This can be represented in terms of an effective temperature.

6. The evolution of a solar mass star has several stages in which the star, having used up the hydrogen fuel in the central region of high temperature, first expands and then sheds gaseous material leaving a small low-temperature white dwarf star of helium composition. This slowly cools by radiation.

7. The behaviour of a high mass star is more spectacular. The central temperature is sufficiently high for hydrogen burning to be followed by helium burning, carbon burning, neon burning, oxygen burning and silicon burning to leave iron as the final ash.

8. The end of thermonuclear burning now leaves an entirely unstable body which explodes as a supernova star. The chaotic conditions allow a wide range of trans-iron chemical nuclei to be formed and, although many are destroyed by further atomic processes, a net quantity of these heavier elements remain to be ejected into interstellar space. This is the origin of the cosmic abundance of the elements.

9. Because these processes occur continuously, the amount of the heavier element component is increasing with time so the cosmic abundance is also changing slowly.

21

The Constitution of Stars

Over 90% of stars in the Universe are on or near the main sequence, like the Sun. Some are slightly more massive and some slightly less massive but looking at the Sun from the Earth is like looking at any typical star from the astronomically near distance of about 149 million kilometres (1 AU). Stars are long lasting sources of energy, often of great strength. They emit electromagnetic radiation over the full spectrum of frequencies from high energy X-rays and γ-rays through to very low energy radio waves. In fact, the spectrum bears the imprint of conditions deep inside which are very close to equilibrium. It must be remembered that recognising a state of thermodynamic equilibrium requires the thermodynamic variables to be measured so that the various critical relationships can be found and checked. All experimental measurements involve errors of some kind, and however small these may be, the equilibrium in a particular case can only be established within a certain error limit. A system with an error within an acceptably small limit must be accepted as being in thermal equilibrium because there is no way of establishing that it is not.[1] Here we investigate the essential properties of stars — both those on the main sequence, like our Sun, and those composed of very dense matter, that is white dwarf and neutron stars. We use the methods of similitude.

I. On the Main Sequence

21.1. A Family of Similar Stars: The Assumptions

Consider a range of stars which all have constitutions that we will now specify. We shall seek those general properties that apply to all the members of the family and differ only by scale.

---

[1]This leads to the concept of local thermodynamic equilibrium whereby equilibrium properties such as temperature and pressure can be applied locally to conditions of nonequilibrium such as for a flowing fluid. The equilibrium is then slightly different from one locality to a neighbouring one.

We make the following six assumptions:

(i) The star is in mechanical equilibrium under its own gravity. Therefore, the weight of material as any level from the centre, r, is supported by the pressure, p, of the material at that level.

(ii) The thermodynamic pressure is dominant throughout the star, the pressure due to radiation being negligible everywhere.

(iii) The material behaves like an ideal gas. The pressure, p, and temperature, T, are related by

$$p = R\frac{\rho T}{\mu} \tag{21.1}$$

where ρ is the material density, μ, is the mean molecular weight of the molecules of the gas, and R is the universal gas constant.

(iv) The transport of energy within the star is entirely by radiation and not by convection. The star is said to be in radiative equilibrium throughout. The material pressure differences at different layers do not contribute to convection. Consequently, the flux of radiation (that is the energy passing perpendicularly through a unit area in unit time) is proportional to the gradient of the radiation pressure and inversely proportional to the obstructive power (called the opacity) of the material to radiation.

(v) The resistance to radiation is measured by the opacity and this arises from interactions between the radiation and the electrons of the material. This form of opacity is called photo-electric. Quantum theory tells us that the opacity per unit mass of material, k, is then related to the density of the material by

$$k = \kappa \frac{\rho}{T^{7/2}} \tag{21.2}$$

where κ is the opacity constant of the material. This relation is often called Kramers' law.

(vi) The energy is generated within the star by thermonuclear processes. This will be specified more precisely in IV.22 but for the moment we can suppose the Bethe empirical approximation to apply. This sets the rate of energy generation per unit mass of material, e, as proportional to the 18th power of the temperature so that

$$e = \varepsilon \rho T^{18} \tag{21.3}$$

where ε is the energy generation constant of the material (very largely hydrogen). The precise power of the temperature is not important but the fact that it is a high power is the crucial thing.[2]

21.2.   Specifying the Family of Stars

The aim is to construct a set of a range of model stars for a series of the four variables μ, ε and κ together with the total mass of the star M. Adding the total radius, R, and the luminosity L, any member of the family is expressed in terms of a standard star. To say the mass and radius of a star are M and R means that the mass is M times the standard and the radius is R times the standard. We can apply this idea to the following assumptions.

(i) The ideal gas law involves the pressure and the temperature. The pressure within a star is the force at the location per unit area and the force is due to gravity. This gives the expression for the pressure.

$$p =\sim G\frac{M^2}{R^4} . \qquad (21.4)$$

Notice that the constant of gravitation has been included here for completeness but its inclusion is in some ways inappropriate because concern is with properties relative to a standard and a constant factor like G is the same for all.

(ii) The density scales as $\rho \sim \frac{M}{R^3}$ so from (21.1) the temperature scales as

$$T \sim \mu\frac{M}{R} . \qquad (21.5)$$

(iii) Using (21.2) and (21.5), the opacity transforms as

$$k \sim \kappa\frac{\frac{M}{R^3}}{(\frac{\mu M}{R})^{7/2}} \sim \frac{\kappa R^{1/2}}{\mu^{7/2} M^{5/2}} . \qquad (21.6)$$

(iv) The radiation pressure is proportional to the fourth power of the temperature T^4 according to Stefan law so the radiation pressure is $1/R$ times the value in the standard , that is

$$p_r = \left(\mu\frac{M}{R}\right)^4 \frac{1}{R} .$$

---

[2]The exact power is determined by the thermonuclear process involved but the 18th power is representative of the range of possibilities.

(v) The obstructive is ρk so that from the expression for the density and (21.6) it follows that

$$\rho k \sim \frac{\kappa}{\mu^{7/2} M^{3/2} R^{5/2}}. \tag{21.7}$$

The radiation flux, f, is (by definition) the radiation pressure gradient divided by the obstructive power so that

$$f \sim \frac{\mu^{15/2} M^{11/2}}{\kappa R^{5/2}}. \tag{21.8}$$

(vi) Finally we can use (21.3) to obtain an expression for the energy production within the star. Using (21.5)

$$e = \frac{\varepsilon \mu^{18} M^{19}}{R^{21}}. \tag{21.9}$$

It is seen that all the characteristics of the family of stars can be expressed as a relationship between the parameters ε, μ, κ, M and R. This is an extraordinary result for an object as apparently complicated as a star. It demonstrates the comment made by Sir Arthur Eddington who was responsible for developing these arguments that a star is, in fact, a very simple object.

21.3.   Some Immediate Conclusions

The arguments so far have been those of simple similitude but have yielded some remarkable results with little effort, other than elementary bookkeeping. One or two consequences can be set down now.

(a) The controlling factors are the mass and radius because they generally enter the expressions with high powers.
(b) Variations in the precise chemical composition are not important. They enter the arguments with very low powers and are almost always very close to unity. It can be noticed that, for stellar compositions of hydrogen with helium, $1/2 < \mu < 2$. Plots of mass and radius will show a small scatter due to differing compositions so the recognition of such scatter will allow an assessment of the chemical differences between members. The obstructive power depends more on the composition.
(c) The radiation flux depends mainly on M, R and μ but is almost independent of the energy generation, ε, of the material. It might have

been expected that ε would have been a controlling factor. The obstruction to radiation photons has been assumed due to electron transitions (photoelectric) and not to the direct interaction with unbound electrons (electron opacity). Both types are present in stars but the former is more important for stars of solar mass or less — the latter applies more for heavier stars.

(d) The temperature within the members of the family will hardly differ. Generally, a smaller M will mean smaller R but with the ratio M/R remaining very much the same.

21.4.   The Luminosity: The Mass–Radius Relation

Further rather remarkable results can still be easily extracted.

(vii) The luminosity of the star, L, is defined as the total flux across the whole area of the star per unit time in the normal direction at every point, that is $f \sim L/R^2$ for the family of stars. (21.8) applies at the surface of the star as well as in the interior so that

$$L \approx \frac{\mu^{15/2} M^{11/2} R^2}{\kappa R^{5/2}} = \frac{\mu^{15/2} M^{11/2}}{\kappa R^{1/2}}. \tag{21.10}$$

The equality sign can properly apply if the units of the variables for the total standard star are chosen to be unity. For instance, if the Sun is taken as standard all stars are compared with the Sun.

(viii) An equivalent expression for the luminosity will also follow from (21.9) which refers to the energy output of the star. There, e refers to the energy production per unit mass so the total energy output must be $eM = L$. Consequently

$$L = eM = \frac{\varepsilon \mu^{18} M^{20}}{R^{21}}. \tag{21.11}$$

The two expressions (21.10) and (21.11) must be identical so that they can be equated to give

$$\frac{\mu^{15/2} M^{11/2}}{\kappa R^{1/2}} = \frac{\varepsilon \mu^{18} M^{20}}{R^{21}}$$

or on rearrangement

$$R^{41} = \kappa^2 \varepsilon^2 \mu^{21} M^{29}.$$

This, according to our model, is the relation that must apply between the mass and the radius if the star is to be in equilibrium with its energy source. It can be simplified within the approximation of eliminating awkward fractions to read as

$$R = \kappa^{1/20} \varepsilon^{1/20} \mu^{1/2} M^{3/4} . \tag{21.12}$$

This is the mass-radius relation for equilibrium implying that $R^4 \propto M^3$. This is a relation found empirically for main sequence stars of about solar mass.

21.5. The Mass and Luminosity Relation

A relation can be deduced for the mass and the luminosity of the star. Inserting (21.12) into (21.11) provides the mass-luminosity relation

$$L = \frac{\mu^7 M^5}{\kappa \varepsilon^{1/40}} . \tag{21.13}$$

The same result would, of course, follow if (21.10) had been used in place of (21.11). It is a very remarkable result. It shows that the luminosity depends very little on the strength of the energy generation whereas it might have been expected to depend directly on it. There is a further surprising result. According to (21.13), if the strength of the energy sources were to decrease the luminosity would *increase*. How can this be? — surely the engine reduces power if you cut the quantity of fuel. The answer can be found in (21.13), (21.5) and (21.3). If the energy source were to weaken slightly the energy flux to the surface would fall. The energy flux then will not be sufficient to hold the star to its initial radius so the star will contract. The effect of contraction is to raise the temperature and so increase the effective energy output. This result is a particular feature of the very high dependence of the energy output on the temperature rather than of the details of the energy output itself. This explains the comment in IV.20 that a star is a self-compensating entity.

21.6. The Central Temperature

The temperature at any point within the star, compared to the standard, is given by (21.5). If T_c is the central temperature of the standard, then

$$T_c = \frac{\mu M}{R}$$

with (21.12) used for R. Then

$$T_c = \frac{\mu M}{R} = \frac{\mu^{1/2}}{\kappa^{1/20}\varepsilon^{1/20}} M^{1/4}$$

which can be expressed in the form

$$\frac{M}{R} = \frac{M^{1/4}}{\mu^{1/2}\kappa^{1/20}\varepsilon^{1/20}} \; . \tag{21.14}$$

This shows that the central temperature varies as $M^{1/4}$. Surprisingly, the central temperature is insensitive to the mass of the star and to the precise chemical composition.

21.7.   The Life Expectancy: Dependence on the Mass

The total energy locked originally in a star will be of order $M\varepsilon$ which is radiated away at a rate given by the luminosity, say (21.13). Dividing the total energy available by the rate of radiating it away will give a measure of the characteristic time, τ, for the radiation to be possible. It is easily seen that

$$\tau \approx \frac{M\varepsilon}{\left(\frac{\mu^7 M^5}{\kappa\varepsilon^{1/40}}\right)} \approx \frac{\kappa\varepsilon}{\mu^7 M^4} \; .$$

It will be seen in IV.22 that the energy source is confined to the central region of the star involving probably no more than 10% of the possible quantity of hydrogen fuel. This means that a more realistic life for the star is $\tau_a \approx 0.1\tau$.

   The inverse fourth power dependence on the mass derived here will not apply to stars very much more massive than the Sun. For very massive stars $\tau \propto 1/M$ is probably a more accurate expression. For stars of about 100 solar masses, the lifetime is likely to be at most 10^6 years. This characteristic time scale could well have been common in the very early Universe involving the first stars.

II.   Very Dense Matter

21.8.   The State of Dense Matter

A solar mass is about 2×10^{30} kg while the hydrogen atoms of which it is composed have a mass of the order of 1.67×10^{-27} kg. The star, therefore, contains the order of $N \approx 2 \times 10^{30}/1.67 \times 10^{-27}$ atoms, that is $N \approx 10^{57}$ atoms. The radius of each atom is of the order 0.5×10^{-8} m and so has a volume $v \approx 4.19 \times (0.5 \times 10^{-8} \text{ m})^3 = 5.2 \times 10^{-25}$ m^3. The total volume of atoms in a rough packing is of order $V \approx Nv = 5 \times 10^{32}$ m^3. For the Sun, $R \approx 7 \times 10^8$ m so that the volume is $R^3 = 1.4 \times 10^{27}$ m$^3 < V$. This means that not all the constituent particles can exist as atoms but a proportion (about 10% in this case) must find a smaller volume. This is achieved by ionization where the electrons vacate the atoms leaving a central nucleus of radius of order 10^{-15} m. The atomic volume is then reduced substantially to $v_p \approx 4.2 \times 10^{-45}$ m^3 per particle and $V \approx Nv_p = 2 \times 10^{12}$ m^3 for the total. A main sequence star contains a central region of some 10% of the volume composed of ionized atoms. By contrast, a white dwarf with radius $\approx 10^7$ m, has a volume $V_w \approx 4.2 \times 10^{21}$ m^3 which is large enough to contain the ionized atoms. The density of matter is then $\rho_w \approx 2 \times 10^{30}/4.2 \times 10^{21} \approx 4.8 \times 10^8$ kg/m^3, or larger than that of a main sequence star by a factor 10^5. Conditions of this type require a different approach and especially the introduction of quantum mechanics.

The atoms will be ionized giving a positive nucleus and negative electrons. Such a collection of charged atomic particles are controlled by two rules.

(i) *It is not possible to know precisely the location of a particle and at the same time know precisely its momentum.* If the position of an object is known to lie at the point x with uncertainty Δx and to have momentum p with uncertainty Δp the Δx and Δp are not independent but are related by the expression

$$\Delta x \Delta p \approx h \tag{21.15}$$

where $h = 6.65 \times 10^{-34}$ Js is the Planck constant of action. This is the Heisenberg Uncertainty Principle.

(ii) *No two electrons, or protons or neutrons, can have the same values of x and p, that is, occupy the same quantum state.* (Strictly speaking, a quantum state will also involve the particle spin which will double the number of possible states but this refinement is neglected here.) Atomic particles which occupy only a single quantum state are called Fermi particles. Particles which can collect together into one state are called bosons.

These two propositions allow the properties of dense matter in bulk to be understood. The analysis can be used for non-relativistic or for relativistic conditions. We consider first non-relativistic motions to calculate an expression for the particle momentum.

21.8.1. Non-relativistic case

According to kinetic theory, the pressure is the result of the momentum of the constituent particles. The pressure P is the product of the number of particles per unit volume n, the mean speed $\langle v \rangle$ and the momentum p. The main contribution to the motion is by the electrons which are substantially less massive than the nuclei. Consequently, the electron pressure is

$$P_e = n_e \langle v \rangle p_e . \tag{21.16}$$

Because the unit volume is involved, $n_e = 1/V_e$ where V_e is the volume per electron. For the x-direction, $\Delta x = V_e^{1/3}$ so that, if $p_x \approx \Delta p_x$ the Heisenberg Uncertainty Principle can be written

$$\Delta p_x \approx \frac{h}{\Delta x} \approx \frac{h}{V_e^{1/3}} \approx h n_e^{1/3} .$$

The mean velocity is re-expressed as $\langle v \rangle = p_e/m_e$ where m_e is the electron mass. The (21.16) becomes

$$P_e = \frac{\alpha h^2 n^{5/3}}{m_e} \tag{21.17}$$

where α is a constant. Exact calculations give $\alpha = 0.0485$. To go further it is necessary to relate n_e to atomic parameters.

There has been no account of the electrical nature of the particles so far. The atoms are electrically neutral so the charge on the nucleus $Z n_p = n_e$ where n_p is the number of ions per unit volume of atomic number Z. The mass density ρ is then given by

$$\rho = A m_p n_p + m_e n_e \approx A m_p n_p$$

because $m_p \gg m_e$. Here A is the atomic number. Therefore

$$n_p = \frac{\rho}{A m_p} \quad \text{and} \quad n_e = Z n_p = \rho \frac{Z}{A m_p} .$$

This expression can be inserted into (21.17) to give the expression for the pressure of the electrons

$$P_e = \alpha \left(\frac{Z}{A} \right)^{5/3} \beta \rho^{5/3} \tag{21.18}$$

with

$$\alpha = 4.85 \times 10^{-2}, \quad \beta = \frac{h^2}{m_e m_p^{5/3}}.$$

(21.18) is an expression for the electron pressure. Notice it does not involve the temperature and so is valid even at the absolute zero of temperature (were it ever possible to reach this state). Such conditions are said to be degenerate and (21.18) is called the degenerate electron pressure. It is proportional to the 5/3 power of the material density. This equation of state is the equivalent for dense matter of the law (21.4) for an ideal gas.

The equation of state (21.18) can be used to find a relation between the mass, M, and radius, R, of the star. This can be achieved by considering the central pressure of the star, P_c. One expression which is analogous to (21.4) is

$$P_c = aG\frac{M^2}{R^4} \tag{21.19}$$

where $a = 0.77$. This can be equated to (21.18) if the central density is used. Then

$$\rho_c = 1.43\frac{M}{R^3}. \tag{21.20}$$

Using (21.18), (21.19) and (21.20) gives

$$R = 0.114 \left(\frac{Z}{A}\right)^{5/3} \frac{\beta}{G} \frac{1}{M^{1/3}}. \tag{21.21}$$

Apparently, $MV = $ constant for the star.

For an electrically uncharged Fermi particle (for instance a neutron) the same formula applies but with $m_e = m_p$. In this case $Z = A$. Then $\beta_1 = \frac{h2}{m_p^{8/3}}$ and

$$R = 0.114\frac{\beta}{G} \frac{1}{M^{1/3}}. \tag{21.22}$$

Again, $MV = $ constant. This agrees with the conclusions of I.4.

21.8.2. Relativistic case

The relativistic case is a little different and leads to a different result. In this case (21.16) applies but now $p_x/m = c$, the speed of light. The result is

$$P = 0.123hcn_e^{4/3}$$

so that $P \propto \rho^{4/3}$. This dependence on the density is different from the non-relativistic case, the material being more easily compressed.

*

Summary

The properties of classes of stars are derived on the basis of similitude.[3] Taking one member as the standard the effects of multiplying the various controlling quantities by a numerical factor is investigated.

1. A star on the main sequence is defined by making six assumptions. These are: the stars is in equilibrium under its own gravity; normal thermodynamic pressure dominates throughout the star; the transport of energy within the star is entirely by radiation; the resistance to the passage of radiation is the photo-electric opacity; the energy generated in the star is through thermonuclear processes and the quantity generated is proportional to some very high power of the temperature.

2. Expressions for the pressure, density, opacity, radiation flux and energy production are derived in terms of the parameters energy generation per unit mass, ε, the mean molecular weight of the material, μ, the opacity constant of the material, κ, the mass of the star, M, and the radius R.

3. Expressions are found for the luminosity, together with the radius–mass relationship and the mass–luminosity relation and for the central temperature. These are all relationships already established from observations.

4. It is concluded that: (a) the controlling factors are the mass and radius; (b) the precise chemical composition is not important; (c) the flux of radiation leaving the surface is only weakly dependent on ε, contrary to what one might have suspected.

5. The properties of very dense stars require quantum mechanics to be understood. The mass–radius relations are derived for a white dwarf star and for a neutron star.

6. It was also found that the life expectancy of a star, from birth until its energy source is used up, is proportional to the inverse fourth power of the mass. Less massive stars can become older than more massive stars.

---

[3] This is closely related to dimensional analysis.

7. The conclusions we have drawn depend entirely on the presumptions defining a main sequence star. Not all stars follow these rules. In particular, it has been supposed that material convection makes a negligible contribution to the total energy transfer within the star. This is approximately true for stars like the Sun (where convection is important only in the upper layers) but is not true for all stars. For these other cases the formulae derived here will need some modification but our concern in what follows is for stars on the main sequence.

22

Stellar Energy Source

The Sun is the central member of the Solar System and must, therefore, be at least as old as the planets. This means that it must have been radiating energy for at least 4.5×10^9 years. The level of radiation need not always have been as great as today (we believe it was less early on) but there must have been significant radiation from early times. At first, an appeal was made to elementary sources of energy, such as the simple contraction of the Sun, but these proved inadequate for one reason or another. The only energy source that proved sufficient, both in magnitude and in time scale, is that based on the atomic transformation of hydrogen into helium, and ultimately transformations to more massive elements up to iron. We consider these energy sources now.

22.1. Isotopes

To understand these processes it is necessary to remember the features of the atomic nucleus. The atomic nucleus is composed of protons and neutrons, held together by the strong nuclear force.[1] The most elementary nucleus is hydrogen, which is composed of just one proton. A single orbiting electron completes the electrically neutral hydrogen atom. There is a modification which has a nucleus consisting of a proton combined with a neutron. This is an example of nuclei with the same number of protons, and so the same electrical charge, but with different numbers of neutrons. They both have the same chemical properties because the chemical nature of the atom is determined by the resultant positive nuclear electric charge. The two atoms are called isotopes of a given chemical element. Every chemical element has

---

[1] This can be represented by gluon particles.

at least one naturally occurring isotope and modern nuclear physics also manufactures isotopes artificially, that do not exist naturally, in nuclear reactors. Some isotopes are stable but others are not and decay into different nuclei with fewer atomic particles and/or fewer positive electric charges. The transformation involves a range of high energy particles not present in the nucleus itself. These include electrons, positrons (positive electrons), pions, and high energy electromagnetic radiations, γ-radiations and X-rays. Electric charge, momentum and energy are all conserved during such processes; energy involves the massless neutrino.[2] Each particle is accompanied by its anti-particle so it will be realised that the atomic processes can be quite complicated.

22.2. The Binding Energy: Fusion and Fission

The different particle aggregates involve different quantities of energy binding the nucleus together and this involves a mass according to the formula of special relativity $E = mc^2$, where E is the energy associated with a mass m and c is the speed of light. The total number of nucleons in the nucleus is the atomic mass number A, or atomic mass, while the number of protons is the atomic number Z. (This equals the number of extra-nuclear orbiting electrons in the atom because the atom is electrically neutral.) The number of neutrons, then, is $(A–Z)$. The plot of the binding energy per nucleon in the nucleus against the atomic mass number A is given in Fig. 22.1. The binding energies per nucleon are generally of the order 8 to 9 Mev $\approx 1.607 \times 10^{-13}$ J. It is seen that the binding energy increases at first from about 1 Mev for hydrogen and quickly reaches a flat maximum in the region of iron ($A = 56$). In fact the region $50 < A < 100$ represents a region of maximum stability for the nucleus because the energy binding the protons and neutrons (collectively called nucleons) together is a maximum. Beyond about $A = 100$ the curve falls off just as it does for elements with $A = 40$ or less. The binding energy is attractive and so, by convention, carries a negative sign (like gravitation). Increasing binding energy, therefore, releases (positive) energy. Increasing A up to $A \approx 57$, means releasing energy, the atoms fusing (fusion): decreasing A from high values down to $A \approx 57$ also releases energy but now the atoms are breaking up

---

[2]It is accepted now that the neutrino might have a mass but it will be very small, probably no more than 10^{-36} kg.

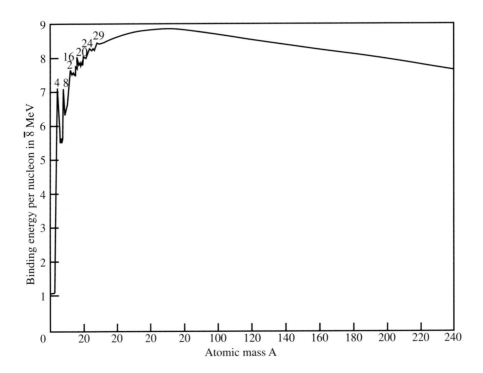

Fig. 22.1.   The binding energy per nucleon over the range of the elements. (1 Mev =
1.607×10^{-13} J.)

(fission). Generally less energy is needed to cause fusion than fission which is
why energy production in nuclear engineering involves fission, starting with
uranium, while the nuclear processes in stars involve fusion, starting from
hydrogen.

There is a question of convenience of units. In dealing with atomic problems it is easier to introduce an atomic mass unit (amu or m_u) rather than
deal always with the very low masses directly. The amu is defined in terms
of electron volts which can be translated into joules, the normal energy unit.
The mass of the carbon nucleus $^{12}C = 12.00$ amu is taken as the definition of mass on this scale. Explicitly, 1 amu $= 931$ MeV, and 1 MeV $=$
1.607×10^{-13} J. Then, 931 MeV $= 1.4969 \times 10^{-10}$ J. The mass itself can
then be deduced by using $E = mc^2$. In this way it is found that 1 amu $=$
1 $m_u = 1.66 \times 10^{-27}$ kg. The mass of the proton is then $m_p = 1.007825$ mu.
while the mass of the neutron is $m_n = 1.008665$ (it is not usual to include
the m_u). The distinction in mass is often irrelevant and then the mass of
the nucleon is taken to be the mass of the proton, m_p.

22.3. Energy from Fusion

Hydrogen is the most abundant element in the Universe and is the starting point of the energy production in stars. Four nucleons (two protons and two neutrons) combine to form one helium nucleus. The He nucleus consists of two protons and two neutrons and the total mass of all the constituents is

$$2 \times (1.007825 + 1.008665) = 4.032980 \text{ amu}.$$

The measured mass of helium is ^{4}He $= 4.002603$ amu, which is less than the masses of the constituents.[3] The difference is

$$4.032980 - 4.002603 = 0.030377 \equiv \Delta m.$$

This is a measure of the binding energy which is released when the nucleus is formed. It amounts to 0.76% of the initial masses and in terms of energy $\Delta m = 4.547 \times 10^{-12}$ J. An alternatively path, starting from hydrogen alone but yielding slightly less energy, is the direct combination of 4 hydrogen nuclei ^{1}H to form one helium nucleus ^{4}He. It is necessary in this process for two protons to be converted into two neutrons through the weak nuclear interaction. The mass difference is now

$$(4 \times 1.007825) - 4.002603 = 0.028697 \text{ amu}.$$

Again, there is a mass loss which is released and is available for use. In relative terms this is 0.72% of the total. (On this basis astronomers have been known to say (incorrectly) that the proton to helium process is only 0.7% efficient because this percentage of the mass vanishes.) In terms of energy 0.0287 amu is equivalent to 4.28×10^{-12} J. Magnitudes such as these may seem small but in fact they describe significant quantities of energy. For very kilogramme of hydrogen, containing $\frac{1}{1.67 \times 10^{-27}} \approx 6 \times 10^{26}$ nuclei, the direct conversion to helium will provide the energy $E \approx \frac{4.28 \times 10^{-12} \times 6 \times 10^{26}}{4} = 6.4 \times 10^{14}$ J. For a body of stellar mass ($\approx 2 \times 10^{30}$ kg) the complete conversion of this amount of hydrogen into helium would provide the enormous energy store of the order 10^{45} J. Such a star will have an observed luminosity of about 10^{36} W so the hydrogen fuel will last for some $\frac{10^{45}}{10^{26}} = 10^{19}$ s $\approx 3 \times 10^{11}$ years if all is burned. If only 10% of the fuel is burned it still gives a life span of 10^{10} years which is rather more than twice the present age of the Earth.

---

[3]It is convenient to identify an element X by its atomic mass number ^{A}X.

The question then is where can such an abundant energy store be found? It requires each hydrogen nucleus to have sufficient kinetic energy to penetrate the mutual repulsion of the positive charges of the others, of the order of 10^{-16} J. This energy of collision must be associated with an environment at a temperature of the order of 10^7 K. The answer to the question, first recognised by Eddington in the 1930s, is at the centre of stars. Take the Sun as a typical star. With mass 2×10^{30} kg and radius 7×10^8 m it has a gravitational energy $E_g \approx GM^2/R = 5.71 \times 10^{41}$ J. There are $N \approx 2 \times 10^{30}/1.67 \times 10^{-27} = 1.2 \times 10^{57}$ hydrogen atoms in its volume: this gives a gravitational energy $\varepsilon \approx 5.7 \times 10^{41}/1.2 \times 10^{57}$ J/proton $= 4.7 \times 10^{-16}$ J. If T is the kinetic temperature, $\varepsilon \approx kT$ where $k = 1.38 \times 10^{-23}$ J/K is the Boltzmann constant. Then, $4.7 \times 10^{-16} = 1.38 \times 10^{-23}$ T so that T $\approx 3 \times 10^7$ K. This environment is entirely suitable for the conversion of hydrogen to helium.

The same arguments allow other elements to form by fusion in stars of greater mass. The central temperature will be correspondingly higher and fusion to form heavier elements will be possible. Thus, helium can "burn" to form carbon and carbon to form oxygen at a temperature of about 1×10^8 K. Carbon can form magnesium, sodium or neon (among other examples) at a temperature of about 5×10^8 K. At 1×10^9 K oxygen can form sulphur, phosphorus or silicon while at 2×10^9 K silicon can burn to form nickel, cobalt and iron. This places the scheme at the maximum binding energy and the fission process becomes endothermic at that stage (energy is absorbed) and so reaches a barrier. These various cases will be considered now in turn.

22.4. The Hydrogen-Helium Process

Four hydrogen nuclei combine to form one helium nucleus. There are two possibilities, one direct and the other indirect involving a catalyst. The scheme, first proposed by Hans Bethe in 1939, is as follows.

22.4.1. The proton-proton chain

The first stage is the combination of two protons (hydrogen nuclei) to form a deuterium nucleus. Using the standard nuclear notation;

$$^1\text{H} + {}^1\text{H} \rightarrow {}^2\text{H} + e^+ + \nu \qquad (22.1)$$

where ^{1}H is the hydrogen nucleus (proton), ^{2}H is the deuterium nucleus, e^+ is a positron (positive electron) while ν is a neutrino. In the next stage hydrogen and deuterium nuclei combine to produce the isotope helium 3

$$^2\text{H} + {}^1\text{H} \rightarrow {}^3\text{He} + \gamma.\qquad(22.2)$$

Two helium 3 nuclei can then combine to yield a helium 4:

$$^3\text{H} + {}^3\text{H} \rightarrow {}^4\text{He} + {}^1\text{H} + {}^1\text{H}.\qquad(22.3)$$

The net result is that six ^1H are used to produce one ^4He nucleus leaving two ^1H: in total, therefore, four ^1H have combined to provide one ^4He.

Once ^4He has been produced other reactions are possible to yield more ^4He. For example (22.1) and (22.2) can be followed by

$$^3\text{He} + {}^4\text{He} \rightarrow {}^7\text{Be} + \gamma.\qquad(22.4)$$

From this point there are two possibilities. Either

$$^7\text{Be} + e^- \rightarrow {}^7\text{Li} + \nu$$

$$^7\text{Li} + {}^1\text{H} \rightarrow {}^4\text{He} + {}^4\text{He},$$

or

$$^7\text{Be} + {}^1\text{H} \rightarrow {}^8\text{B} + \nu$$

$$^8\text{B} \rightarrow {}^8\text{Be} + e^+ + \nu$$

$$^8\text{Be} \rightarrow {}^4\text{He} + {}^4\text{He}.$$

In each case 4 hydrogen nuclei have been combined to produce one helium 4 nucleus. This cycle can proceed with pure hydrogen as the initial material and incomplete cycles will provide a range of nuclei.

22.4.2. The carbon-nitrogen cycle

An alternative cycle applies at the slightly higher temperature of 5×10^7 K but requires ^{12}C to be present initially, as catalyst, and ^{14}N is formed as an intermediary. The cycle is as follows:

$$^{12}\text{C} + {}^1\text{H} \rightarrow {}^{13}\text{N} + \gamma$$

$$^{13}\text{N} \rightarrow {}^{13}\text{C} + e^+ + \nu$$

$$^{13}\text{C} + {}^1\text{H} \rightarrow {}^{14}\text{N} + \gamma$$

$$^{14}\text{N} + {}^1\text{H} \rightarrow {}^{15}\text{O} + \gamma$$

$$^{15}\text{O} \rightarrow {}^{15}\text{N} + e^+ + \nu$$

$$^{15}\text{N} + {}^1\text{H} \rightarrow {}^{12}\text{C} + {}^4\text{He}.$$

Again it is seen that four ^{1}H nuclei combine to form one ^{4}He nucleus as before. The catalyst ^{12}C is transformed during the process but is restored at the end.

This is the main set of reactions but subsidiary chains are also possible. For instance, the last step may be expanded to involve nuclei of oxygen and flourine before reaching helium. These several reactions are possible if the initial gas is not pure hydrogen.

22.5.   Reactions at Higher Temperatures

Other reactions become possible when the temperature is raised and consequently the heavier nuclei can strike each other with sufficient energy to penetrate each others' nucleus. The main reactions given here are associated with increasingly high central temperatures. Other reactions also occur but those shown here are the dominant ones:

(a) Helium burning — temperatures of order 10^8 K
A triple helium process creates carbon which becomes oxygen

$$3 \, ^4\text{He} \rightarrow \, ^{12}\text{C} + \gamma$$
$$^{12}\text{C} + \, ^4\text{He} \rightarrow \, ^{16}\text{O} + \gamma.$$

Like all nuclear processes, there are alternative roots. Thus

$$^4\text{He} + \, ^4\text{He} \rightarrow \, ^8\text{Be} + \gamma$$
$$^8\text{Be} + \, ^4\text{He} \rightarrow \, ^{12}\text{C}^* + \gamma.$$

The star on the carbon denotes it to be in an excited state. This state was originally predicted by Hoyle and later observed. It represents a "resonance" reaction at temperatures of about 1×10^8 K.

(b) Carbon burning — temperatures of order 5×10^8 K

$$^{12}\text{C} + \, ^{12}\text{C} \rightarrow \, ^{24}\text{Mg} + \gamma$$
$$^{12}\text{C} + \, ^{12}\text{C} \rightarrow \, ^{23}\text{Na} + \, ^1\text{p}$$
$$^{12}\text{C} + \, ^{12}\text{C} \rightarrow \, ^{20}\text{Ne} + \, ^4\text{He}.$$

The "burning" of carbon is seen to produce Mg, Na and Ne.

(c) Oxygen burning — temperatures of order 10^9 K

$$^{16}\text{O} + {}^{16}\text{O} \rightarrow {}^{32}\text{S} + \gamma$$

$$^{16}\text{O} + {}^{16}\text{O} \rightarrow {}^{31}\text{P} + {}^{1}\text{p}$$

$$^{16}\text{O} + {}^{16}\text{O} \rightarrow {}^{31}\text{S} + {}^{1}\text{n}$$

$$^{16}\text{O} + {}^{16}\text{O} \rightarrow {}^{28}\text{Si} + {}^{4}\text{He}.$$

Here a range of elements is produced including helium itself although there will be no helium in the initial material.

(d) Silicon burning — temperatures of order 3.5×10^9 K

A wide range of reactions occurs here which is in the regions of the maximum of the binding energy curve.

$$^{28}\text{Si} + \gamma \rightarrow 7 \, {}^{4}\text{He}$$

$$^{28}\text{Si} + 7 \, {}^{4}\text{He} \rightarrow {}^{56}\text{Ni}$$

alternatively

$$^{28}\text{Si} + {}^{28}\text{Si} \rightarrow {}^{56}\text{Ni}$$

$$^{56}\text{Ni} \rightarrow {}^{56}\text{Co} + e^+ + \nu$$

$$^{56}\text{Co} \rightarrow {}^{56}\text{Fe} + e^+ + \nu.$$

This produces iron, which is at the end of the chain of fusion processes. In these ways the stars produce the elements heavier than hydrogen and helium up to and including iron. It can be noticed that all the nuclei produced this way have mass numbers that are multiples of 4. This stresses the importance of the α-particle structure ^{4}He in the composition of these nuclei.

22.6.   The Escape of Radiation from a Star

It is necessary to say a few words about the escape of the energy from the star once it has been released in the thermonuclear reactions at the centre. The energy is in the form of photons of electromagnetic radiation. For a temperature T, $E = h\nu = kT$ where ν is the frequency of the radiation, $h = 6.67 \times 10^{-34}$ Js is the Planck constant and $k = 1.38 \times 10^{-23}$ J/K is the Boltzmann constant. If $T = 10^7$ K, then it follows that $\nu \approx 10^{18}$ s^{-1} which is in the X-ray region of the spectrum. The central region of the star contains X-ray photons.

The photons are scattered by the surrounding atoms and execute a Brownian (random) motion, moving outwards from the centre. The characteristic feature of the motion is that the direction of the photon after a collision is not correlated with the initial direction. The important feature of this random process is that the distance travelled after N collisions (called steps) is proportional to $N^{1/2}$, the square root of the number of steps and not the number of steps itself This implies that the distance travelled away from a starting point is quite small. For instance, if such a random walk has a step length of 1 metre and 10^6 steps are taken, the distance travelled is 10^3 metres and not 10^6.

Each collision of this type was found experimentally by Compton to reduce the frequency of the photon in a characteristic way. The effect is called the Compton effect. Continual collisions reduce the initial frequency of the radiation until it reaches the surface when the mean frequency will be in the optical region. Being a random process different photons will have made different paths through the star and suffered different numbers of collisions. Although in the mean the number is large there will be some that have made fewer collisions on the journey. These will still possess the high energy characteristic of the interior. Others will have made more collisions than the average and will have been reduced to a very low frequency. This spread of path lengths, encompassing X-rays at one end and radio waves at the other, forms a characteristic spread of energies and is called a black-body spectrum. For the Sun, the surface temperature is about 5800 K which corresponds to a mean frequency of $\nu \approx 1.2 \times 10^{14}$ s^{-1}.

The random motion of the photons in a star has steps of different length although the average step length is probably about 1 mm. The photons will take of the order of 10^7 years to pass from the centre to the surface. The density of material is low at the surface and the mean free path for collisions is now very large. Under these conditions the photons are essentially independent and do not collide. This region of free flight is thin and marks a clean surface to the star as seen by observers from outside.

22.7.   Synthesizing the Heavier Elements: r- and s- processes

Although the heavier elements can be made through the fusion process up to and including iron there are heavier elements than iron in the Universe. The question arises how these heavier elements are formed remembering that the natural processes involve fission at the higher atomic mass numbers, and so the reduction of atomic mass. The answer has been found to involve the catastrophic supernova explosions of the more massive stars. The fusion of elements involves high temperatures which give the nuclei high speeds and so sufficient energy to overcome the electric repulsion of the like charges when they collide. This condition allows nuclei to be formed up to iron in stars. Nuclei of higher mass cannot be formed by this high speed collision approach but can be formed through the activity of neutrons. These have no electric charge and so can enter a nucleus with relative ease even at the lower temperatures. The requirement, then, is for a copious supply of neutrons and it is here that the conditions in a supernova explosion allow the synthesis of heavy elements.

The neutron is a radioactive particle even when free, decaying to a proton with the emission of an electron with a half-life of about 20 minutes. It is even less stable when placed inside a nucleus so the introduction of a neutron to the nucleus results in a proton being added and an electron ejected.[4] This process could be repeated through the periodic table, producing continually heavier nuclei. The nuclear stability can be accounted for by adding neutrons that remain stable within the nucleus. This process is effective if neutrons can be added slowly, the nucleus being given time to accommodate to the new occupancy. This is the slow process or s-process of nuclear synthesis by neutron capture and beta decay.

The situation is different if the neutron source is more copious. Then, neutrons can be captured before an electron emission has been possible so that the new nuclei will be likely to eject the stable α-particles as part of the stabilization process. The fast process is called the r-process (r for rapid). The result will be a range of atoms forming rather different isotopes than for the s-process.

The s- and r- processes are able to account for the general features of the abundance of the elements except for the lighter elements such as lithium, beryllium and boron. One explanation for their occurrence is the result of cosmic ray collisions with atoms of the interstellar medium. There is considerable activity, and success, in attempting to understand the abundance of the chemical elements by these methods.

*

Summary

The details of several thermonuclear fission energy sources for stars are given. These are associated with a hierarchy of temperatures starting with about 1.5×10^7 K which applies to the centre of the Sun.

---

[4]An electron has insufficient energy to remain in the nucleus.

1. The energy holding the nucleons within a nucleus together differs from one element to another. The lowest is for hydrogen but this increases with the number of nucleons up to $A \approx 55$ where it reaches an energy of about 8.7 Mev ($\approx 1.5 \times 10^{-13}$ J).

2. Energy of binding is released by fusion when A increases up to $A \approx 55$ but is released by fission as A decreases from $A = 92$ down to $A \approx 55$.

3. The release of energy by fission requires lower energies to proceed than by fusion. Fission energies can be made available on Earth in the nuclear reactor, used for the production of domestic power. Energies for fission are found in stellar conditions.

4. The simplest process is the conversion of hydrogen into helium using the proton-proton chain or the carbon-nitrogen cycle. The latter requires the slightly higher temperature of 5×10^7 K. These reactions also produce berylium, boron and lithium.

5. Higher temperatures produce further reactions. Helium burning to produce carbon, oxygen and berylium occurs at a temperature of about 10^8 K; carbon burning to produce neon, magnesium and sodium proceeds at a temperature of about 5×10^8 K; oxygen burning to produce sulphur, phosphorus, silicon and helium itself requires a temperature of 10^9 K; silicon burning to produce nickel, cobalt and iron requires a temperature of the order of 3.5×10^9 K.

6. The production of elements heavier than iron requires very special conditions and energies and these are found in supernova events. The crucial event is the copious production of neutrons which can enter nuclei easily because they do not carry an electric charge. Once in a nucleus some neutrons will decay into protons hence synthesizing a heavier nucleus. A range of heavy nuclei are formed in this way but many are unstable. Many remain, however, providing the heavier elements found in nature.

7. If the Universe had a beginning and this was dominated by hydrogen the first stars will have had to use the proton-proton chain to produce energy.

23

The Sun and Its Interior

The Sun is typical of over 90% of the stars in the Universe and it is there for us to observe only 149 million kilometres away (which is 1 astronomical unit). It is composed very largely of hydrogen and derives its energy from thermonuclear processes which convert hydrogen nuclei into helium nuclei,

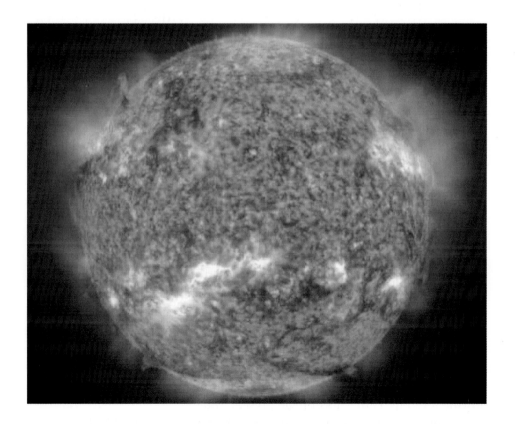

Fig. 23.1. The Sun — the typical main sequence star. (NASA/ESA: SOHO)

Table 23.1. Some bulk data for the Sun with representative values for calculations.

Property	Magnitude	Representative values
Mass (kg) — M_S	1.989×10^{30}	2×10^{30}
Radius (km) — R_S	6.95510×10^5	7×10^5
Volume (m^3) — V_S	1.412×10^{27}	1.4×10^{27}
Mean density (kg/m^3) — ρ_S	1,409	1.4
Surface temp. (K) — T_S	5,780	5,800
Luminosity (J/s) — L_S	3.854×10^{26}	4×10^{26}
Solar constant (W/m^2)	1,368	1,400
Composition (no. atoms)	H (92.1%), He (7.8%)	H (100%)

considered in Chapter 22. The great brightness of the Sun makes its surface very difficult to observe in detail viewed from telescopes on the Earth's surface (indirectly, of course). The situation has been changed substantially over recent years by the introduction of space vehicles able to observe the Sun in space, beyond the atmosphere, and so to obtain a clear image such as that of Fig. 23.1.

23.1. Internal Conditions

No direct measurements can be made of the interior of the Sun but theoretical studies have provided a consistent picture which is believed to be reliable. The volume of the Sun is $V = 1.4 \times 10^{27}$ m^3 while it contains $N \approx 2 \times 10^{30}/1.67 \times 10^{-27} = 1.3 \times 10^{57}$ atoms. Each atom, therefore, has an associated volume $v = 1.4 \times 10^{27}/1.3 \times 10^{57} \approx 10^{-30}$ m^3. This corresponds to a linear dimension $d \approx 10^{-10}$ m. The radius of a neutral atom is of the order 10^{-8} m but the radius of the nucleus is of order 10^{-15} m. It seems that the atoms in the Sun generally are ionized and that they are loosely packed. It follows that the radius is sustained by radiation pressure rather than simple compression, at least over the larger part of the volume. Consequently, the mechanical equilibrium involves the balance between the compressing force of self-gravity of the mass and the opposing force of radiation. The radius will increase with the temperature. If there were no heat source, the radius would be that appropriate to simple atomic compression which would provide a radius 1/10th less of the present value.

The major concentration of mass is in the central region. The heat source is equally restricted there. Half the solar mass is contained within the central

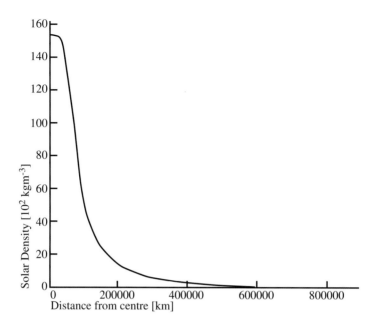

Fig. 23.2. The theoretical distribution of density with distance from the centre of the Sun.

sphere of radius 0.25 R_S which is 0.015 V_S — rather more than 99% of the energy generation occurs here. The profile of the density is shown in Fig. 23.2. It shows a central density of $\rho_c = 1.5 \times 10^5$ kg/m^3 (more than 15 times the density of iron) and a surface density of about 1×10^{-4} kg/m^3 (which is not a bad laboratory vacuum). With this density distribution half the solar mass is contained within the sphere with radius 0.5 R_S. The corresponding temperature profile has a central value of about 1.5×10^7 K and a visible surface value of 5800 K. This is shown in Fig. 23.3.

The radiation produced at the centre of the Sun is in the form of γ-rays and X-rays, together with neutrinos. The neutrinos hardly interact with matter so they escape virtually unaffected into space. The photon radiation shows the opposite effect. Radiation photons are inelastically scattered by free electrons, according to the rules of Compton scattering. The scattering causes the mean radiation frequency to decrease, the frequency reducing to optical levels at the surface. The mean free path of the photons (the distance between collisions) is very short, perhaps a few millimetres. The collisions are random, so the photons execute a Brownian motion on their way to the surface. Even at the speed of light, a photon takes the order of a million years to pass from the central region to the surface, giving an average

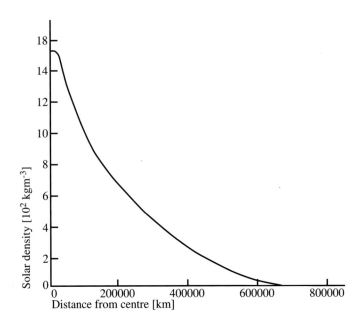

Fig. 23.3. The theoretical distribution of internal temperature in the Sun.

speed of about 2×10^{-5} m/s making some 10^{12} encounters in the process. This extended time scale provides a stability to the radiation process. The thermonuclear processes are themselves subject to short-term random effects so the amount of radiation varies in a corresponding way. The long delay in the emission of photon energy allows the Sun to maintain a mean equilibrium which transcends these thermal fluctuations.

The high densities near the centre make radiation the main mode of energy transfer there. This mode extends out to about 0.85 R_S (perhaps at distance 1.5×10^5 km from the centre) where the density becomes low. Beyond that the temperature has fallen dramatically, to the level of about 10^4 K where atoms can form even though they are highly ionized. Photons can be absorbed by the outer electrons in an atom, leading to a significant increase in the opacity to radiation at this level. One consequence here is a blockage in the passage of heat passing to the surface. If heat transfer by radiation is blocked the transfer mechanism changes to the less efficient method of material convection. The result is natural convection appearing from that point to the surface.[1] It is thought that there is more than one

---

[1] Convection was not included in the analysis of a main sequence star developed in IV.21. This is a good and sufficient approximation because the region of convection had little effect on the equilibrium of the star.

layer of convection. A layer of giant cells is coated with a layer of super granular cells and this is itself covered by smaller cells which form the visible surface. Each layer of cells is about a factor 10 smaller than the layer below. The surface layer has cells about 2,000 km deep and 1,000 km across.

The overall correctness of this picture is being confirmed by current observations in helio-seismology. Here the natural vibration of the Sun due to internal motions is revealed at the surface by small oscillations. The modes present come from different layers of the Sun — the longer wavelengths probing deeper than the shorter wavelengths. These studies can reveal information about the conditions deep inside. The measurements are difficult and time-consuming but promise to yield important information in the future. It may be possible to probe distant stars in the same way, through small variations of luminosity, in due course.

23.2. The Surface: The Photosphere (not to be viewed either *directly* or through optical instruments)

The mean surface temperature is 5800 K. It radiates in wavelengths from the longest to the shortest but those in the broadly visible range (between the infrared and ultra-violet) the emission is very similar to that of a black body. A solar spectral distribution is shown in Fig. 23.4. There, the dotted line is the theoretical Planck curve for a black body at 5800 K. There are components in the radio and X-ray regions but these are not black body — more will be said of them later.

The visible surface is sharp and apparently smooth, as if the Sun were a liquid sphere. This is seen in Fig. 23.5. In fact viewed in detail it is a ragged gaseous sphere which merges imperceptibly into the surrounding space. A more dynamic view of the surface is shown in Fig. 24.4, of IV.24. How is this difference to be accounted for? The passage of photons through the Sun has been seen to be by Compton scattering up to a thin layer in the outer regions. Beyond that the transport is by direct trajectories, the photons behaving as independent projectiles that move away from the Sun. The surface of the transition region between these two regimes is sharp and it is this very narrow spherical region that is seen as the surface of the whole Sun. That the surface is not a simple layer is shown by the phenomenon of limb darkening. The edges of the Sun are observed to be darker than the centre. This is shown in Fig. 23.6. The light that is seen by the observer from the edges travels through more gas (length a) than that from the centre

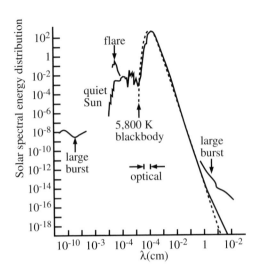

Fig. 23.4. The energy spectrum for radiation from the Sun. The dotted line shows a black body spectrum for the temperature 5800 K. Other regions do not follow the black body form.

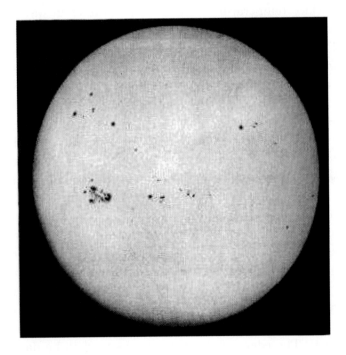

Fig. 23.5. A low power telescopic image of the solar surface. (Mount Wilson Observatory)

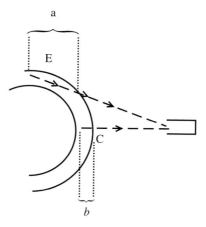

Fig. 23.6. The radiation from the centre of the body (point C) passes through less atmosphere than that from point E on the limb.

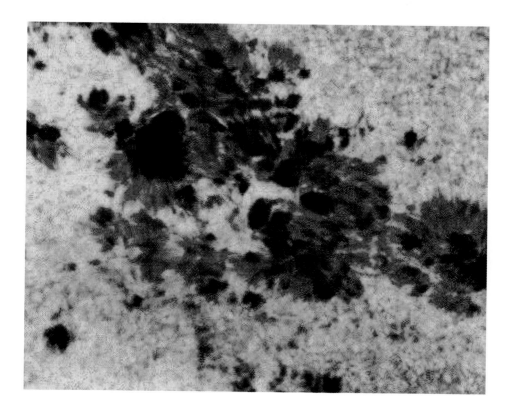

Fig. 23.7. A large group of sun spots. (National Solar observatory)

(length *b*). This means the edge radiation suffers greater absorption than that from the centre so that more light is received from the centre than from the edges. It results in that the Sun is seen as a sphere in perspective in the sky and not as a "flat" plate.

A major feature of the solar surface is sunspots (see Fig. 23.5 for a general view), which appear as dark spots (with the area of each spot comparable to the size of the Earth or greater) which may be a few or many to form a group. A sunspot group is shown in detail in Fig. 23.7. Although such groups are a major feature, there are periods when there are no spots at all. A closer view of an individual spot shows it to have internal structure and to be closely related to the surface granulation. Individual granules have a characteristic length of a few thousand kilometres. The granules change with time indicating an active surface. Each granule is edged in black showing a lower temperature: apparently these are the convective cells anticipated in the last Section. These features are shown in Fig. 23.8

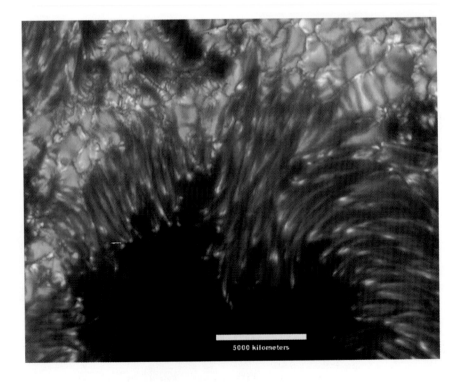

5000 kilometers

Fig. 23.8.   Details of the granulation at the edge of a sun spot. (Royal Swedish Academy of Sciences)

which refers to the boundary of a sunspot. The scale of the region is shown by the white bar which represents 5,000 km, or rather more than 80% of the radius of the Earth. The measurement of temperatures in the region show that the blackness of the spot is due to a lower temperature than the mean for the granules. The reduction is several hundred degrees K. The lower tememprature results in a reduction of radiation and, although the spot is actually bright, it appears dark in comparison with the higher temperature surrounding surface. The filaments are distorted by the dark spot. Although they are several thousand kilometres long, they are only 100 kilometres wide. As a matter of fact it is only recently that such details have been seen, such is the development of modern telescope.

The solar surface area covered by sunspots varies over an approximate eleven-year period, passing from sun spot minimum to maximum and then back to minimum. The surface can even be devoid of activity for a short period. The first signs begin, symmetrically north and south, at about $\pm 30°$ latitude, the activity moving quickly to lower latitudes (nearer the equator) and to some extent north/southwards up to about latitude $\pm 50°$. After about 2 years the maximum occurs and much of the middle surface is covered by groups of spots, such as those shown in Fig. 23.7. The surface coverage drops as the maximum is passed until the minimum occurs again after about another 9 years. Then the cycle starts again. The intensity of sun spot coverage varied from cycle to cycle as does the duration of the cycle. The series of cycles can be viewed as a "butterfly" diagram as was first suggested by A. E. Maunder in 1904. This is shown in Fig. 23.9.

Although the diagram covers the years 1954 to 1979, the phenomenon has been known for several centuries, in fact since the telescope was first applied to astronomy in the early 17th century. There was a period during the 18th century when no sunspots were reported from anywhere in the world. It is possible to see the larger sunspots with the naked eye under appropriate conditions (for instance, on a sunny, winter, misty day). No reports from mediaeval Europe or earlier that any mark has been seen on the Sun suggests they were not there. Extensive records have come from ancient China but no reports of anything resembling sunspots have been found. This period when sunspots were absent from the surface is called the Maunder minimum after the man who first recognised it.

The conditions near the surface have been explored very recently by the Michelson Doppler Imager (MDI) carried aboard the SOHO automatic space solar observatory and which detects the motion of material in the surface region. It has been found that the regions surrounding and below

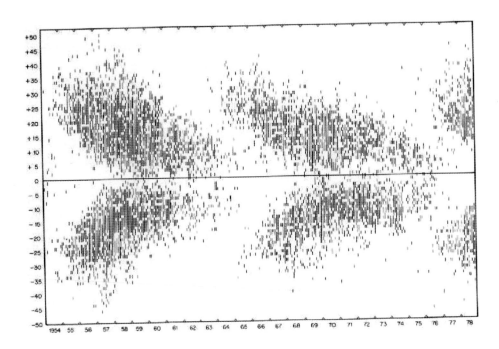

Fig. 23.9.   The Maunder "butterfly diagram" showing the area coverage of sunspots for the years 1953 to 1979. (Mount Wilson Observatory)

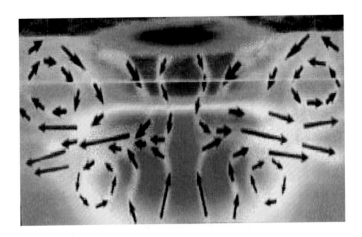

Fig. 23.10.   The inferred fluid motions below a sunspot. (MDI, SOHOH: NASA, ESA)

a sunspot is one of wide fluid flows. The inferred pattern of motions is shown in Fig. 23.10. Eddies apparently lie around the spot, sucking solar material downwards and so making a region where internal convection is not taking material upwards there. The result is that the region immediately under the spot is at a lower temperature than the surrounding surface explaining why the spot appears black relative to the surroundings. The precise details involve magnetic forces.

It was realised in the early 20th century that sunspots are the seat of substantial magnetic fields. Spectroscopic analysis able to detect the Zeeman effect has shown that the spots in one hemisphere have a particular magnetic polarity and those in the other the opposite polarity. The polarity of a hemisphere reverses with each cycle. This means that, although the visible period of the sunspot cycle is 11 years, the full magnetic period of the cycle is 22 years. Magnetic lines of force are found to emerge from one side of the spot and re-enter the solar surface on the other. It became clear that magnetic effects on the Sun are an important characteristic of its internal behaviour.

23.3. Solar Rotation

It is found from observations of sun spots that the rotation of the Sun decreases with latitude. Thus, the (sidereal) rotation period at the equator is 25.7 days but is 28 days at latitude 30°, 37 days at latitude 60° and 33.4 days at latitude 75°. The Michelson Doppler Imager carried aboard the SOHO craft allows the solar vibrations to be detected and so the rotation periods deep inside. This is shown schematically in Fig. 23.11. It is found that the rotation periods, which differ so much on the surface, move together with depth until they settle towards a common value at a little below 0.7 the solar radius. This transition region extends over a 20,000 km ring of matter involving shear of the material. Much below, at about 0.33 solar radius, the different curves combine to a single period of about 28 days. This period refers to the core which would seem to rotate as a solid body.

The rotation periods are obtained from the observational data by mathematical analyses and unexpectedly a residue of data remained once the mean rotation was extracted. This turned out to be a jet stream system circulating the poles not unlike the jet streams which circulate in the Earth's atmosphere. The jets are at latitude 75° and at a depth about 40,000 km below the photosphere. The jet streams, both north and south, move some 10% faster than the surrounding gas. They are perhaps 30,000 km wide. There

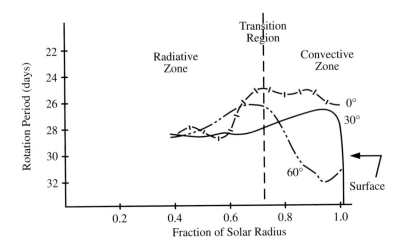

Fig. 23.11.   Showing the rotation periods at three latitudes deduced from measurements using the Michelson Doppler Imager. (After SOHO, NASA/ESA)

are also alternating fast and slow bands some 20,000 km below the surface. These bands are not fixed in location but moves from higher latitudes towards the equator over the period of the sunspot cycle. The boundaries between the high and low speed current bands are regions where the temperature is slightly higher than the local material. They are also regions associated with sun spot groups. The movement of the ionized gas carries magnetic fields with it and so acts to concentrate them. The movements of fluids and of magnetic fields are intimately connected within the Sun. The SOHO orbiter is providing new in-depth details of unexpected features of the Sun. One special facility now is the ability to investigate the Sun over a range of wavelengths and simultaneously, which is not possible from the Earth's surface due to the interference by the atmosphere. Data in the ultraviolet are proving especially interesting.

*

Summary

1. The Sun is a large fluid ball with a central temperature a little more than 1.5×10^7 K and a surface temperature of 5800 K.

2. The mean gradient of the temperature is, then, about 0.02 K/m. It would be difficult to recognize so small a gradient in everyday living so

the internal region can be supposed to be in local thermodynamic equilibrium everywhere and the interior in thermodynamic equilibrium.

3. The solar density has a value of about 1.5×10^5 kg/m^3 at the centre. This falls off very quickly with distance from the centre and has fallen to about 10^4 kg/m^3, about 30% out from the centre.

4. The visible surface (the photosphere) radiates as a black body over the central wavelengths. It has the appearance of a smooth spherical surface but the passage of radiation through the atmosphere causes limb darkening, making it appear like a three-dimensional sphere rather than a flat disc.

5. The surface carries sunspots seen as dark marks on the surface. The characteristic length of the spots is several thousand kilometres. They tend to appear in groups and the area of the Sun covered goes through an eleven year cycle, minimum to maximum and return to minimum.

6. They first appear at higher latitudes, north and south, and the area covered moves down to the equator before the spots disappear. This behaviour can be represented by the butterfly diagram of Maunder.

7. The fluid beneath a spot carries considerable convective activity.

8. Sunspots are associated with strong magnetic fields. The polarity is different in each hemisphere and changes from one sunspot cycle to the next. The period of the magnetic sunspot cycle is, therefore, about twenty-two years rather than the eleven for the sunspot numbers.

9. The Sun does not rotate as a rigid body in its outer regions but the central region does have a single rotation period as would a solid body.

24

Solar Emissions of Particles:
The Solar Wind

Although the Sun is a gaseous body we saw that it appears to have a "solid" surface because the mean free path of photons is very short, up to that radius from the centre after which the atoms are effectively free, moving in straight line trajectories. This region beyond that of random motion can be taken to be the "atmosphere" and it is the topic now. It is a complicated region as is seen from Fig. 24.1. This shows a single sunspot region photographed on 2001 March at three different electromagnetic wave lengths. The appearance of each is very different. The top picture is taken in visible light shows well formed dark spots against a bright, smooth, sharp photosphere. The characteristic dimension of the group is of the order 120,000 km. The profile of the Sun is well defined at the left and right horizons at this wavelength. The middle picture is in extreme ultraviolet radiation. Now the bright active regions, not geometrically well defined, stand out against a darker background. The background itself has more detail and the surface profile is less sharp.

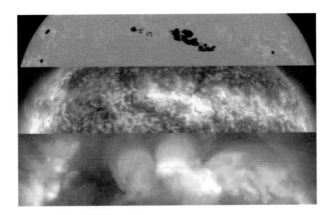

Fig. 24.1. A group of sunspots viewed in three wavelengths. (SOHO — MDI/EIT Consortium, Yohkoh/SXT Proect)

The regions above the sunspots are strong emitters of ultraviolet radiation. The surface granules are also emitters. The lower picture is in X-rays and shows a diffuse spread of gas surrounding the active regions. The profile of the Sun has been effectively lost here. There are large cloud-like regions where X-rays are emitted.

Is all this different activity taking place at the same level in the atmosphere? No — the three panels show radiation from three different heights above the optically visible surface. The surface itself is the photosphere and, as we have seen, has a mean temperature of a little below 5,800 K. That is the first picture of the sunspots. Above that is a gaseous layer called the chromosphere with temperatures in the range 10^4–10^5 K. This is the ultraviolet region. Above that is the solar corona with a mean temperature of a few times 10^6 K which gives emissions in the X-ray region. It must be admitted that these high temperatures are rather technical because the amount density of matter is very low and the heat content quite small. The effect on the atoms, however, is precisely that for the temperatures involved which are measured by levels of ionization, even though few atoms may be involved.

24.1. Above the Surface: The Chromosphere and Corona

Figure 24.1 shows that the photosphere is enclosed by a thin atmosphere involving high temperature with the emission of high energy radiation. This is the chromosphere. This relatively thin region is encased in a more extensive region called the corona. Before observations with spacecraft were possible, which have allowed full photographs of the Sun such as Figs. 24.1 and 24.5 to be obtained, details of the outer regions enclosing the visible surface could only be detected during a short time (a very few minutes) of a solar eclipse by the Moon. One excellent modern example of a traditional eclipse is shown in Fig. 24.2, taken from the High Altitude Observatory on Mauna Kea in Hawaii in 1991. It is seen that the region away from the Sun, called the corona, is highly ordered. The equatorial plane is well marked by an extended gaseous region while the polar regions show distinctive plumes away into interplanetary space. The fluid lines remind one of the paths of magnetic lines of force and it is known that magnetism does, indeed, play a large part in the dynamics of the corona. In fact, the paths of the streamers show a certain similarity to the (distorted) lines of force of a simple dipole magnet. The first thoughts were that the Sun is the source of such a magnetic structure and that the ionized material in the atmosphere was responding to

Fig. 24.2.   An excellent picture of the outer layers (corona) of the Sun seen during the solar eclipse of 1991. (High Altitude Observatory, Hawaii)

a magnetic force. This early surmise was proved correct locally but entirely wrong on a solar scale.

Actually, it was realised during the 19th century and developed in the 20th that limited details of the chromosphere can, in fact, be detected from the Earth's surface without a lunar eclipse at all by using a special telescope technique. A disc of size equal to that of the image of the Moon in the telescope at the focal point is placed there to obscure the image of the bright Sun and so form an artificial eclipse. A cross section of the chromospere can be studied continuously during daylight in visible light. This method gave details of the chromosphere although not the overall picture seen during an eclipse. The method has, however, had a renaissance recently. The automatic solar probe SOHO obtains photographs such as Fig. 23.1 directly but these give little detail of chromospheric structure. By using the blocking ring details become very clear as in Fig. 24.3. The enormous arched emissions of hydrogen gas are clear but so are the masses of gas shot off from the surface (the so-called mass ejections) at 8 and 2 o'clock. These represent a major interaction between the Sun and interplanetary space.

Fig. 24.3.   A view of the Sun from SOHO in which a central ring cuts out the main glare of the Sun. (SOHO: NASA, ESA Consortium)

Fig. 24.4.   The loops around sunspots on a quiet Sun, in ultraviolet light. (NASA, Trace)

The surface of the sun shows many features involving filaments and arched structures. One example, of the Sun on a quiet day, is shown in Fig. 24.4 taken in ultraviolet light. The surface is one of turmoil. The bright region towards the horizon is a sunspot group with glowing gases flowing out of it. The temperature here is about 10^6 K compared to the black regions where the temperature is some three orders of magnitude lower. It is surprising that the apparently flat solar surface is, in fact, more like a sea in turmoil.

More disturbed conditions give rise to prominences. These are huge arches of gas extending enormous distances out into space. One example is shown in Fig. 24.5 and another in Fig. 24.6. The first shows a single loop of enormous proportions while the second shows two enormous arches rising above the photosphere. These ejections often contain millions of tons of gas and can rise to a height of several hundred thousand kilometres. They can last anywhere between a few hours and a few weeks. Some reappear after a solar rotation, 27 days later — indeed, the average lifetime is about a month although some last twice or three times longer. This is several solar rotations. The pictures here show the prominences from the side. Seen from

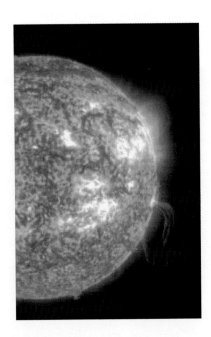

Fig. 24.5. A simple prominence taken in extreme ultraviolet helium light by the SOHO Imaging Telescope, 2001, February 12 during sunspot maximum. (SOHO–EIT Consortium, ESA, NASA)

Fig. 24.6. An arched prominence taken in Hα light in 1997. (Big Bear Observatory, Caltech)

above they form a line known as flocculae. The very large sources of energy driving prominences have still not been positively identified but they are believed to involve magnetic fields.

A prominence is a source of considerable energy. High energy electrons and protons are believed to spiral through the arch maintained by the magnetic field joining the two "feet", of opposite polarity, at each end. One effect is the release of high energy radio waves. The impact of electrons on the photospheric materials at the negative foot causes the production of hard X-rays and γ-rays. The impacts at the positive foot cause the production of neutrons and soft X-rays. Since prominences are associated with sunspots, the different intensities of sunspot numbers during a cycle provide different effects on the atmosphere.

24.2. Magnetism

Solar magnetism is a very difficult subject. The electrical conductivity, σ, of the material is high and the length scale, L, is large so the product σL^2 is enormous. The time for a magnetic field to decay in the solar plasma is proportional to this product which means the field does not decay. Consequently, the magnetic field is said to be frozen into the material and fluid and field move together. This is thought to be the basic mechanism driving many of the material motions at the surface.

The Sun does not possess a visible general intrinsic field so there is no intrinsic field component beyond the photosphere. The fields that are observed weave out of and into the main body and there is no single region where the field may be supposed to be formed. It is often thought that one possible region of origin could be movements of stress below the convective region but this is still to be demonstrated. One obvious primordial source is the magnification of the initial weak field which permeated the material that collapsed initially to form the Sun. The Ulysses spacecraft established that the lines of magnetic force are distributed quite evenly in latitude over the solar surface and is not concentrated in the regions of high latitude, including the polar regions, as previously believed. This new distribution is shown in Fig. 24.7. The field may look like a dipole field at first glance but more careful scrutiny shows that it is anything but that. The polar lines of force are open exposing the solar surface to space. This is also the case in the equatorial latitudes and particle emissions are possible particularly from those regions as will be seen in the next Section. The details of the solar field change with the position in the sunspot cycle. At sunspot minimum the field is simpler and more static. It becomes more complicated and variable as the cycle progresses until it has a quite complex form at maximum. There are still many obscurities in solar magnetism, which remains a subject of active research.

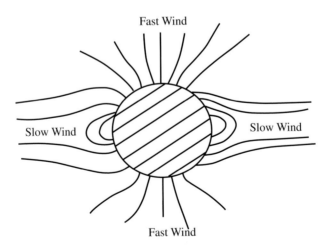

Fig. 24.7.   Showing the open magnetic field lines at the equator and at the two poles.

24.3. The Solar Wind

The very high temperatures of the corona would be expected to lead to the emission of particles into space as the corona continually expands. This was suspected early on. Such a "wind" was postulated explicitly in the 1950s to explain the fact that comet tails point away from the Sun, whether the comet is approaching the Sun or moving away from it. The tail acts as a sort of flag in the wind. It was the first space mission to Venus in the 1970s that established the actual presence of a substantial general emission of particles from the Sun. It became clear that the emission is variable in its intensity and could be viewed as a sort of space weather. The emission is called the solar wind. It has now been studied in detail for several decades and is found to pervade the full range of the Solar System. The solar wind emerges from holes in the corona (coronal holes) where the magnetic field is peeled back to open directly to interplanetary space. The activity varies with the phase of the sunspot cycle because the structure of the magnetic environment changes with different phases of the cycle. The simplest situation is shown in Fig. 24.8, referring to sunspot minimum, where coronal holes are seen clearly at the north and south poles. There is one active region close to the equator.

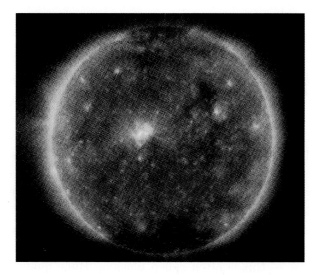

Fig. 24.8. The Sun at sunspot minimum on August 22, 1996 photographed in extreme ultraviolet light at 195 Å (1.95×10^{-4} m) using ionized iron corresponding to a temperature of 1.5×10^{6} K. The corona holes at the north and south poles are clearly evident. (SOHO, NASA–ESA)

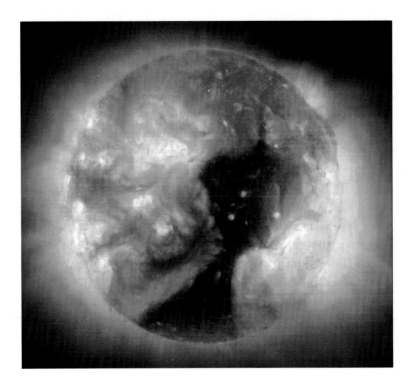

Fig. 24.9. The large coronal hole observed 2002 January extending from the south pole well into the northern hemisphere during a period of high sunspot activity. (SOHO; NASA–ESA)

In this case, the poles alone give direct contact with interplanetary space. Figure 24.9 shows (in the far ultraviolet) the other extreme, at sunspot maximum, where a gigantic coronal hole can be seen covering the southern pole, much of the southern hemisphere and penetrating into the northern hemisphere. This provides a massive opportunity for electrons and nuclei to leave the Sun at high speeds. There are also eight or ten active regions with sunspots providing extensive activity. The Sun in this state has a major effect on the ionospheres of the planets.

The ejected material is a mixture of magnetic lines of force and electrically charged ions, mainly hydrogen but also including all the components of the solar abundance — this is often called plasma. Huge quantities of plasma can be ejected in extreme conditions called coronal mass ejections (CEMs). Two essentially simultaneous CEMs are seen in Fig. 24.3. The scale is shown by the inclusion of an image of the Sun in the central ring. It is clear that the ejections have a substantial volume and can reach out very nearly a solar radius before moving away into interplanetary space.

The solar wind has two components, one slow and the other fast. The slow component emerges from equatorial latitudes while the fast component emanates from the two polar regions. The faster component has a speed of about 800 km/s while the slower component has a speed of about 450 km/s. The slower speed can be understood as the natural expansion of the hot corona and is now known to originate in the equatorial latitudes. The high speed component emanates from the polar coronal holes but the energy source for these high speed particles is still not known. The mean density of the solar wind at the Sun is about 2.0×10^8 atomic mass units/m^3 (1 amu $= 1.66 \times 10^{-27}$ kg). This has been reduced to about 6.0×10^7 amu/m^3 at the radius of Mercury, 3×10^5 at the distance of Jupiter and down to some 5×10^3 at the distance of Pluto.

24.4. Present and Future Variability

It has usually been assumed in the past that the Sun's output of radiation is essentially constant but recent measurements from space vehicles and improved theories of stellar structure suggest this is, in fact, not so.

Theory shows that a main sequence star of about solar mass begins its life with a luminosity of about 80% of the Sun's present luminosity. This slowly increases with age, closely linearly, until the luminosity has become 130% higher that the initial value after about 10^{10} years. At this age the star leaves the main sequence (upwards on the H-R diagram) and the luminosity increases. The luminosity then rises to about 1,200 times its present value, the radius expanding to the orbit of Venus or beyond. This is a first red giant phase. Further variations follow until, after a series of helium shell flashes associated with helium burning, the star becomes a white dwarf star after about 12×10^9 years, having shed some 45% of its initial mass by the ejection of gases. The star then remains planet sized, slowly losing its heat by radiation. For the Sun, its present age is probably a little less than 5×10^9 years and it is often said to be middle-aged.

Although it is obviously not possible to check such calculations in detail, it is known that the solar output is subject to some variability. The Sun has been observed by space vehicles (and especially SOHO) in detail over at least two solar sunspot cycles and it is established now that the luminosity varies throughout the cycle, increasing at sunspot maximum over sunspot minimum by about 0.5%. Somewhat controversial is the indication that the last sunspot cycle has involved a greater luminosity than the one before. The effect on the planets is too small to quantify and it has till now to be firmly established. Theoretically, it would be expected to be so but more observations are needed in the future to make any effect precise and quantitative. The effect on the Earth's climate is not known.

*

Summary

1. The Sun is a truly boiling cauldron with an atmosphere (the corona and the chromosphere) showing great activity.

2. The surface is covered in filaments and large arches of gas, called prominences, the largest of which rise several hundred thousand kilometers above the surface.

3. These high structures would be of great relevance to a companion with a very small semi-major axis.

4. The Sun is not the seat of an ordered intrinsic magnetic field. The wide range of fields that are observed are not ordered but are associated with various surface features.

5. The Sun is the source of a strong stream of ejected atomic particles, called the solar wind. The composition of the solar wind is obviously the same as the particle distribution in the surface region which is the cosmic abundance of the elements.

6. The property of the Sun is found also to occur in other stars, and often much more strongly. The intensity of the stream is believed to become lower with the age of the star.

7. The solar wind passes through the solar system and beyond, until it encounters the interstellar plasma, perhaps some 100 AU downstream. This meeting is called the heliopause and marks the edge of the solar system which is, in this way, enclosed in a large magnetic bubble.

8. The intensity of the radiation from the Sun has been found to show very small variations on several time scales. The effect of these variations on the climate of the Earth is not yet known.

Part V

EXOPLANETS

Bodies of planetary size and mass around other stars with special reference to solar type stars.

25

A Planetary System from Afar: The Solar System

We can see the main members of the Solar System from Earth, perhaps using telescopic aids, but would find such observations increasingly difficult if we were to move steadily further away from the Sun, into space. Ultimately, as we got very far away, the angle of the planets subtended at the eye would be so small that we would be able to discern very little of the System. This is the situation we face when we attempt to find other planetary systems similar to our own. It is important to ask, therefore, what observations could be made from far away to see a planetary system and, at least, what evidence would allow some indications of the existence of the planetary system itself?

25.1. Observing the Motion of the Central Star

For the Solar System, the Sun itself would be visible from large distances as a point of light. Using photon detection methods of great sensitivity its radiation would be detectable for several hundred parsecs but the accompanying planets would not be seen directly. Their presence can be confirmed from their gravitational effects on the Sun itself moving about the common centre of mass. This will be the case for other solar-type stars.

Suppose the star has a single companion. The star and planet orbit each other in two linked elliptical paths with a common focus and the same eccentricity, as shown in Fig. 25.1. The shape (ellipticity) of each orbit is the same as are the times for one orbit but the linear scale of the orbits, and the speeds of the bodies following them, are inversely proportional to the mass of the body. In the diagram, the heavier star follows the solid line while the less massive companion follows the dotted curve. This is the case for the Solar System, the Sun moving in response to the gravitational pull of each planet.

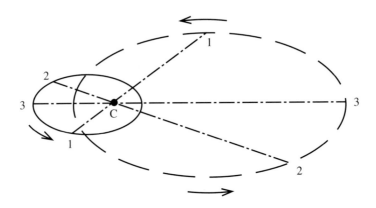

Fig. 25.1.   The paths of two masses orbiting under gravity with a common focus C acting
as the centre of the force. The greater mass follows the solid orbit while the smaller mass
follows the dashed orbit. The corresponding motions 1, 2, 3 are shown in each orbit.

Table 25.1.   The effect of three planets on the motion of the Sun about the
centre of mass of the Solar System. The associated solar orbital period is given
in column 3, the mean solar orbital radius in column 4 and the mean solar
orbital speed in column 5.

Planet	Eccentricity of planet orbit	Period of orbit planet (ys)	Semi-major axis of Sun (km)	Speed of Sun in orbit (m/s)
Jupiter	0.0483	11.86	7.4×10^5	12.4
Saturn	0.0560	29.46	4.1×10^5	2.7
Earth	0.0167	1.00	4.5×10^2	0.09

Data from the Solar System provide an example of the magnitudes that
can be involved. The situation is simpler here because the orbits are very
closely circular. The centre of mass of the System is displaced from the
centre of the Sun by about 7.7×10^8 m which is about 1.1 solar radii. This
dispalcement is very largely the effect of the planet Jupiter alone. The Sun
follows an elliptic path with small eccentricity about the centre of mass, as
described in I.3. The full solar motion is the superposition of the motions
caused by each of the individual planets even though the effect of Jupiter
dominates those of all the other planets. The effects of Jupiter, Saturn
and Earth on the Sun are listed in Table 25.1. The second column lists
the eccentricities of the planetary orbit which is also the eccentricity of the
corresponding solar orbit. The third column lists the orbital period of the
planet which is also the orbital period of the Sun due to this interaction.
The third column lists the consequential semi-major axis of the Sun (very

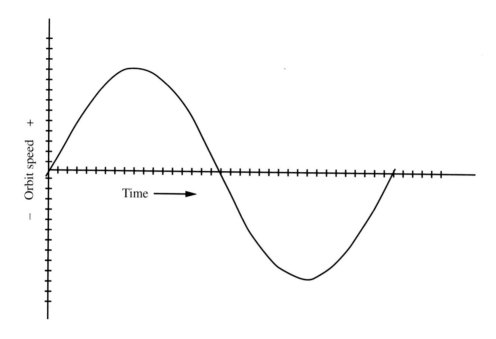

Fig. 25.2. The orbit speed-time curve (in arbitrary units) for a body in a circular orbit about the mass centre. The orbital speed is constant.

nearly circular in practice) while the last column lists the speed of the Sun in its orbit due to each planetary perturbation.

It is seen from the Table that the speed of the Sun's orbital motion due to the planetary perturbation is small, but by no means negligible. As it happens, such speeds can be detected by Doppler-shift measurements if they are not too small. At the present time speeds beyond about 3 m/s can be measured with confidence but new techniques are about to reduce this to 1 m/s. This will mean that companions with masses as low as 10^{25} kg and above are open to investigation. The motion of a single planet in a circular orbit will provide a single sinusoidal speed-time curve for the orbit of the star about the common centre of mass of the form shown in Fig. 25.2. The curve here is a sine wave. The amplitude is arbitrary since it depends in practice on the ratio of the mass of the companion to that of the star.

The speed in the orbit is not constant if the orbit is elliptical. This will be the most common case and the variations of speed through the orbit are accounted for by Kepler's laws of planetary motion. Remember here that the companion moves more quickly when it is nearer to the central star (perihelion) but slower when it is further away (slowest at aphelion). The effect is to distort the simple shape of the sine wave to the form shown in

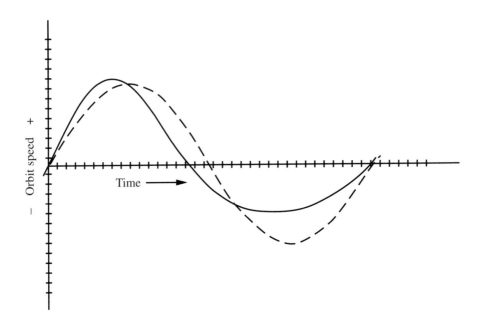

Fig. 25.3.   The orbit-time curve for an eccentric orbit (solid curve).   The curve for a circular orbit is shown dotted for comparison.

Fig. 25.3.   The sinusoidal form is shown as dashed for comparison.   The setting of the zero of time is arbitrary and is chosen here to be that of the uniform speed.   The speed is seen to increase at first and subsequently to slow to speed up again as perihelion is approached.   Obviously, accurate measurements are required if a precise measurement of the ellipticity of the orbit is to be obtained this way.

The Solar System has two especially large planets (Jupiter and Saturn) so this case is of interest.   One example of the simultaneous motions of two companions of unequal masses is shown in Fig. 25.4.   The orbits are presumed circular and the smaller mass is supposed to orbit outside the more massive companion, with an orbital period twice that of the larger mass.   The profile is similar to that of a damped wave during one combined period but the profile is repeated with the period of the larger orbital time Just as the composite curve is constructed by adding the two separate curves so the observed profile can be decomposed into its component parts.   This certainly requires accuracy of measurement but it also must involve a sufficient time of observation for at least a significant fraction of the orbital time of the longest member to be measured.

This is of extreme relevance to observations of exo-planets.   The first such companion was discovered in 1995 and observations have been undertaken,

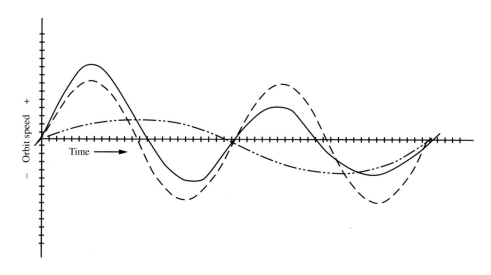

Fig. 25.4.   The orbit-time curve for two companions in orbit.

in increasing number, since then. The longest observed complete period, therefore, is about 8 years. For the Solar System, this is barely 0.7 of the orbital period of Jupiter (11.87 years) and rather less than 1/3 that of Saturn (29.42 years). Over an eight-year period it would be possible to derive a reasonably accurate orbit for Jupiter and to show evidence for the existence of Saturn but probably not an accurate Saturn orbit. The situation is more encouraging for planets with a much smaller semi-major axis, and so substantially shorter orbital periods. To this extent, at least, any survey of exo-planets conducted so far will be deficient in information about the companions with larger semi-major axes quite independently of the inability to find the smaller companions.

The general study of the observed orbits must be based on Fourier analysis and for this the accuracy of the data is of great importance. The study is in its infancy and will require observing time to develop. The accuracy of the method as a terrestrial detection method is limited by the random motions of the Earth's turbulent atmosphere. This will be overcome by making observations from space vehicles. The turbulent conditions at the surface of the star under investigation also set a first limit on the lowest mass that can be observed. Generally, it will not be possible to observe bodies much below about the Uranus/Neptune mass using the Doppler method.

The motions of the star-companion system are controlled by the conservation of energy and angular momentum. The application of the requirements

of these constraints allows the combined mass of the two bodies to be found
as well as the orbital details. If the mass of the star is known from other
sources, the mass of the companion can be deduced. Unfortunately, there
are not quite enough observed variables to allow the system to be specified
completely and in particular the inclination of the orbit of the companion remains unknown. This means that the absolute motion in the orbit
cannot be found but only the component along the direction of sight of the
observer. This provides a minimum mass for the companion through the
quantity $M \sin i$ where i is the inclination of the orbit and M is the actual
mass of the companion. It might be expected that the companion will orbit
near the equatorial plane of the central star ($i \approx 90°$) in which case the minimum mass is not likely to differ greatly from the actual mass. If a density
is presumed for the companion a tentative radius can be assigned to it. This
is no more than an indication because it was seen in I.4 that the companion
can be expected to have one of a range of compositions in principle. It might
be expected in practice that the companion will be a hydrogen body but this
must be checked by observation.

25.2.   The Case of a Transit

The very special case when the observed orbit of the companion crosses the
face of the central star as observed has special significance. This is said to
be a transit of the companion. The extreme case is the solar eclipse due
to the Moon, which is strictly a transit but one where the solar radiation
is stopped from reaching the Earth. The transit of the planet Venus is the
same in principle but very different in practice because its apparent size is
minute in comparison with that of the Sun. It was of importance in the past
because the angles of parallax of the transit can be used to determine the
actual distance of Venus from the Earth and the Sun.[1]

The importance of a transit is that the radiation of the star is reduced by
an amount proportional to the area of the companion disc and this reduction
can be measured. Assuming the companion is effectively black, as a planet
would be, the area causing the reduction of radiation can be calculated
with some accuracy. This is the cross-section area of the companion and a
radius for it then follows, assuming the companion to be essentially spherical.
If, further, the mass of the star is known alternatively, for instance through

---

[1]Captain James Cook famously observed the transit of Venus from the Pacific Ocean in
the 18th century.

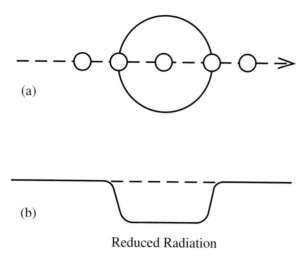

(a)

(b)

Reduced Radiation

Fig. 25.5. Schematic representation of a transit of a star by its companion as viewed from Earth. (a) The companion moves across the stellar face; (b) the intensity of the radiation decreases.

stellar theory, the observed motion again allows the actual mass of the companion to be found. For a transit, the inclination of the orbit of the companion is known and the dynamical structure of the system is consequently known completely. With the radius and the known, the density follows at once and the probably composition of the companion can be deduced. The technique is illustrated in Fig. 25.5. The planet appears as a black dot passing in front of the star. The approach requires measurements of the intensity of radiation (techniques of photometry) of great accuracy if it is to give reliable results. To gain some idea of the magnitudes involved, the passage of Jupiter across the Sun reduces the solar radiation received on Earth by about 1% while a body of Earth size will reduce the solar radiation by about 0.01%. These are small amounts but are measurable using modern apparatus.

Transits of exo-planets observed from the Earth will always be rare but partial transits will be less so. At present three systems are known with orbital inclinations close to 90°. These are the companion of the star HD 209458, the companion of the star OGLE-TR-56 and the companion of the star BD-10 3166. Finding partial transits will increase the number of events and the optical gravitational lensing experiment OGLE is designed to find full and partial transits. It has had some success and the third transit example above was discovered this way. Observations from a space vehicle will be likely to have more success. The European Space Agency (ESA) has accepted a mission, the Eddington mission, designed to survey about 500,000 stars. The detectors will be designed to have sufficient sensitivity to measure the small change of stellar brightness associated with the transit of a body of the size of the Earth. The launch date has not been announced but it is likely to be around 2007 or soon after.

25.3. Polarimetry

Light reflected by a rigid surface is polarised so that the electric vector is parallel to the surface at the point of reflection. In the present terms, the light reflected by a companion will be polarised in a way depending on the nature of the surface. The light from the star, on the other hand, is black-body radiation and is not polarised. Observations of the star system can allow the polarised light to be traced and so the reflecting body identified. Polarimeters are apparatus designed to detect polarised radiation and are being developed for automatic measurements in space. Such equipment is expected to fly by the end of the decade.

25.4. Nulling Interferometry

A more immediate possibility is to use the inverse of normal interferometry, where two beams are superposed to enhance the crests and troughs, to give instead a zero beam. By arranging the crests of one wave from a star to lie over the troughs of another wave from the star, the result is that the light beams cancel each other and the star disappears. The light reflected from a planet, however, follows a different path through the optical system and will not vanish. The presence of the planet will be evident in the field of the eyepiece.

The first steps are being taken to apply this method in practice. The observations will first be made from the Earth's surface using a large telescope (for instance, the 8-metre telescopes of ESO's Very Large Telescope) combining the beams with the ESA's ground based European nulling interferometer experiment (GENIE). The aim eventually is to have such instrumentation on an automatic spacecraft and ESA's planned Darwin mission is to be the ultimate equipment for "seeing" planets. NASA is also working to construct a Space Interferometry Mission (SIM). The aim is to resolve bodies separated by as little as 0.01 arc second. A number of telescopes will be mounted on a 10 m arm able to be moved to keep a light path stable within 10 Å (1×10^{-9} m). It is planned that SIM should be in orbit by 2008. These missions should, in principle and perhaps also in practice, be able to detect the least massive of planets, even those with masses below that of Earth.

25.5.   Astrometry

This study is to measure precisely the position of a star in real time and so detect its motion directly. The method requires extreme instrumentation and must be performed outside the Earth's atmosphere to achieve sufficient accuracy. ESA is designing a mission along these lines, to be launched during the next decade and to be named Gaia. Only bodies of Jupiter mass and above will be able to be detected by this method in the foreseeable future.

25.6.   Direct Imaging — White Dwarf Stars

The glare of the central star is less obtrusive for a white dwarf system than for a main sequence star system. The reflected intensity of a planet at optical wavelengths will be a factor 10^8 lower than the radiation intensity of the star so the companion cannot be seen. The situation is different for a white dwarf star in infrared wavelengths. Now, because the temperature of the star is low, the radiation of the star is not critically larger than that of the planet, perhaps only greater by a factor 10^3. It then becomes possible to view the two directly. Accurate infrared measurements cannot be made from Earth because of the infrared interference there but measurements can be made in space. Various test arrangements are being devised for later application in automatic probes.

25.7.   Conclusion

A new era was opened in the last decade of the last century with the discovery of companions of planetary size orbiting stars. The methods of detection are as yet crude but nevertheless are giving interesting results. Only the most massive companions can be detected at the present time but work is underway to improve the equipment to allow less massive objects to be discovered. An essential part of these developments must be observations from automatic space probes. The ultimate aim must be to map the surface details of the companions so that bodies comparable to our Earth can be identified, if they do, indeed, exist. This is a very grand enterprise which may begin to bear fruit in a decade or so from now.

*

Summary

1. The appearance of the Solar System for an observer far away is speculated upon. This acts as a guide to the appearance of other companion systems view from the Earth.

2. The various methods for detecting such systems are described, both of those used now and those that are planned for the future.

Post Script

Very recently NASA's Spitzer Space Telescope has detected directly the infrared light from the two known exo-planetary bodies HD 209458b and TrES-1, previously found by transit methods. The measurements confirm that each planet is indeed a hot Jupiter with surface temperatures of the order of 1,000 K. The future use of different infrared wavelengths offers the possibility of studying the planets' atmospheric compositions and wind structures. This is the first direct detection of planets and confirms the validity of the methods used before in the visible wavelengths. Conditions here were particularly favourable for observations. The small semi-major axes of the orbits leading to high surface temperatures causes the planets to emit heat strongly and so to be sources of comparable strength to the central star itself.

26

Observed Exo-Planet Systems

The first exo-planets were discovered in 1991 by Wolszczan and Frail, around the millisecond pulsar star PSR B1257+12. The first exo-planet around a solar type star was discovered in 1995 by Michel Mayer and Didier Queloz of the Geneva Astronomical Observatory. Since then many more companions to solar type stars have been recognised and more are being discovered month by month. At the present time, 136 planetary systems have been found involving 154 planets: 14 of these are multiple planet systems. What are the characteristics of these systems? How do they compare with the Solar System? That is the subject now.[1]

26.1. Pulser Systems

The pulsar PSR B1257+12 has been found to have three Earth mass companions and one Moon like mass in orbit about it. There is a possible 5th member with perhaps cometary mass and a period of about 3 years. The data are collected in Table 26.1. The small values of the semi-major axes for the inner three companions are strange when it is realised that they orbit a pulsar. They are all within the orbit of Venus and yet it would be expected that the initial star would have expanded to beyond 2 AU (beyond the orbit of Mars) before it collapsed to form the pulsar. The three companions would have been absorbed and should have disappeared. The fourth planet would not have been affected. Why these three planets are there is, consequently, a mystery. Their masses are what might be expected for the core of Jupiter like planets or heavier. This would be understandable if the hydrogen covering had been blown away by the supernova event. The problem of the relatively small semi-major axes remains.

---

[1]An up-to-date list of exo-planets is found on the internet address
http://www.obspm.fr/encycl/catalog.html

Table 26.1.   The orbital data for the companions of the pulsar PSR
B1256+12.

Companions of PSR B1257+12				
$M \sin i$ (M_E)	0.015	3.4	2.8	≈ 100
Semi-major axis (AU)	0.19	0.36	0.47	40
Period (days)	25.34	66.54	98.22	6.21×10^4
Eccentricity	0.0	0.0182	0.0264	?

A second pulsar is also known, PSR B1620–26, though its properties have
not been as accurately defined as in the previous pulsar. The details for the
companion are

$M \sin i -$ $1.2 < M \sin i < 6.7$ ($M_J = 1.898 \times 10^{27}$ kg)

Semi-major axis (AU)   10–64 AU

Orbital period   61.8–389 years

Eccentricity   0–0

26.2.   A Companion to a Solar-Type Star

The first planetary companion around a solar type star was discovered in
1995 using Doppler measurements of the orbital speed of the star. If ν_s is
the frequency of the radiation emitted by the source and ν_o is the frequency
measured by the observer, the theory gives the relation

$$\frac{\nu_o}{\nu_s} = \frac{1}{1 \pm \frac{\nu}{c}},$$

where ν is the magnitude of the speed of the source relative to the stationary
observer and c (= 2.99×10^8 m/s) is the speed of light in a vacuum. The
positive sign is used if the source is moving away from the observer ($\nu_o < \nu_s$
so the light is reddened) while the negative sign is used if the source is
moving towards the observer ($\nu_o > \nu_s$ so the light is more blue). Applying
this theory to the motion of a distant object in an orbit means that the
perihelion and aphelion motions are measured, one moving away from the
observer on Earth and the other moving towards the observer. The plot that
results of the inferred speeds against time is shown in Fig. 26.1 which is a
repeat of the first measurements of the first companion found, that of the
star 51 Pegasus. The initial discovery was made by Meyer and Queloz. The
discovery was confirmed very soon after by Marcy and Butler whose curve is

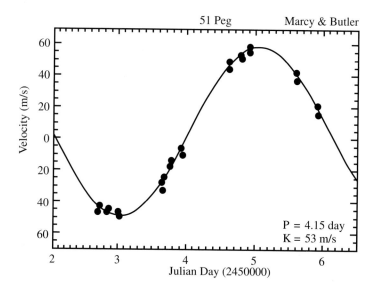

Spectroscopic type

Fig. 26.1. The velocity-time data for 51 Pegasus. The discovery by Meyer and Queloz is confirmed by Marcy and Butler.

shown. The observed data profile is seen to be fitted closely by a sinusoidal curve, suggesting that the orbit is closely circular. The speed of the star is seen to vary between $+53$ to -53 m/s with a period of 4.15 days. This has more recently been refined to 4.23 days. The inclination of the orbit is unknown. It was deduced that the minimum mass of the companion is $0.46\ M_J = 8.73 \times 10^{26}$ kg. The semi-major axis is 0.0512 AU $= 7.63 \times 10^9$ m which is about 25 times the stellar radius of about 3.1×10^8 m. The orbital eccentricity is 0.013 so the orbit is nearly circular.

26.3. Stellar Transits

There are three situations where the inferred data can be checked directly. This is the transit where the companion passes directly across the stellar surface between the observer and the star. The Doppler motion of the star can be observed as usual so the minimum mass is determined together with the orbital speed and semi-major axis. There is more information now because the transit allows (a) the inclination of the orbit to be deduced and so the mass of the companion can be given precisely and (b) the radius of

Table 26.2.   The data for the three transit star systems detected so far.

Comanion to Star	Mass (M_J)	Radius (R_J)	Mean density (kg/m^3)	Semi-major axis (AU)	Orbit period (days)	Eccen. n.	Inclin. i	Stellar mass (M_S)
HD 209458b	0.64	1.4	250	0.045	3.53	0.0	86.1	1.05
Ogle-TR-56b	0.90	1.30	573	0.023	1.2	–	86.2	1.04
Ogle-TR-3b	0.50	0.7 (?)	≈ 250	0.022	1.19	–		≈ 1.0

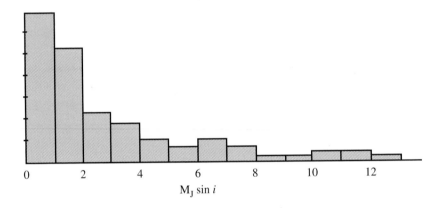

Fig. 26.2.   The distribution of $M_J \sin i$ in units of the Jupiter mass.

the companion to be found observationally. This provides direct evidence for the size of the body involved and so of its density. The data for the three stars are collected in Table 26.2. In each case a single companion has been found. The two systems are each based on a star closely similar to the solar mass. The mass of one companion is 1.4 times that of the other. The radii are similar but the mean density of the more massive companion is 2.3 times that of the other two. Both semi-major axes are very small, no more than a factor 10 greater than the stellar radius.

26.4.   A Survey of the Measurements

The 154 companions found so far can be considered as a group of objects. The distribution of the masses is shown in Fig. 26.2. If we make the reasonable assumption that the orbital inclinations are broadly the same then $M_J \sin i$ essentially measures M_J. The majority of bodies have about a Jupiter mass: indeed, some 30% of bodies have a mass not greater than

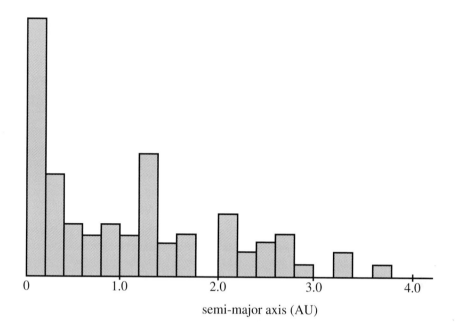

Fig. 26.3.   The distribution of the semi-major axes in AU.

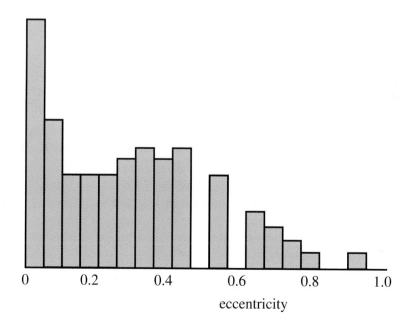

Fig. 26.4.   The distribution of eccentricities.

Table 26.3.   Twelve multiple companion systems about solar type stars. Two are triple systems while ten are double systems.

Companion to Star	$M \sin i$ (M_J)	Semi-major axis (AU)	Period (days)	Eccentricity	Distance (pc)	Stellar mass (M_S)
ν Andromeda	0.69	0.059	4.6170	0.012	13.7	1.3
b, c, d	1.19	0.829	241.5	0.28		
	3.75	2.53	3.518	0.27		
55Cnc	0.84	0.11	14.65	0.02	13.4	1.03
b, c, d	0.21 (?)	0.24 (?)	44.28 (?)	0.34 (?)		
	4.05	5.9	14.69 yrs	0.16		
HD 38529	0.78	0.129	14.31	0.29	42.3	1.39
b, c	12.70	3.68	5.96 yrs	0.36		
Gleise 876	0.56	0.13	30.1	0.12	4.72	0.32
b, c	1.98	0.21	61.02	0.27		
HD 74156	1.56	0.276	51.61	0.649	64.56	1.05
b, c	> 7.5	4.47	6.301 yrs	0.395		
HD 168443	7.7	0.29	58.116	0.529	≈ 33	1.01
b, c	16.9	2.85	4.766 yrs	0.228		
HD 37124	0.75	0.54	152.4	0.10	≈ 33	0.91
b, c	1.2	2.5	4.096 yrs	0.69		
HD 82943	0.88	0.73	221.6	0.54	27.46	1.05
b, c	1.63	1.16	1.218 yrs	0.41		
HD 12661	2.30	0.83	263.6	0.096	37.16	1.07
b, c	1.57	2.56	3.957 yrs	< 0.1		
HD 160691	1.7	1.5	638	0.31	15.3	1.08
b, c	1?	2.3?	3.562 yrs	0.8?		
47 Ursa	2.41	2.10	3.00	0.096	13.3	1.03
Majoris b, c	0.76	3.73	7.11	< 0.1		
ε Eridani	0.86	3.3	6.85 yrs	0.608	15.9	0.9
b, c	0.1?	40?	280 yrs?	0.3?		

about 1.898×10^{27} kg and perhaps nearly 60% have masses less than twice this value.

The sizes of the orbits are shown schematically in Fig. 26.3 which plots the semi-major axes up to 4 AU. About 25% have a semi-major axis below 0.2 while relatively few have values in excess of 2.0. This means that many companions are close to their primary.

The distribution of eccentricities is plotted schematically in Fig. 26.4. Some 19% are seen to have eccentricities below 0.05 while 30% have values below 0.1. Surprisingly, over 50% of companions have an eccentricity in excess of 0.4. In the Solar System these are eccentricities appropriate to comets. These high eccentricities mean that the orbital distance at perihelion is very much less than at aphelion.

26.5. Multiple Companion Systems

There are fourteen systems found so far with more than one companion: twelve are listed in Table 26.3. Two contain three companions while the

Table 26.4.   16 single companion with an eccentricity less than 0.1.

Companion to Star	Eccentricity y	Semi-major axis (AU)	$M \sin i$ (M_J)	Period (days)	Distance (pc)	Stellar mass (M_S)
HD 83443b	0.08	0.038	0.35	2.986	43.54	0.79
HD 179949b	0.05	0.045	.84	3.093	27.0	1.24
HD 187123b	0.03	0.042	0.52	3.097	50	1.06
τ Bootesb	0.018	0.046	3.87	3.487	15	1.3
HD 209458b	0.0	0.045	0.69	3.525	47	1.05
HD 76700b	0.0	0.049	0.197	3.971	59.7	1.0
51 Pegasusb	0.013	0.0512	0.46	4.23	14.7	
HD 168746b	0.081	0.065	0.23	6.430	43.12	0.92
HD 130322b	0.048	0.088	1.08	10.724	30	0.79
Gl 86b	0.046	0.11	4.0	15.78	11.0	0.79
HD 195019b	0.05	0.14	3.43	18.3	≈ 20.0	1.02
ρ CrBb	0.05	0.23	1.07	39.845	16.7	0.95
HD 28185b	0.06	1.0	5.6	385	39.4	0.99
HD 27442b	0.02	1.18	1.43	423	18.1	1.2
HD 23079b	0.06	1.48	2.54	627.3	34.8	1.1
HD 4208b	0.04	1.69	0.81	829.0	33.9	0.93

remainder contain two. The members cover a wide range of masses, a very wide range of semi-major axes, a wide range of distances from the Earth but a comparatively narrow range of stellar masses. The eccentricities range from 0.012 to nearly 0.65.

26.6.   Small Eccentricities

The planets of the Solar System are characterised by a small orbital eccentricity and it is of interest to pinpoint similar systems elsewhere. It turns out that 16 systems have been found so far with an eccentricity below 0.1 (Table 26.4). Two involve three companions while the remainder involve two. The distances from Earth range from rather more than 13 pc up to nearly 65 pc.

26.7.   Systems with a Large Semi-Major Axis

The Solar System has the major planets with orbital radii in excess of 5 AU but an exo-system with such a large value is not yet known (see V.25). A companion of Jupiter mass or greater will attract comets and meteors away from the inner regions and so protect any smaller companions that might be there. For the Solar System such protection could come for large bodies beyond about the orbits of the asteroids and this would set an inner distance of about 2 AU. In Table 26.5 is collected data for the systems with a

Table 26.5.   Twenty-nine systems with a semi-major axis in excess of 2 AU.

Companion to Star	$M \sin i$ (M_J)	Semi-major axis (AU)	Period (years)	Eccentricity	Distance (pc)	Stellar mass (M_S)
HD 10697b	6.59	2.0	1083	0.12		
HD 213240b	4.5	2.03	951	0.45		
HD 114729b	0.9	2.08	1136	0.33		
γ Cephei	1.76	2.1	903	0.2		
HD 2039b	5.1	2.2	1190	0.69		
HD 190228b	4.99	2.31	1127	0.43		
HD 136118b	11.9	2.335	1209.6	0.366		
HD 50554b	4.9	2.38	1209.6	0.366		
HD 196050b	3.0	2.5	1289	0.28		

Table 26.5. (*Continued*).

Companion to Star	$M \sin i$ (M_J)	Semi-major axis (AU)	Period (years)	Eccentricity	Distance (pc)	Stellar mass (M_S)
HD 37124b	0.75	0.54	152.4	0.10	≈ 33	0.91
	1.2	2.5	4.096 yrs	0.69		
ν Andromeda b, c, d	0.69	0.059	4.6170	0.012	13.7	1.3
	1.19	0.829	241.5	0.28		
	3.75	2.53	3.518	0.27		
HD 12661 b, c	2.30	0.83	263.6	0.096	37.16	1.07
	1.57	2.56	3.957 yrs	< 0.1		
HD 216435b	1.23	2.6	1326	0.14		
HD 30177b	7.7	2.6	1620	0.22		
HD 106252b	6.81	2.61	1500	0.54		
HD 33636b	7.8	2.7	1620.54	0.41		
HD 216437b	2.1	2.7	1294	0.34		
HD 23596b	7.19	2.72	1558	0.314		
14 Herb	4.90	2.83	1730.461	0.37		
HD 168443 b, c	7.7	0.29	58.116	0.529	≈ 33	1.01
	16.9	2.85	4.766 yrs	0.228		
HD 72659b	2.55	3.24	2185	0.18		
ε Eridanib	0.86	3.3	6.85 yrs	0.608	15.9	0.9
	0.1?	40?	280 yrs?	0.3?		
HD 39091b	10.37	3.34	2083	0.62		
HD 38529b	0.78	0.129	14.31	0.29	42.3	1.39
	12.70	3.68	5.96 yrs	0.36		
47 Ursa Majorisb	2.41	2.10	3.00	0.096	13.3	1.03
	0.76	3.73	7.11	< 0.1		
Gl 777Ab	1.15	3.65	2613	0.0		
HD 74156b	1.56	0.276	51.61	0.649	64.56	1.05
	> 7.5	4.47	6.301 yrs	0.395		
55Cnc b, c, d	0.84	0.11	14.65	0.02	13.4	1.03
	0.21 (?)	0.24 (?)	44.28 (?)	0.34 (?)		
	4.05	5.9	14.69 yrs	0.16		

Table 26.6.   Two companions with a small eccentricity and semi-major axis in excess of
2 AU.

Star	$M \sin i$ (M_J)	Semi-major axis (AU)	Period (years)	Eccentricity	Distance (pc)	Stellar mass (M_S)
47 Ursa	2.41	2.10	3.00	0.096	13.3	1.03
Majoris b, c	0.76	3.73	7.11	< 0.1		
55Cnc	0.84	0.11	14.65	0.02	13.4	1.03
b, c, d	0.21 (?)	0.24 (?)	44.28 (?)	0.34 (?)		
	4.05	5.9	14.69 yrs	0.16		

semi-major axis of at least that value. There are 29 entries, some involving
a single companion but eight with more than one companion.

26.8.   Small Eccentricity and Larger Semi-major Axis

To mimic the Solar System it would be required that the semi-major axis
should be greater than at least 2 AU and the eccentricity be less that 0.05.
We refer to Table 26.5. It is clear that very few systems satisfy the requirement. Indeed, only two systems discovered so far are sufficient. These are
listed in Table 26.6.

These two systems are the most likely to mimic solar system orbits. There
is no indication yet that they contain planetary bodies of Earth size.

*

Summary

1. Details are given of the companions to the pulsar systems PSR B1257+12
   and PSR B1620–26. These were the first systems to be discovered.

2. The first companion to a solar type star was Pegasus 51b using the
   Doppler approach. Details of the system are given.

3. Data for two systems discovered by transit methods are given.

4. The distribution of the masses of the companions, the semi-major axes
   and the eccentricities are presented.

5. Data are given for the companions found so far.

6. Two systems are selected for which the companion has (i) a Jupiter mass,
   (ii) a closely circular orbit and (iii) a semi-major axis in excess of 2 AU.

27

Assessing the Observational Data

The present list of exo-planets is biased by observational constraints. On the one hand, the accuracy of current measurements restricts the lower mass that can be found. On the other hand, the observations have not yet been made over sufficient time to be able to follow the full orbital details. Nevertheless, some details can be regarded as established and will take their place in any final analysis.

27.1. Firm Characteristics

There are two characteristics to notice, which show that many of the exo-planetary systems found so far differ basically from the Solar System. These are:

(1) There are many companions with very small semi-major axes, often less than 0.10 AU ($= 1.5 \times 10^{10}$ m). The smallest value so far is 0.024 AU $= 3.6 \times 10^9$ m which is only about ten times the radius of the central star. The interaction between the companion and the star will be strong at all times but especially at perihelion.
(2) There is a good proportion of systems with a large eccentricity for the orbit. This leads to a substantial difference between the star-companion distance at perihelion and at aphelion. Such orbits are cometry in the Solar System.

These features must be represented in any complete analysis of the exo-planetary systems although the present proportions of these features will surely be different as more systems are discovered.

At the present time, the features that can be expected to appear in the future are:

(c) More companions with larger semi-major axes, and so longer orbital
periods. The longer time periods to be expected are those in excess of
ten years but the observations have only been conducted over the last
eight years. Those stars which appear at present without companions
(and there are over a hundred of them) could well be found to have
companions with larger orbits later.

(d) These observations could also increase the number of multiple companion systems. It will be interesting to find whether multiple systems
are, in fact, a common occurrence or the relative rarity that they now
seem to be.

27.2.   More Massive Companions

The stellar systems considered in V.26 involve companions with a minimum
mass below $13M_J$ but higher masses have been observed. A list of twelve
such systems is given in Table 27.1. The pattern here appears very much
like that for lower mass companions. Many semi-major axes are small and
eccentricities large, just as before.

It is usual to take minimum masses up to $13M_J$ in the normal statistics
because masses beyond this limit are accepted as being brown dwarfs if the

Table 27.1.   Twelve systems with a companion with a minimum mass in excess of $13M_J$.

Star	Minimum mass (M_J)	Semi-major axis (AU)	Orbit Eccentricity
HD 162020b	13.73	0.072	0.28
HD 110833b	17	0.8 (?)	0.28
BD-04782b	21	0.7	0.28
HD 11275b	35	0.35	0.16
HD 98230b	37	0.06	0.0
HD 11445b	39	0.9	0.54
Gl 229	40	–	–
HD 29587b	40	2.5	0.00
HD 140913b	46	0.54	0.61
HD 283750b	50	0.04	0.02
HD 09707b	54	–	0.95
HD 217580b	60	≈ 1	0.52

composition is overwhelmingly hydrogen. A hydrogen body of mass $13M_J = 2.47 \times 10^{28}$ kg will have a radius of about 9.4×10^7 m giving a mean density $\rho \approx 7.2 \times 10^3$ kg/m^3. This is slightly larger than that of Earth. Beyond this mass the radius will increase and the density will decrease as the mass is increased, to the brown dwarf level. The composition may not, however, be of hydrogen. It could, at least in principle, be a silicate or a ferrous body. For a silicate body ($Z = 15$, $A = 30$) the radius $R \approx 2.07 \times 10^7$ m for $M(\mathrm{sil}) = 2.467 \times 10^{27}$ kg. This gives the density $\rho(\mathrm{sil}) \approx 6.65 \times 10^5$ kg/m^3. Alternatively, for a ferrous body ($Z = 28$, $A = 56$) giving the radius $R(\mathrm{fer}) \approx 1.78 \times 10^7$ m with a density $\rho(\mathrm{fer}) \approx 1 \times 10^6$ kg/m^3. Such bodies have not yet been detected.

27.3. A Special Case: Transit Systems

The system HD 209458b is special because the companion is seen to transit the star and so the dynamical structure can be found without ambiguity. The orbital eccentricity is essentially zero and the semi-major axis is small, $a = 0.045$ AU $= 6.71 \times 10^9$ m. The mass of the companion is $0.69 \times [\sin 86°.1]M_J = 1.31 \times 10^{27}$ kg and the radius $R_p = 1.4\,R_J = 1.4 \times 7.19 \times 10^7 = 1.01 \times 10^8$ m. The distance between the Sun and the companion, R_{sc}, is 67 times the radius of the companion, R_p. The mass of the star is $1.05\,M_{\mathrm{sol}} = 1.05 \times 2 \times 10^{30} = 2.10 \times 10^{30}$ kg.

Consider a small mass m at the surface of the companion. The force holding this to the companion, F_c, is given by

$$F_c = \frac{GM_c m}{R_P^2} = 8.56 \text{ mN}.$$

The force acting on the elements due to the star is given by

$$F_s = \frac{GM_s m}{(6.71 \times 10^9 - 1.01 \times 10^8)^2} = 3.21 \text{ m}.$$

The ratio of the forces is $F_c/F_s = 8.56$ m/3.21 m $= 2.7$ in favour of the companion. It seems that $F_c/F_s > 1$ here so the hold of the companion on its atmosphere is, apparently, nearly three times that of the star. Consequently, the atmosphere will tend to stay with the planet. The margin is small, however, and the atmosphere of the companion will be distended by the proximity to the star. This conclusion is confirmed by the low density of the companion. The companion will be subject to intense radiation from the star and mass ejections will cause substantial ionisation in the companion atmosphere. This is a situation unknown in the Solar System.

27.4.   Small Semi-Major Axes: Rôle of Eccentricity

The last example shows how tenuous is the hold of a companion on its atmosphere when the distance between the two bodies is uncomfortably small. The situation can be investigated as follows.

Suppose the star of mass M_s is at the focus of the eclipse with eccentricity e and semi-major axis a. The companion has mass M_P and radius R_P. The perihelion point is P distance $a(1 + e)$ from the star while the aphelion point is A, distance $a(1 + e)$ from the star. The star and companion come nearest together at perihelion. Then, the force due to the planet holding an atmospheric particle of mass m into the atmosphere is

$$F_P = \frac{GM_P m}{R_P^2}$$

while the force due to the star acting to remove the mass is

$$F_S = \frac{GM_s m}{[a(1 - e) - R_P]^2} .$$

The atmosphere is held by the companion if $F_P/F_S > 1$. Forming the ratio this condition is found to be satisfied if

$$e < 1 - \frac{R_P}{a}\left[1 + \left(\frac{M_S}{M_P}\right)^{1/2}\right].$$

This condition provides a lower limit to the flattening of the ellipse which is necessary because the flatter the ellipse the closer the star and companion come together at perihelion.

It is interesting to study the interaction for a case of high eccentricity. For this purpose consider the star HD 74156b with the parameters $a = 0.276 = 4.11 \times 10^{10}$ m and $e = 0.649$ with $M \sin i = 1.56\ M_J = 2.96 \times 10^{27}$ kg. For the star $M_S = 1.05 \times 2 \times 10^{30} = 2.10 \times 10^{30}$ kg. We may take $R_P \approx 6 \times 10^7$ m. The arguments above lead to the condition $e < 0.959$ and the inferred $e = 0.649$ certainly satisfies this constraint. It seems that even in this case the companion atmosphere is not stolen by the central star. The most extreme test is for the star HD 80606. Here the inferred eccentricity has the extreme value $e = 0.927$. The calculation gives $e < 0.943$ which again is just satisfied by the observations. The conditions on the surface of the companion will, nevertheless, be extreme especially during sun spot maximum on the star.

The effect of the eccentricity, in specifying the semi-minor axis of the orbit, specifies how closely the star and companion meet during the perihelion region of the orbit. The limit is when the distance between the bodies is the sum of their radii, which will provide a value of e close to 0.99. No case of this closeness has been observed so far. Very close encounters coupled with intense stellar activity will allow the star to extract some companion atmosphere at each perihelion. Continual very close encounters could denude the companion of its atmosphere all together if the value of e is large enough. This could yield a silicate surface and could be evidence for it. No case has been discovered so far.

27.5. The Future?

Predicting the future is for thought experiments but not for observational studies. It will be interesting to see whether Solar System like structures are found, presumably from systems where nothing has been detected so far because long orbital periods will be involved. It is certainly the case, however, that systems which depart significantly from the Solar System structure have been confirmed and these may indeed be the most common form.

The Jupiter-like companions with semi-major axes less than 3 AU are certainly different from our System and, by the observation of transit companions, are confirmed as genuine. Semi-major axes of greater value may be found to be common eventually when the observations have been under way for a longer period. Where there are large planetary bodies there can be expected to be small ones. It can be expected that Earth sized companions will be found eventually although this can certainly not be guaranteed.

*

Summary

1. The exo-planetary systems detected so far are generally unlike the solar System.

2. In general there are many systems with small and very small semi-major axes, and large eccentricities

3. This may be because the short history of these studies (only some 8 years so far) has not allowed Jupiter mass companion with large semi-major axes to be detected.

4. Comments are made of future prospects.

Part VI

EXO-BIOLOGY

An exploration of the possibilities of life other than that on Earth and some thoughts on the role animate matter might play in an inanimate Universe.

28

Life on Earth

Living organisms are everywhere on Earth. The most elementary forms are found in all environments, hot, cold, high altitude, under the surface, in highly radioactive areas, in high acidity and in high alkalinity. Advanced life has inhabited the seas and oceans and covered the surface. It is interesting to discover how this has come about as one example in the Universe. Life on Earth is the only life yet known and the nature and speed of the evolution, and the constraints on life here, is the best guide we have to that of any other animate ecosystems that might exist.

28.1. Early Life

The Earth is 4,550 million years old. The earliest rock yet found is 3,860 million years, or some 700 million years younger. What happened during this period is not known directly but can be inferred from general evidence within the Solar System. Every ancient surface bears evidence to a severe and continuous bombardment by a very wide range of planetesimals in the period soon after its formation. This will also be a period of hot bodies since heat of formation of the planets will still have been substantial. An impact will also provide heat, since the kinetic energy of the impacting body will be converted completely to heat and perhaps light on impact. It is very likely that the surface of the Earth will have been largely molten at this time, not only through impacts but also due to large volcanic outpourings of lava for the interior. The gases that were released dominated the atmosphere which was not oxidising at this time but contained substantial quantities of CO_2 and other gases. There will have been very little condensed water on the surface, if any, but quantities of steam rising from rain accumulating on the hot surface. Electric storms will have been everywhere and anything less like the Earth today would be difficult to imagine. It certainly would seem an unlikely venue for the formation of living organisms yet this seems to have occurred.

The early, plastic, surface in this so-called Archaen period will have be recycled very quickly and examples of the very earliest surface materials have never been found. This is not the case for the lunar surface where the earliest material has been found. The earliest known examples of elementary living material appear to have been mature so it is probable that living material is older than that found so far.[1] Theoretical calculations suggest that the immediate surface began to solidify to a depth of perhaps 200 km after about 200 million years. The surface will probably have solidified to a significant depth by some 4,000 million years ago, although volcanic activity persisted, holding a surface temperature much higher than it is at present. It was in this inferno that the most elementary living organisms seem to have appeared and spread with some vigour. It is worth noticing that the radiation from the Sun was less in the past than now. At the time of the formation of the Earth, its radiation was no more than 70% of today's intensity.

Whether the earliest life formed spontaneously on Earth or whether it was "seeded" from outside is not yet possible to tell. It certainly could have been carried from space in meteorites and could have withstood the rigours of re-entry into our then dense atmosphere held firmly inside the meteoritic material. It is possible that living material did not arrive here already formed but that the precursors came from outside. It is known that comets contain organic materials that could react to form amino acids, which are the building blocks of proteins that are essential to life. For instance, it has been found that the comet Hale-Bopp contains, apart from water, also ammonia, formaldehyde and hydrogen cyanide. There is the third possibility that life formed on Earth from material already here. It is clear that there was a time very long ago when living material did not exist here because the Earth itself didn't exist. Presuming that the Universe had an origin at a finite time in the past it is clear that living material must have evolved as part of the developing Universe. This would not follow if some form of steady state universe were valid, with an infinite past history. On the other hand, our Galaxy will not have had an infinite past history and it does include living materials so there must have been an origin for it. Whatever its origin, early life was of the most elementary form. It also follows, if the various

---

[1]The traces of carbon found so far have no apparent animate form and could possibly not be associated with living materials. This is generally not thought to be the case but if it were the first evidence of irrefutably living material would be about 1,000 million years younger. The ratio of heavy to lighter carbon, however, supports the early carbon as being animate.

components of the Universe have a finite (though perhaps, by our standards, an enormously long) life that living materials will have an ending, whatever the ultimate state of evolution may be. All living materials are mortal.

28.2.   The Characteristics of Early Life

The early conditions of the Earth involved substantial temperatures. Life then must have been thermophilic, or heat loving. Life diversified for the first 3,000 million years but stayed about the size of modern archa and streptococcus. Only later did it increase in size. The earliest evidence of life on Earth is a small fossil carbon ball embedded in rock, 3,860 million years old and found in the oldest Greenland rock. This ball will not have been the initial development and the Greenland rock is not the earliest rock on Earth. It is possible that the earliest life, before DNA (deoxyribonucleic acid) and proteins had evolved, consisted entirely of RNA (ribonucleic acid). This is a single strand structure which is not truly alive because it is not able to control its own reproduction in the way that the double helix DNA does. The early RNA could have been floating on the warm shallow seas and lagoons. The formation of DNA molecules was a vital step in the evolution of life. The DNA quickly usurped the more specialist roles in living and at some unknown time the DNA was able to surround itself with a protective membrane. These were the first cells. The three main branches of life, then as now, developed from these early cells.

The known thermophiles divide into three main root sources, Archaea, Bacteria and Eucarya (Eukaryotes). These appear to have branched out from an unknown initial life form. These three roots have members as follows, very roughly in evolutionary order:

Archaea: The existence of these organisms was unknown until the 1970s. They are single-celled life forms, resembling bacteria in some ways but quite distinct from them in their DNA composition. They include members that inhabit some of the most extreme environments — near rift vents in the oceans at temperatures above 100°C and in extreme acidic or alkaline environments. They are also associated with biological systems themselves. For instance, they inhabit such regions as the digestive tracts of cows, termites and some marine creatures, where they generate methane. They also occur in oxygen-free environments such as marsh mud, ocean depths and even in petroleum deposits. The early conditions on Earth seem to have involved such gases as CO_2, CH_4 and NH_3.

Bacteria: Although bacteria are best known for causing diseases, this aspect is almost a trivial component of their existence. Some require oxygen for their activity, others are anaerobic and cannot exist in the presence of oxygen while a third group thrive in oxygen but can manage without it. The Earth's ecosystem is heavily dependent on bacteria that break down rotting organic material, recycling CO_2 into the atmosphere so enabling new plants to grow. Other bacteria, some within plant roots, fix nitrogen into nitrates and nitrites that make nitrogen available to plants.

Eucarya: These are bodies made of cells with a nucleus. Examples are slime moulds, algae and various types of fungi. These are the early life forms leading to plants and animals and, ultimately, *homo sapiens*. The more complex of these creatures require a component of oxygen in the atmospheric gas.

This evolutionary pattern is associated with the way that the environment developed at the Earth's surface. At the beginning there was a reducing atmosphere, rich in hydrogen compounds and also at very high temperatures. Later oxygen became a component of the atmosphere, possibly due to the photo-dissociation of water and the loss of hydrogen, and later, as plants developed, through photosynthesis. At the same time there was an increase in oxygen so the overall temperature of the Earth's surface was reducing.

The distinctive nature of the DNA in archaea and bacteria may contain an important message. If the DNA in bacteria could have evolved from that of archaea, or if they both evolved from some common ancestor, then this simply indicates that as conditions changed so evolutionary changes also took place to adapt to the new conditions — a Darwinian explanation. On the other hand if there is no clear evolutionary pathway from archaen to bacterial DNA, or the possibility of a common source, then this may indicate two distinct origins. This would promote the suggestion that the generation of life is not an unlikely phenomenon and may arise in a variety of forms to thrive in whatever environment happens to be available. This is an example of adaptive evolution.

Most of these elementary life forms still survive today, though perhaps in a slightly modified form. If there was just one initial common form that started everything then this has probably not survived. It appears that all the elementary life forms thrive in hot conditions. Geothermal hot springs (at temperatures of 60–70°C) and rims of active volcanic cones all contain these elementary living forms. The first widespread living communities were probably stromatolite reefs very much like those that exist today in the hyper-saline waters of Shark Bay, Australia. The reef occupies the top three or four inches of the surface of the water, growing into layers. The top layer is an oxygen producing photosynthetic cyanobacteria. The middle region is composed of bacteria that can tolerate some oxygen and sunlight but not at full intensity. The bottom layer involves bacteria that thrive in the complete absence of oxygen and sunlight. The whole complex traps sediments in the water and binds particularly with calcium carbonate ($CaCO_2$). The oxygen-loving organisms continually rise to the top forcing the others down so there is continual movement. There may be a thousand million individual members of a colony living in symbiotic relationships with each other.

28.3. Oxygen in the Atmosphere

The early organisms, like all those today, can be viewed as particular types of mechanical engines. Each takes in nutrients and rejects waste. In the Earth's early days various gases were produced but not oxygen. The evolution of cyanobacteria changed this situation. They produced oxygen which poisoned other forms giving them the advantage of the best photosynthetic sites. The main advantage of oxygen is in the efficiency of the production of energy. The oxygen-rejecting, or anaerobic, bacteria ferment sugar to form energy whereas the oxygen-loving, or aerobic, bacteria utilise the special combustion properties of oxygen. The result is that the former obtain 2 units of energy by the fermentation of 1 molecule of sugar, the latter obtain 36 units. The evolutionary change to oxygen and photosynthesis was inevitable. It is possible that the mitochondria which burn oxygen and sugar in the cells of most eukaryotic are the direct descendents of the earliest oxygen breathing bacteria. Eukarya require oxygen because the energy production mechanisms requires it. For plants, the chloroplasts which allow photosynthetic process to be active may well have evolved from the earliest cyanobacteria that are now contained within plant cells.

It took nearly 1,000 million years for oxygen to accumulate in the atmosphere in significant proportions. The reason is the existence of dissolved iron in the early seas. The oxygen was produced under water and combined with iron to form the oxide. When essentially all the free iron had been oxidized the oxygen produced began to enter the atmosphere. Larger organisms, such as Gyrpania, were able to evolve and the move to an oxygen-rich atmosphere advance rapidly. The availability of oxygen in the atmosphere provided an efficient source of energy able to power larger land creatures that were to evolve later. Without that source, the evolution of large and complex creatures would not have been possible. Intelligent creatures require a prolific source of energy which can only be provided by an efficient source.

28.4.   The Evolutionary Sequence

The evolutionary sequence is becoming clear and is shown schematically in Fig. 28.1. The relative time-scales for the development of different life forms, shown in the figure, are just estimates since it is not possible to put precise dates on the important events. It will be most graphic to look at the various

Time Ago (million years)	Living Forms
4000-2000	Single cell creatures probably largely RNA based
2000	First eukaryotes: cell structure with nucleus: DNA well established
1100	Sexual reproduction: multiplication of eukaryotes forms. Photosynthesis and the appearance of free oxygen
600	Multicelled animals: worms and sponges
550	Lobites, clams – many invertebrates; the first vertebrates
400	The emergence of life from the sea to the land: Early land plants, fish invade land
350	Seed bearing tree ferns: early amphibians
280	Early reptiles 3 or 4 metres long
200	Beginning of large dinosaurs
180	The first birds
65	End of the dinosaurs; vertebrates get smaller; mammals
4	First ancestors of modern man
0.1	Homo sapiens
	THE FUTURE?

Fig. 28.1.   A highly simplified chronology of life forms on Earth so far.

stages compressed in time to something within our experience. If the age of the Earth is taken to be 24 hours, the first 12 hours involve single-celled creatures only. The first oxygen, later to benefit larger organisms, is seen at about 11 o'clock, just before noon. Eukaryotes appear at noon. Sexual reproduction, so important for the evolution of advanced creatures, appears much later, at about 6 o'clock in the evening. Multi-celled animals make an appearance at 21.00 hours. The first ancestors to modern man enter the picture about 22.40 and *homo sapiens* appears about 1.2 seconds before midnight. It is seen that the evolutionary pattern was slow to emerge from the very elementary forms but eventually accelerated once the elementary forms were well established. It is astonishing how quickly mankind has developed and put a stamp on the Earth. Mankind's grasp of the Earth is increasing apace and it can only be wondered what the next 24 hours, or even 24 seconds, will bring.

Although life on Earth has been characterized by a mean evolution towards more advanced forms, evolution has occurred through a broadening process in which many elementary life forms have persisted during the development of the more advanced forms. The microbial life forms still account for perhaps half of the inventory of the life forms on Earth.

Two separate life systems exist on Earth. One is the familiar form living on the surface region of the Earth. The other has been discovered only over the last thirty years living over the deep high temperature ocean "vents" of the mid-ocean ridge structure that circumvents the oceans. As far as is known, the chemical composition of the two forms is the same and the structures are similar. But each has developed quite independently with different heat sources (one solar and the other geothermal) to provide energy to their different environments. For instance, in the deep ocean there are blind shrimp-like creatures, eyes not having evolved because there is no sunlight at that depth in the ocean. The detailed study of the mid-ocean ridge animate systems, several kilometres below the surface of the sea, is still in its infancy.

The appearance of the mid-ocean ridge life on Earth has led to the possibility that that the initial living organisms might have formed under water. The early Earth will have had warm shallow surface water and sub-surface life might have developed in such waters. If this were so, it might also have developed independently below icy surfaces of satellites such as Europa (Section 7.3.2), one of Jupiter's large satellites. The environment deep under the ice could well be warm and congenial to at least elementary living organisms. The eventual exploration of Europa will be of enormous interest.

28.5.   The Movement of Continents

One factor which has affected the evolution of living things has been the movement of the continents over geological history. The location of the continents has been found partly from geomagnetic data using remanent magnetism to locate the magnetic poles, and partly from geological arguments. The earliest reliable data date from about 500 million years ago and this is shown in Fig. 28.1a. Other figures show the distributions 325, 175 and 50 million years ago. It is seen that some modern features have emerged during the last 50 million years and especially the location of India as part of Asia and the isolation and movement of Australasia. The Atlantic Ocean has yet to open fully and Britain and Scandinavia have yet to form. Such

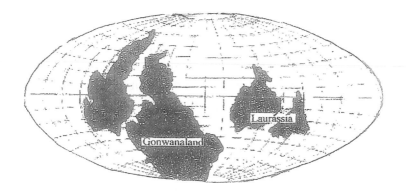

Fig. 28.1a.   The distribution of the continents 500 million years ago.

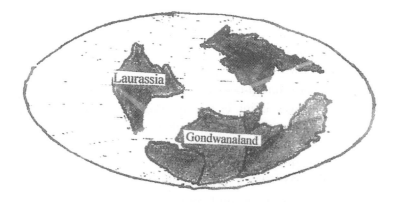

Fig. 28.1b.   The distribution of the continents 325 million years ago.

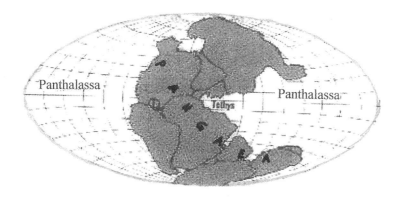

Fig. 28.1c. The distribution of the continents 175 million years ago.

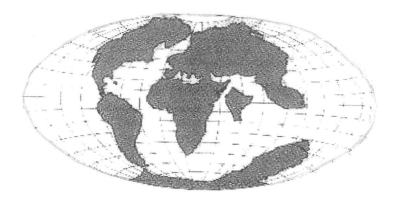

Fig. 28.1d. The distribution of the continents 50 million years ago. They are approaching the modern form.

changes of land mass affect the matter living on it or in the seas around it in crucial ways. Fragmented land masses lead to isolated evolution: the volume of the surrounding waters is increased allowing new development of ocean life. What causes these movements of continental mass is not yet clear. The assumption that it is associated with internal mantle plumes provides many other problems.

28.6. Life on the Atlantic mid-Ocean Ridge

There is a second ecosystem on Earth quite separate from that on the surface or in the oceans. It was discovered quite by chance some 30 years ago when the deep oceans were beginning to be explored, to a large extent for military purposes. The thermal vents of the mid-Atlantic Ridge were surprisingly found to be populated by complex life forms evolved for life in that environment. The vents are ejecting very hot superheated water with minerals in black plumes which give the vents the name smoky chimneys. Around this hot water are giant sea worms, crabs and shrimp-like creatures. All are blind because sunlight does penetrate to the depths of the Atlantic ocean some 3 kilometres down. Bacteria are present and the driving energy appears to be sulphates. This ecosystem has evolved without the help of solar energy, the independent source instead being the heat arriving from the mantle through the vents system. As one vent becomes dormant the whole system moves, unaccountably, to another.

The basis of the life there is DNA, as for other creatures. Presumably this living activity and that of the rest of the Earth evolved independently. The age of the mid-ocean ridge system is not known. It presumably dates back to the earliest evidence of continental drift, rather before 500 million years ago. The relation between the mid-ocean ridge system and the rest of living creatures remains to be understood. It is clear, however, that ecosystems can arise and thrive without the active help of the Sun as an energy source.

28.7.   Changes of Climate

One of the most important influences on the evolution of living materials is climate. Increased heat can be tolerated more easily than falling temperatures. There is a general climate change due to such astronomical factors as changes of the tilt of the Earth, changes in the orbit around the Sun and variations of the radiation output of the Sun and other effects. Changes of the land mass distribution have their own effects. The interiors of large land masses tend to be hot in summer and cold and damp in winter. The ocean currents affect the climate near the coasts. Clouds have a strong effect, preventing evaporation over the sea and hopefully providing evaporation over the land. Clouds also reflect incoming solar radiation reducing the surface temperature.

The global climate on Earth has changed considerably over the eons. For instance, during the time of the dinosaurs the average temperature was some 14°C hotter than now. By contrast, 100,000 years ago the Earth was plunged into a frozen ice-age period from which it is slowly recovering now.

28.8.   Some Final Comments

We have considered this very brief guide of the development of life on Earth as an indication of the processes that have molded life here. Life on exo-planets, should it exist, would be influenced equally by such random processes there.

The essential feature for evolution is the change in continent/ocean areas/volumes. Fragmentation means more restrictions and separations which enhance the chances of evolutionary change. They might also restrict the food supply and affect the climate in beneficial/adverse ways. Movement of land from hot to colder regions will also have its effects. One deficiency at the present time is an almost complete lack of information about climate in geological periods. Intense research is being conducted at the present time to explore this vast area. This is using a range of indicators such as fish ear bones (they show rings like tree rings dependent on temperature and other factors), gas ratios (and especially the relative abundances of the oxygen isotopes O^{16} and O^{18}) and other factors. Rock geology is very important in discovering hot (desert) periods and cold (snowball Earth) periods. All these factors affect life because they affect food and the environment.

The complexity of the effects of climate are beginning to be clear over the last 11,000 years. This is an infinitesimal period in geological time but it nevertheless indicates some of the effects. It appears that warmer conditions are associated with the rise of the Roman Empire, with the crossing of the Alps by Hannibal and the Viking journeys including the colonization of Greenland. The effect of a warm, stable climate is very clear from history. The present cultural scene arose from developments in the warmer regions of the Mediterranean and the Near East. It does not follow, however, that continued heating would be entirely beneficial.

It is often forgotten that the Earth is in the process of emerging from the most recent of its periodic ice ages and the globe is warming. Some 10% of the continental land surface is still covered by a worldwide system of glacier ice. This is about one-third of the glacial area covered in the past at the height of the ice age when the mean global temperature was probably about 5°C colder than now. The area of continental glaciation is still shrinking and the global warming will likely eliminate it altogether. The levels of the oceans will consequently rise above their present values. Many of the major cities over the Earth lie on or near the low lying coastal regions, having arisen from commerce, and these areas will be under threat as the mean gloabl temperature rises further.

There are many plans to eliminate global warming but these do not always take account of the major causes. Moving out of an ice age is one — the mean temperature must rise but by how much? The very presence of an atmosphere causes surface warming (by some 25°C for the Earth) and this will be enhanced by the particular composition. Flora and fauna naturally emit molecules which act to increase the temperature (greenhouse gases such as methane gas and carbon dioxide). There is a variation of the solar insulation due to astronomical factors, such as variations of the solar radiation and variations of the Earth's orbit. These are greenhouse factors that must be there. There is an undoubted enhanced greenhouse contribution due to human industrial activities though it is very difficult to assess its actual contribution. Over much of the Earth's history the mean temperature has been 10–15°C higher than now. The Earth is, in many ways, moving towards its more normal temperature. Can Homo sapiens adapt to these new conditions quickly enough?

From the point of view of our present enquiry it is very clearly demonstrated that environmental changes have had an enormous effect on the evolution of living materials. It can be concluded that the most elementary organisms are almost impossible to eradicate and would probably appear almost whatever the physical conditions. More advance life is different. On Earth it took 4,000 million years to pass beyond the bacterial stage to small macroscopic life. Once that hurdle had been crossed the pace of evolution increased and it took only a period of 500 million years to develop intelligent creatures and ultimately *homo sapiens*. The evolutionary processes still go on and it would seem unlikely that *homo sapiens* is the ultimate living creature on Earth.

It is, however, not safe to predict the future from the past. In the past the environment has dictated the behaviour of intelligent life but now the environment itself is subject to some measure of control. This must affect the direction of future evolution at least in some small measure. It is an interesting question to ask whether, if advanced life were to be eliminated from the Earth, it would evolve again and in the same general way.

*

Summary

1. The Earth is the stage for the development of animate materials and its physical properties control the creatures living on it.

2. The physical state is controlled partly by internal mechanisms and partly by conditions imposed from outside.

3. The surface life forms appear to have developed quite independently of life forms deep on the mid-ocean ridges. This means that the surface evolution has been driven by solar energy whereas that under water has been based independently on heat from within the Earth.

4. One major influence has been the slow movement of the continents.

5. The evolutionary sequence continues and we seem to be living through a period of extinctions now. There is no reason to suppose that *homo sapiens* is the final product of evolution.

29

What Makes a Planet Habitable?

Habitability depends on what it is that has to live there. The requirements for elementary life are different from those for advanced life which may be different from those for technologically capable life. These cases must be considered separately.

29.1. An Overall Requirement

Living material is a collection of complicated molecules held together by the binding forces between the atoms that constitute it. Its essential feature is self-replication and the ability to adapt to its environment according to the Darwinian laws. This will generally require the intercession of catalysts of some kind to accelerate the various chemical processes so, expressed another way, the body can be defined as being autocatalytic. To comply with Darwinian requirements it must also be modular, that is it is constructed from modules. DNA is modular. If any one component of a module is changed the descendent molecules are changed only in that one module. A living system must build up to maturity and maintain itself in a steady state. This is not an equilibrium state since there are processes of decay and renewal active side by side. The system requires an intake of matter and energy and, for steady state conditions, the ejection of lower grade matter and energy. In this special respect it acts as a machine.[1]

There are two linked aspects to living things. One controlling the metabolism, the other being the information controlling the development of the living thing and guiding its evolutionary development. The information is contained in the cell. In what follows we will be interested particularly in the conditions necessary for continued metabolism in its most general form

---

[1]Attempts have been made to apply the second law of thermodynamics to this side of its actions.

and will not be concerned directly with evolutionary matters. We might notice, however, that the flux of matter generally includes that from other living things. The earliest life must have been self-contained in that it was adequately supported by simpler minerals and heat. In a real sense the earliest microscopic creatures might be said not to have been alive in the sense that a virus is not alive.

The living organism generally requires a stable environment to within a certain measure, determined by its complexity. Low complexity organisms can usually survive harsher conditions of stability than the more complex ones. Judging from the example of the Earth, microbial organisms are easily formed even under the extreme conditions of the early Earth but more complex forms require much more nurturing. Thus, it took 4,000 million years for post-microbial life to appear on Earth. Once it appeared evolution proceeded very fast providing intelligent creatures after only a further few hundred million years. It is not yet clear whether the time before the appearance was an incubatory period requiring a stable environment for the full extent or whether the advanced life appeared because the environment had become stable.

Living material may suffer damage from a number of causes, the more advanced life being more vulnerable than elementary life. One major danger is being forced to absorb energy, in one form or another, above a certain limit. This can be in many forms. For advanced life, for instance, it may be the absorption of kinetic energy due to a fall or the absorption of radiation (such as X-rays or ultra-violet rays and so on). This will disrupt the molecules of which it is made, or they may be disrupted by a cut or some similar misfortune. Microbial life is able to withstand extreme cold by hibernation for considerable periods but advanced life generally cannot do this except in certain small degrees. The more elementary advanced life is able to renew body parts.

Generally a heat source is necessary to maintain the body at a temperature where chemical reactions can proceed with appropriate speed and efficiency. This might be intrinsic to the planetary body on which the life forms or it might alternatively be supplied by a nearby star. For a constant heat supply the planet must be orbiting the star.

29.2. Atomic Constraints: Binding Energies

All living things are composed of atoms arranged into molecules and the stability of the whole organism is compromised by an instability of the

atoms and molecules of which it is composed. The interactions between
the atoms are short ranged and join together to form molecules from which
living materials are constructed. The energies of interaction between atoms
and molecules are, therefore, of importance. The simplest atom is hydrogen
consisting of one proton and one orbiting electron.

The equilibrium radius for the hydrogen atom in its lowest energy state
(the so-called ground state) is the Bohr radius a_o with $a_o \approx 5 \times 10^{-11}$ m. This
is the characteristic atomic size. The energy associated with the electron in
its ground state orbit is called the Rydberg energy, Ry $= 2.2 \times 10^{-18}$ J. We
might notice that the atomic properties (as opposed to the nuclear properties) are non-relativistic. To see this compare Ry with the rest energy of the
electron, $E_e = m_e c^2$ where $m_e = 9.1 \times 10^{-31}$ kg is the mass of the electron
at rest (the rest mass) and $c = 3 \times 10^8$ m/s is the speed of light in a vacuum.
It follows that $E_e = 8.2 \times 10^{-14}$ J. The ratio Ry$/E_e = 2.7 \times 10^{-5} \ll 1.$[2]

The temperature, T, associated with a given energy, E, is given to sufficient generality by the relation $E = kT$ where $k = 1.38 \times 10^{-23}$ J/K is the
Boltzman constant. It follows that an energy of 2.2×10^{-18} J is equivalent to
a temperature of $T \approx 2 \times 10^5$ K. A temperature of this magnitude will cause
hydrogen atoms to begin to break apart. A photon of this temperature has
a frequency of about 3×10^{15} s^{-1} and a wavelength of about 1×10^{-7} m.
This is just in the ultraviolet part of the spectrum. Radiation of this or a
higher frequency will disrupt the hydrogen atom. Heavier atoms can withstand correspondingly higher frequencies before being broken up but are
vulnerable to ultraviolet radiation. Living material contains hydrogen and
so is sensitive to irradiation by ultraviolet light. The material is generally
opaque to these frequencies and it is the outer barrier (the skin or hide) that
is particularly affected. Higher frequency radiation is more penetrating and
more disruptive throughout the total volume of the body. An environment
will be hospitable to life only if such radiation is either absent or present in
minute amounts. It is true, however, that a small quantity of hostile radiation can be a source of mutations to add variety to the evolutionary pattern
of the material.

The energies of the electric forces holding atoms together, the binding
energies, are much smaller. These generally arise from a sharing of electrons
(the so-called valency bonding) either directly or indirectly. A weak bonding
is also given by transient dipoles called van der Waals forces. The valency
bonding is associated with energies in the general range 10^{-2}–10^{-3} Ry, that is

---

[2]A reader familiar with spectroscopy will recognise the quantity $(2\text{Ry}/E_e)^{1/2}$ as the fine
structure constant $\alpha \approx 1/137.18$.

10^{-20}–10^{-21} J. This energy is small enough to fall in the optical and infrared regions of the electromagnetic spectrum. In terms of living materials, radiation of such energies can raise the temperature of fluids associated with cells. Again the skin or hide acts as a general barrier.

29.3. Stellar Radiation

A main sequence star provides a long term and stable energy source. The characteristics of such stars are believed to be well understood. The energy source is the thermonuclear conversion of hydrogen into helium in the central core of the star. The temperature here is of the order of 1.5×10^7 K and the energy of the photons of is of order 2.1×10^{-16} J or 10^2 Ry. Then, the frequency $\nu = 3.2 \times 10^{17}$ s^{-1} which lies in the X-ray region of the spectrum. It has been seen previously that the photons move towards the surface, along the (very small) decreasing temperature gradient and lose energy as they go due to collisions with electrons. The emerging radiation will, consequently, cover the whole range of frequencies although the average photon will suffer a number of collisions to make its final frequency in the optical or near infra-red regions. The result is that the star tends to radiate all frequencies as a black body at an effective temperature T_e. There is a maximum frequency $\nu_{\max}$ of the spectrum, unique to the temperature T_e, which can be calculated from Wien's radiation law. This is the condition that $T_e/\nu_{\max} = 9.7 \times 10^{-12}$ Ks. It follows that the higher the black body temperature, the higher the maximum frequency. The maximum frequency is an important parameter in our discussion.[3]

29.4. Heat from the Central Star

These conclusions have an immediate bearing on the particular type of star that is suitable as a source of heat to support living materials. Remembering the previous conclusions, the maximum frequency received by the surface should be below the ultraviolet region but the heat received on the planet should leave water above its freezing point at least over much of the surface (perhaps excluding the polar regions).

The energy emitted by a black body is proportional to the fourth power of its temperature. This is a very severe law: if the temperature is reduced

---

[3]This simple picture can be complicated in practice by various absorption and emission processes at the stellar surface.

by one order of magnitude (that is by a factor 10) the emitted radiation is reduced by a factor 10^4 (that is by 10,000). The same is true in reverse, if the temperature of the black body is increased by a factor 10 the intensity of radiation in the ultraviolet and X-ray regions is now very much enhanced. It is clear that a prescribed range of surface temperatures on the planet places a very tight limit on the black-body temperatures of the star that are acceptable. It was seen in IV.21 that a relation exists between surface temperature (now the effective temperature) of the star and its mass. In placing a limit on the effective temperature we are also placing limits on the acceptable stellar mass. This limit arises in another connection.

The lifetime of a star, during which it remains on the main sequence, depends on its mass. The more massive the star, the more quickly it uses up its thermonuclear fuel so that more massive stars have a shorter time on the main sequence than less massive stars. Theory suggests that, if the mass is not too large, the lifetime is inversely proportional to the fourth power of the mass. For very heavy stars the dependence is much stronger. For a main sequence star of stellar mass the lifetime is a few times 10^{10} years. If the experience of life on Earth is a guide, there is an upper limit to the mass of a star that is to act as a centre for a habitable planet. It should be in the range of one or two solar masses, or perhaps about $1 - 5 \times 10^{30}$ kg. There is, of course, a lower limit because the energy emitted (expressed through the effective temperature) also decreases with the mass. There is anyway a lower limit to the mass of a star determined by the central temperature and so the mass, which must be sufficiently large for thermonuclear processes to take place at all. This lower limit is about 1/10 solar mass, or about 2×10^{29} kg. It is likely that the mass of a star able to support advanced life will straddle the solar mass by about an order of magnitude. We can say, then, that the mass M will satisfy a criterion of the type

$$\approx 5 \times 10^{29} < M < \approx 5 \times 10^{30} \text{ kg}.$$

The range of suitable stars is seen to be narrow. It is interesting, however, that a high proportion of the stars in the sky lie in this mass range.

29.5. The Role of an Atmosphere: Planetary Mass

Even a low mass star emits considerable quantities of ultraviolet and X-radiation so it might seem that living material will not be able to survive a stellar source of radiation. There is a second line of defence — the planetary atmosphere. Among other things, this presents a protective layer for the surface but its effect depends on its composition. One effect is to filter out certain of the incoming radiation; another effect is to restrict the wavelengths of the radiation able to escape to space. An example of the first effect is the presence of ozone in the Earth's atmosphere which prevents much of the incident ultraviolet light from reaching the surface. The Earth's atmosphere also excludes much of the infrared radiation from entering, giving only a thin optical window to the outside. An example of the second effect is the presence of CO_2 in the atmosphere which causes global warming — the effect is extreme on Venus where the warming has become very great leading to high surface temperatures.

The presence of an atmosphere of a particular composition sets limits to the mass, M_p, of a planetary body itself. The atmosphere is held by the gravitational force between the planet and each of its constituent molecules. For an undisturbed atmosphere the escape velocity, v_c, is that for which the kinetic energy of a representative atmospheric molecule has precisely the same magnitude as the gravitational potential energy in the gravitational field of the planet. Then

$$v_c = \frac{2GM_p}{R_p + z},$$

where R_p is the radius of the planet and z the altitude of the atmospheric molecule. In general, $R_p \gg z$ so we can write $(R_p + z) \approx R_p$. The speed of the molecule depends on the temperature, T_g, of the gas according to $v_c \approx \sqrt{T_g}$ so the ability of a planet to contain an atmosphere depends on the gas temperature. This is determined by the luminosity of the star and by the semi-major axis of the orbit. Unfortunately all the atmospheric molecules have not got the same mean speed. Some molecules will move faster than average and others will move more slowly. The slow ones make no problem but the fast ones do. This distribution of speeds is according to the Maxwell distribution which maintains a constant fraction of its members with speeds above the critical for retention by gravity. These fast molecules escape into space, the gas redistributing itself to retain its Maxell "tail". Having a continuing number of molecules beyond the critical for retention means that the atmosphere gradually is lost. The crucial measure is the time scale over which the atmospheric pressure declines. Only the very largest planets, such as Jupiter and Saturn, will retain their atmosphere essentially without loss. Lower masses will retain their atmosphere for times of order 10^9 years or more. It turns out that the smallest mass able to retain a stable atmosphere for a significant time in excess of 10^9 years is of the order 5×10^{23} kg or about 1/10 the mass of the Earth.

There is an upper limit. It is likely that advanced living materials will not thrive in a hydrogen atmosphere so the planet mass must not be so high as to retain such components. This sets an upper limit of about 5×10^{25} kg or about 10 times the Earth's mass. This is not the only criterion for the maximum mass.

29.6.   The Role of Water

An essential requirement of any advanced living materials is the ability of chemical agents to mix and interact and to allow rejected waste materials to escape. For this reason a liquid environment is crucial. On Earth this vital liquid is water. It is interesting to ask why water has this importance. Is life likely to have adapted to water because it was here or is the presence of water a necessary requirement for life promoting life?

Is the basis of DNA important? A number of liquids not incompatible with the solar abundance are immediate candidates from a purely physical point of view. These are listed in Table 29.1 with the appropriate melting and boiling points, all at atmospheric pressure. Another possibility is hydrogen cyanide. Carbon dioxide passes directly from solid to vapour and so is not relevant.

The selection between the various options must be based on the chemical criteria for the rate of chemical reactions. This must be viewed in relation to the temperature of the planetary surface. Radioactive heating alone will raise the surface of a body of Earth mass to an equilibrium temperature of about 100 K In order to facilitate the chemical reactions, the (required) atmosphere must raise the temperature further to about 300 K. From Table 29.1, only water and ammonia cover this range but the boiling

Table 29.1.   The melting and boiling data for several liquids.

Liquid	Melting Point (K)	Boiling Point (K)
hydrogen	13.9	20.3
oxygen	54.3	90.13
water	273	373
carbon monoxide	67	83
methane	111.7	134.7
nitrogen	63.1	77.3
ammonia	239.8	283.8

point of ammonia is low. In fact, only water appears to cover the required range. This conclusion is supported by the critical point data. Thermal stability is provided by the simultaneous existence of the three phases. Changing temperature will change the proportions of each phase present at the time. Increasing temperature increases the vapour in the atmosphere while a falling temperature increases the liquid water. The result of the three phases present at the same time is to provide a thermal stability. For water these conditions apply in a band at the distance of the Earth from the Sun. The same situation could apply for ammonia between the distances of Mars and Jupiter; for methane this could apply in a thin region at about the orbit of Neptune. Insofar as the Sun is a typical star these conclusions and distances must be typical for all solar type stars. The physical properties of water are supplemented by its chemical solvent properties which singles it out as a unique fluid for the production of living materials. Its wide universal occurrence may be significant.

29.7. Surface Features: Body Size and Scaling

It is interesting to see whether any knowledge can be gleaned about the controls on the size and behaviour of living materials on planets. Once again, these problems are best tackled from an atomic point of view. It will be found that the overall size and structure of living material are determined by the interaction energies between atoms and molecules.

There are about 10^{27} atoms in a body of the size of a human being and the binding energy for all the atoms is about 10^6 J. A body will come to great harm if it is forced to absorb this amount of energy. Actually even 1% of this amount would provide a major perturbation. Most living materials have very much the same mean density which is rather less than that of water, so the size of a body is a measure of its mass. One important constraint on the size must be the quantity of energy it can safely absorb during its normal activities. As an example, walking about the surface will involve bumping into obstacles from time to time or falling, perhaps even down a hole. Falling will involve gravitational energy and this must be small in comparison with the binding energy. A height of about 30 m is indicated by these arguments. This is about the characteristic size of the largest creatures that have lived on Earth. At the other extreme, small creatures with a mass of order 10^{-2} kg could have a size of about 10^{-2} m. This is in the range of insect sizes on Earth. The importance of these effects will also depend on the texture of the surface. The maximum size of the living materials will be greater the smaller the planet because gravity is smaller. The pull of gravity may be less but the strength of the atomic forces remains the same.

It is interesting to consider the effect of scale on the design of living creatures. Much of science fiction is based on this but, as we will see, it is not possible to convert small creatures into large ones by simply scaling up their size. If a body is to be mobile on a solid surface it must have legs of some kind (snakes are a special example but they also are quadruples and have their size limits as well). The weight of the body is ultimately supported by the ankle cross sections, one half holding the weight while the other half moves the body forward (or back). The cross-sectional area of the ankles varies as the square of a characteristic size, say, R^2. The weight of the body is determined by its volume in a given gravitational field and so varies as R^3. The "weight/support" ratio therefore varies as $R^3/R^2 = R$. This ratio increases with increasing size — if the creature were to grow substantially larger its would become weaker and ultimately couldn't support itself.[4] The sizes of living creatures, therefore, depend partly on the strength of gravity and partly on the nature of the terrain.[5] Smaller planets, with lower gravity, would be able to support larger creatures than larger planets with higher gravity.

*

Summary

1. Living material cannot live exactly anywhere. The more advanced the life, the more demanding are its requirements.

2. At the minimum, it requires constant heat to provide a stable environment with a temperature appropriate for the efficient activity of the chemical reactions necessary to maintain life.

3. The constant heat source requires the planet to be in an orbit round the star.

4. The spectrum of the radiation from the star must not contain too high a proportion of high energy wavelengths. It must also provide sufficient intensity in the infrared wavelengths. This places a constraint on the surface temperature of the star.

---

[4]The science fiction accounts of gigantic spiders, as one example, could not occur. Even the multiple legs would not be able to support the creature which would sink to the ground under its own weight.
[5]There are other factors such as size in relation to available food supplies.

5. There is a relationship between the surface temperature and the stellar mass. The spectrum requirement restricts the mass of the star to be of broadly solar type, within a range of perhaps 0.5 to 2.0.

6. Protection of the surface of the planet from harmful solar radiations requires a planetary atmosphere of appropriate thickness. This itself places a constraint on the mass of the planet, and particularly a lower limit because the atmosphere must be held by gravity over the lifetime of the body.

7. There is a relation between the maximum size a living creature can attain and the gravity of the planet.

8. These arguments have been based on atomic properties and binding energies and an attempt has been made not to be biased by our experience of living creatures on Earth now. Have we succeeded?

30

An Anthropic Universe?

Some earlier arguments have explored the characteristics of environments that might be expected to support extra-terrestrial life in the Universe. Extra-solar systems that might possibly satisfy these requirements have also been identified. Two systems were found in particular that might be suitable for further exploration when observational techniques make this possible. In the meantime it is important to ask why we should expect other life forms to be present in the Universe anyway and what role animate matter could possibly play in an inanimate world. These are the most basic of questions that can be asked and, to gain perspectives, must involve properties of the Universe itself.

30.1. Describing the Physical World

Let us start from the very beginning and ask the age old question: why is the world as it is? Although this question has arisen from time immemorial, the answers have always been inconclusive and have not always been of the same form. In earlier times the answer revolved about the belief that this is the way God made it. Why He should have thought of this plan was beyond discussion but this is what He provided. The spread of physical conditions provided a very wide habitat for mankind that was generally congenial. It is rather as if the birds were able to say "the air was put there for us to fly in — isn't it lucky that it is there. And isn't it lucky that it has a composition we can all breathe?" Not everything, of course, was favourable to living things. What difficulties there were, were ascribed to human weakness and folly. This was corrected by overseeing angels, such as the Greek Furies (the three Eumenides, Alecto, Tisiphone and Megaera, called the Kindly Ones by those likely to receive the correction). All these constructions were devised to make some sense of the Universe we inhabit, within our understanding. All civilisations, however simple, had their distinct story of the origin of all things.[1] Good opposed evil in the guise of the actions of semi-human gods, vying with each other. These ideas are still current although we now often give them different names and actions.[2]

These perspectives began to have a change of character after the discoveries by Copernicus, Kepler, Galileo and Newton which introduced the then new scientific method. It became clear, in the 17th century, that there are underlying rules (called laws if they were invariably followed) showing the Universe to be a physically controlled entity. It began to seem possible to attempt to penetrate the underlying logic of existence. The work of Gregor Mendel and of Charles Darwin and of others, in the 19th century, showed that the development of living material involves rules and can itself be included in this endeavour. The undertaking became more insistent with the discovery of the atomic and sub-atomic particles, starting with the discovery of the electron by J. J. Thomson in 1897. This sub-layer has strange magnitudes for the masses and electric charges and other properties. Why should the masses and charges have the magnitudes that they have? Are they related and if so how? The multiplicity of atomic and sub-atomic particles has become somewhat extreme and the story isn't yet ended. The question raised by the ancient Greek thinkers of what underlies the Universe has still not been solved.[3] Each underlay discovered so far is associated with an energy greater than that for the layer above. Will our knowledge be curtailed when we are unable to summon up yet greater energy for exploration? Do the layers in principle extend for ever? Is knowledge open ended in that you can have as much as you can afford? How much will forever lie hidden? This we will never know. Arguments of symmetry have shown an ability to penetrate the structure of things to some extent. Many of our descriptions of the physical world have the common root of continuity.

More recently it has been realised that the presence of life itself raises very special questions, whether it is universal or has just happened on Earth. If it has arisen in only one place (here on Earth) the difficulties are the same as if it had arisen everywhere, as we shall see.

---

[1] Indeed, a civilisation can be characterised by its particular account of the formation of the Universe.

[2] Richard Feynman, the American physicist, used to say that in medieval times it was thought that the planets moved round the Sun by being pushed along the orbit by angels. The discovery of the law of gravitation taught us that the angels do not push along the orbit but instead towards the Sun. The direction of pushing has changed but our understanding of what does the pushing remains obscure.

[3] The modern approach involves the quantum vacuum but it has many components. Its properties are not understood. The question of what the vacuum actually is, is not considered.

30.2.   Consequences of the Strength of the Forces of Nature

The great forces of nature number four: gravitational, the strong nuclear force, the weak nuclear force and the electromagnetic force.[4] Planetary problems involve all four forces one way or another. The application involves the following peculiarities:

1. The balance between gravity and electromagnetism observed now is such that the electromagnetic force is about 10^{39} times weaker than gravity. If this ratio had been substantially less the stars would have been many millions of times less massive and would have burned millions of times faster. The planetary bodies would either not have formed (if gravity had been weaker) or would have been strangely different (had gravity been stronger).

2. The Universe is full of hydrogen which is the fuel for stellar radiation. The existence of hydrogen results from the fact that the nuclear weak force is 10^{28} times the strength of gravity. Had it been only a little less weak hydrogen would not have formed but helium would have. A world without hydrogen would be a strange one indeed: helium is a largely inert element so there would be virtually no chemistry. One immediate consequence would have been no water. There could be no life.

3. If the strong nuclear force (which holds nuclei together) had been just a few percent stronger it would not have been possible for protons to form. Without protons there could be no atoms. Without atoms the Universe would be composed of radiation.

4. If the strong nuclear force were only a few percent weaker, the atoms that might form would not be stable enough to allow stable matter to become established. The Universe would at best be a weak jelly.

5. The masses of the proton and the neutron are slightly different, the difference being very nearly twice the mass of the electron. If this mass difference had had another magnitude then either all the protons would have become neutrons or all the neutrons would have become protons. Either way, there could have been no atomic nuclei.

6. The range of natural chemical elements, with atomic masses from 1 to 92, is formed in supernova explosions. The synthesis of the various elements depends on very special factors but the case of carbon is exceptional. Carbon is the basis of living materials on Earth and can be synthesised only because of the special ratio between the strong force and

[4]New forces may be discovered in the future, probably at the sub-atomic level.

the electromagnetic force. This ratio allows carbon to reach an excited state of 7.65 Mev (1.224×10^{-12} J) at a temperature of about 1.5×10^7 K at the centre of the star. These conditions allow a resonance between ^{4}He, ^{8}Be to produce ^{12}C. The binding occurs during a period of 10^{-17} s. This is very short on our everyday time scale but long on the nuclear time scale of 10^{-23} s, which is about the time necessary for light to traverse a nucleus.

7. Just as carbon is necessary for life so is water. This is a very strange substance in having an extraordinary crystal structure which gives a greater mean separation of atoms in the solid form (ice) than in the liquid form. The reason is well understood in the structure of the H_2O molecule but the hydrogen, oxygen linkage is unusual. Indeed, workers in the theory of liquids will admit that water is not a liquid from the point of view of their theories. It is more closely an amorphous solid. The result is the vision of icebergs floating in the oceans with the solid (ice) being less dense than the liquid.[5] There is, indeed, a minimum density at 4 K. The effect of this unique property is for a volume of water to freeze from the top rather than from the bottom. Ice rises because its density is lower than water, giving a layer of ice at the top and water at close to freezing at the bottom. Living things in the water can, therefore, survive very cold weather because there will always be some liquid lower down. Were this not so, life could not have survived the cold periods on Earth.

30.3.   The Beginnings

It is important to know whether the Universe had a beginning or not. If it did, life must have had an origin in the past: if it did not, life need not have had one. Conditions in the Universe deep in the past can only be guessed for there is no direct information. There is one piece of observation which applies now and that is the Hubble red shift. In the 1920s, Edwin Hubble and his colleagues (using the then new 100-inch Hooker telescope at Mt. Wilson, California) found that the spectral lines emitted by galaxies are shifted towards the red. There is one exception — the Andromeda galaxy[6] M. 31,

---

[5]These matters were considered earlier in I.5.

[6]M 31 was the 31st entry in the list of hazy patches in the sky listed by Messier in his catalogue of objects. These were designed to allow the observer to avoid these objects when searching for comets. Most of the Messier objects, in fact, turned out later to be galaxies. Messier's designations are still used today.

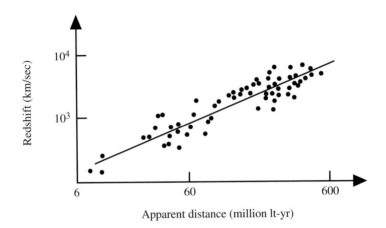

Fig. 28.1a.   A plot of the redshift observed in the spectrum of a galaxy against its apparent distance. The straight line is called Hubble's law.

which shows a blue shift. The spectral shifts were interpreted as Doppler shifts meaning that all the galaxies (other than that in the constellation Andromeda[7]) are moving away from us. The speed of the movement has only recently been found with accuracy and is about 65 km/s/megaparsec ($= 2.10 \times 10^{-18}$ m/s). A simple plot of the Hubble law is shown in Fig. 28.1a. The distance scale extends out to about 6×10^{8} light years ($\approx 6 \times 10^{13}$ km). The law has been found to apply to all distances measured so far. The mean rate of expansion is very low. The slope of the line is called the Hubble constant, $\mathscr{H}$. Theory reveals that $\mathscr{H}$ is not actually a constant of nature but a parameter which depends on the size of the Universe and its motion at any epoch.

If the galaxies are moving apart they were closer together in the past and there must have been a time, if the expansion has continued unabated in the past, when all the material in the Universe was indefinitely close together. The dimensions of $\mathscr{H}$ are inverse time. Its inverse gives a characteristic time for the "age" of the visible Universe, $\tau = 1/\mathscr{H}$.

The Hubble law is generally accepted now as evidence for an initial phase of the Universe where matter was highly compressed. The expansion is presumed to have been caused by an explosion of some kind, causing the matter (then in the form of energy) to move apart. This hypothesis is called the Big Bang. Considerable detail has been attached to this concept but it still remains hypothetical. Tracing the expansion back the "age of the

[7]The movement of M 31 towards our Galaxy is one example of random motion among clusters of galaxies.

Universe" is the characteristic time τ since the expansion began.[8] The best current value is $\tau \approx 13.6 \times 10^9$ years which is very slightly more than 3 times the age of the Solar System. The hypothesis has been developed into a theory by the application of general relativity, developed initially by a number of workers, including Einstein, Eddington, de Sitter, and Lemaitre.

There has been considerable uncertainty about the future behaviour of the Universe. The theory allows three possibilities: (i) the Universe will expand forever with increasing speed; (ii) it will expand but come to rest after an indefinitely long time; or (iii) it will reach a maximum size and contract back to a point again. Recent measurements show the expansion to be increasing and suggest the density of matter and energy have a critical value which describes the possibility (ii). It seems the Universe will expand forever, becoming indefinitely dilute.

An alternative theory to the Big Bang is the statement that there was no beginning and that the Universe has existed for an infinite time in the past and will do so for an infinite time in the future. The most recent exponents of this model are Hoyle, Bondi and Gold. Every location in the Universe must remain essentially unchanged and will be like every other location. This must involve some production of matter to compensate for the movement of matter away through the Hubble expansion. There is, then, a continuous creation of matter replacing that locally that has moved away due to the Hubble expansion. This approach is not generally favoured by the majority of cosmologists at the present time although it has certain theoretical advantages. Future development could be centred on an appropriate time dependence for the creation of matter, perhaps large quantities being produced periodically rather than either as a continuous drizzle throughout time or as an initial once for all creation (big bang).

There is one further piece of observational information. The whole Universe is bathed in a radiation of microwave frequencies having a black body (thermalised thermodynamic equilibrium) form. Its temperature is measured accurately as 2.728 K. This places it in the general frequency range $\nu \approx 2.1 \times 10^{10}$ s^{-1}. This temperature actually forms the lowest natural temperature in the Universe. Local laboratory cryostats achieving a lower temperature are colder than the Universe! The presence of this background radiation was actually first observed by McKellar in 1941 from the sharp interstellar absorption lines due to CN. They were identified soon after by Adams. The observations were not very precise and were associated with

---

[8]There is no knowledge of conditions before the explosion nor what caused it. The modern world has its mysteries and mythologies.

a background temperature in the range $T = 1.8$ K and $T = 3.4$ K. This radiation was rediscovered in 1963 by Penzias and Wilson with more accurate apparatus and the temperature was given as 2.73 K, well within the range suggested by Adams. The result in 1963 came at a time of intense activity on the Big Bang theory and the radiation was immediately accepted as evidence for the de-coupling of matter and radiation when the temperature fell to a value of about 3×10^3 K. The expansion of the radiation with the Universe would reduce this temperature to about 2.73 K today.

There is ambiguity in the interpretation of the radiation, as has been stressed by Hoyle and Burbidge. The conversion of hydrogen into helium in stars provides the radiation we see everywhere. This radiation is absorbed by interstellar medium and re-emitted and in this way is rapidly thermalised. Using estimates of the quantity of helium in the Universe, which is presumed to have originated from hydrogen, it is found that the frequency (and so the temperature) is of the same order of magnitude as the observed cosmic background radiation.

The theories about the origin of the Universe have been viewed in a little more detail because they control the background to thoughts on exo-biology. If the Universe has had a beginning at some time in the past life must have originated after that and the search for the way it formed is legitimate. If the Universe did not have a beginning in the past there is no reason to suppose that we can ever find out how it started — indeed, did it ever start? A universe with an infinite history in the past must present us with insurmountable difficulties in understanding its basic structure. It is usual to presume that there was a finite beginning but it is safest to admit that the details are, at present, entirely hypothetical.

30.4. The Size of Our Universe

Even though it may not be clear what the origin of the Universe was, its size can be deduced with some certainty. The important quantity is the characteristic time deduced from the value of the Hubble constant. The outer ridge of the expansion will travel with the speed of light, c, so the size, $d(\text{now})$, will be the age (in seconds) multiplied by c, that is

$$d(\text{now}) \approx 13.7 \times 10^9 \times 3.17 \times 10^7 \times 2.99 \times 10^8 = 1.299 \times 10^{26} \text{ m}.$$

This is an unimaginally large distance.

It must be realised that this does not cover the whole Universe but only that portion which can be observed from Earth. The speed of galaxies increases with distance and the redshift moves the frequency to zero when the galaxy reaches the speed of light — the galaxy has disappeared from our view. There must be more of the Universe beyond our horizon but we will never be able to observe it.[9] The nature of matter cannot have changed as it passes beyond our horizon so we presumably know what the Universe outside is like even though it is not possible to estimate how much there is of it. This is one of a number of major uncertainties about the Universe.

It is possible to gain some estimate of the mean density of matter. It is estimated that an average galaxy (if there be such a thing) contains some 5×10^9 solar masses that is about 1×10^{40} kg. It is also estimated that there are some 10^{11} galaxies visible within our observable horizon giving the total mass of the order of 1×10^{51} kg. The volume of the visible Universe is $V = \frac{4\pi}{2}d^3 \approx 2.2 \times 10^{78}$ m^3. A very rough estimate of the mean density of the visible matter within our observable region follows as $\rho_m \approx \frac{1 \times 10^{51}}{2.2 \times 10^{78}} \approx 4 \times 10^{-28}$ kg/m^3. It must be stressed that this is a theoretical estimate and not a direct measurement. It will change with any new estimates of the age of the Universe. It is not the end of the story. It is known that other forms of mass are also present, particularly dark matter and dark energy, so the more likely value of the density in terms of matter[10] is $\approx 10^{-26}$ kg/m^3.

Much of this background remains obscure. The nature of the dark matter remains unknown except that it interacts gravitationally with ordinary matter. The dark energy appears to be involved with the underlying vacuum. It seems to be linked to the cosmological constant of general relativity and is the driving force for the expansion of the Universe.

30.5. Model Universes: Anthropic Principles

So far the discussion has not involved the presence of life although it has been seen that the Universe observed now has a rare set of physical properties congenial to the presence of living bodies. The conclusion that the Universe was in some way designed for the development of advance life may be difficult to avoid.

---

[9]A galaxy that could be observed at the edge last year will disappear over the edge this year and disappear from view. It will not have ceased to exist for its own stars but will have to us.

[10]Mass, m, and energy, E, are interchangeable through the special relativity formula $E = mc^2$.

One suggestion involves a statistical set of universes covering all the possible combinations of values of the physical constants. Our Universe would be but one of an infinite number of others, of which we cannot be aware. The example here is the phase space of statistical mechanics perhaps expressed in a quantum form and related to the many-worlds concept. These Universes cover the full range of physical conditions from a Universe composed of a thin medium of gases through to a Universe which is a solid clump of energy. Life cannot develop in almost all others which makes our Universe unique in allowing it to happen. Without life the Universe cannot be observed and so is unknown. There is no possibility of ever verifying this statistical array and it is of little interest to science.

The fine balance between forces clearly evident in our Universe represents an altogether unlikely combination. This has led to a number of statements of an Anthropic Principle.

A **Weak Anthropic Principle** is permissive and will state that (a) the structure of the Universe does not exclude the possibility of carbon based life developing at some sites within it and (b) the Universe is old enough for this to have happened. Such a statement does not exclude the possibility of living material but does not insist that it must form.

A **Strong Anthropic Principle** is authoritarian and states that the properties of the Universe are such that living materials will develop within the Universe given sufficient time and that evolutionary forces will allow this to develop into advanced technological forms.

The time necessary for this to happen is a separate question. Judged from the experience on Earth, it must require some 5,000 million years to reach an advanced stage of development. This is rather less than one-third the characteristic time of the Universe (quoted as its age). There is an important consequence. Assuming a Big Bang, stars and galaxies are likely to have formed after about 1,000 million years and a second or third generation of stars would have been necessary to provide the chemical elements heavier than hydrogen and helium needed to provide planets and living materials. The technological societies could, therefore, have formed after about 6,000 million years from the beginning. The radius of the Universe then was $d(\text{initial}) \approx 5.69 \times 10^{25}$ m or little more than one-fourth the present value. Nevertheless the distances between galaxies will have been large and the distances between stars in the galaxies will have been very little different from now. Even from the beginning exo-systems would have been isolated and links between them probably non-existent. Isolation may be an inherent characteristic of living colonies.

30.6.   Information and the Universe

If the anthropic principle does apply it is interesting to ask why it should. What purpose could animate material have in the Universe at large? One possible purpose was suggested by the present author long ago. It revolves around information.

The universe is a closed system (a black hole in the classical definition[11]) so nothing has been lost or gained since its formation. This means that all the information for its complete evolution was with it at the moment of its formation. This was hidden information or virtual information. It would appear as the evolution proceeds. Virtual information will be converted into actual information in the evolutionary process. In this way, virtual information is lost as the material Universe evolves.

A second container of information is life itself. The cell contains all the information for the production of new life and the evolution of complexity can be regarded as the production of information. Presuming the Universe to have had a beginning, this animate information will have begun after the Universe was formed. A balance can be envisaged between the lost inanimate information as the universe evolves and the gained animate information as life develops. Remembering the links between information and entropy this balance can be seen as a conservation of entropy. It is known that the universes provided by general relativity, including expanding universes, evolve at constant entropy (isentropic) and here is a detailed example of it. This is as yet a conjecture but it can be raised to testable science only when the two concepts of animate and non-animate information have been given a quantitative form. This has not yet been done satisfactorily.

This hypothesis amounts to a strong anthropic statement because it implies that living material is required to allow the non-animate material to evolve. Such a relationship between animate and non-animate materials would make our Universe unique and would give living materials a central place in its structure. This would be in keeping with the central assumption of modern science that the Earth and its various associations are neither central nor unique in the Universe. The fact that the same physical principles apply throughout the known Universe, so that the same developments of the physical universe can be expected to arise

---

[11]General relativity sets the condition relating mass M and "radius" R for a black hole as $M/R > 6.7 \times 10^{26}$ kg/m. Taking $R \approx 1.3 \times 10^{26}$ m found in 30.4 this gives $M > 8.7 \times 10^{52}$ kg which agrees well with the estimate of the mass found there.

everywhere, is a reason for presuming the existence of extra-terrestrial life forms. The argument has not been developed into a quantitative statement because a precise definition of information in living materials has yet to be developed.

30.7.   Extra-terrestrial Visitations

If the Universe is designed for the development of intelligent, technically advanced, life the question raised long ago by the Italian physicist Enrico Fermi becomes very pertinent; "if they are there why have they not been here?" At first sight this sounds very reasonable until it is turned round: "we are here, why have we not been there?" There are many reasons but they all revolve around a present lack of technology on the one side and psychology on the other. We list some thoughts now from our point of view. These problems would beset extra-terrestrials as well.[12]

1. We are not yet technically capable of moving very far from Earth. There have been six short visits to the Moon, 250,000 miles away, but we are not yet able to visit Mars which is also more or less on our doorstep. Our most advanced technology is very little older than 50 years and is still improving rapidly. It will take time to develop further. The next human visitation will be to the planet Mars but not for some years.

2. We have not yet identified another planetary body which would be capable of maintaining advanced life. The search is on but will take more time yet. Once discovered, such a body would give a target for a visitation by us and for some form of communication.

3. The nearest star is some 4 light years away and target stars for visitations are likely to be up to ten times further away. The two systems where living materials could be located are each about 40 light years away. This is a huge distance for an expedition.

4. Propulsion systems for long interstellar journeys, such as ion drives, are still only in their infancy. There will be little possibility of travelling even at a significant fraction of the speed of light. There is a limit to the acceleration that can be achieved and the acceleration that can be withstood by proposed spacecraft and the occupants. There must, of course, be deceleration at the other end. Perhaps 1/10 the speed

---

[12]It has been suggested (as a joke?) that sensible aliens would not wish to have anything to do with us.

of light is a realistic average speed to aim for. The nearest star then becomes 40 years away, or 80 years for a return journey, while at present a likely system is 400 years away (800 years with return).

5. The average life span of extra-terrestrials is, of course, unknown but it is very unlikely to be as long as this. To travel the 40 light years there and back would require an expedition covering a number of generations of people. An expedition returning to Earth this year (2003) would have set off before the year 1203 AD. Birth and death, and the upbringing of generations on the voyage would present major sociological problems. Technology would also have developed enormously over this time.

6. The logistics of such an expedition are now very clear. Everything needed for the expedition must be carried — all the things needed to support a community in complete isolation. There are no corner shops, supermarkets or hospitals in space. The craft needed for such an enterprise would be huge. This would pose a major problem for engines, which would require fuel to be carried for the whole expedition time. Saving fuel would be a major constraint on the use of the propulsive power and so on acceleration. It would be logical to wait until the rate of advance in technology has slowed.

7. It has taken some 4,500 million years for our technology to develop to the stage where we recognised these various problems as beyond us at this stage. Can we expect extra-terrestrials in our galaxy to be better placed and more advanced than us. Their star is likely to be of approximately the same age as the Sun.

8. The vast distances between advanced life forms is more than likely to be the essential answer to the Fermi paradox, should the extra-terrestrials be there.

9. For these various reasons it would seem most likely that advanced life will congregate in isolated colonies. There may, eventually, be some indication that other colonies exist but there is very little chance of members of different colonies actually meeting. In the very long term information may pass by communication at a distance but this is by no means clear. Direct contact would give no obvious genetic advantage. These various thoughts suggest that trips between extra-terrestrial neighbours are most unlikely although it would be an enormous advance if their presence could be detected.[13]

---

[13]It would follow that UFOs involving extra-terrestrials cannot exist. What is actually observed during such sightings is not clear.

10. It is interesting to speculate on the form that extra-terrestrials might take, presuming they exist. Because the rules of physics are universal and the same evolutionary rules apply (of adaptive radiation) everywhere, it could follow that nature has chosen the same solutions to evolutionary problems everywhere. This would suggest a broad similarity of life forms everywhere, modified by local circumstances such as the strength of gravity.

*

Summary

1. It has been realised for some time that the physical constants measured in the laboratory have special values which cannot be explained on rational grounds.

2. Even small changes would have provided a Universe where animate materials could not have formed.

3. It is suggested that this is not chance but that the Universe involves a relationship between inanimate and animate materials.

4. It has not proved possible to find absolute properties of the Universe but only relative ones. It is possible only to describe what is observed and to relate one observation to another. The actual physical nature of things remains hidden.

5. Exploration is a many ways similar to ordering pictures without knowing more about them.

6. The size and nature of the Universe may isolate individual colonies of animate materials one from another so they will have no usefulness in communicating together. It may be possible to do little more ultimately than find the presence of other life forms indirectly. It would still be possible for information to pass between them on a lengthy time scale.

7. Animate material at a location is as ephemeral as the inanimate material there though the time scale may be long. It is a mysterious Universe.

8. Might nature have reached similar solutions to evolutionary problems everywhere? Just as stars and cosmic bodies have a universal form, might life forms also be similar?

Epilogue

Nothing has been said so far about the origin of the Solar System. The Galaxy is accepted as having a finite age so an origin for the Sun and planets must also be accepted. The same argument applies to the other stars which have been observed to have companions in orbit about them.

Before exo-planet systems had been discovered it was generally presumed that, if they were found, their structure would mimic that of the Solar System. It was confidently expected that all planetary systems would follow a single pattern and that the different systems would show different stages in the evolution of a single class of object. The distribution and size of the planetary companions might differ in detail but the general pattern observed for the Solar System would appear everywhere. There had been difficulties in developing a theory for the origin of the Solar System and it was hoped that exo-systems would give clear guidance in solving this problem and the problem was unresolved. It was expected that the different observed systems could be labelled in much the same way as Hubble had attempted to do for spiral and elliptic galaxies a half a century before.[1] This would be a step towards understanding the origin of planetary systems. It has been seen in the previous pages, however, that the observed exo-planetary systems are not obviously the same as our Solar System. In particular, the massive companions with low semi-major axes and high eccentricities present great difficulties. Such massive objects can hardly have formed near the central star and there is no obvious mechanism for drawing them in. The whole matter is very much more complicated than had been thought. The theories already developed would have to be extended to account for the new observations. It is still accepted that there is a single mechanism.

---

[1]In the 1930s, Edwin Hubble constructed a sequence of pictures of galaxies showing spiral and elliptic galaxies in a sequence of openness in an attempt to describe their evolution. The sequence is of only historical importance now.

Two different theories had been developed for the origin of planetary systems, one more popular than the other, which is more recent. It may be that both theories describe processes that are important and that the answer is a combination of the two.

(a) The Solar Nebular Theory

This dates back to Laplace and to Kant. In essence it proposes that large molecular gas clouds of cosmic composition (and therefore very largely molecular hydrogen) condense under the appropriate circumstances to form a large central condensation (the star) encased in a cosmic dust which itself condenses and accretes to form planets and satellites. The particular features depend on the temperature reached during the condensation process and the distribution of temperature away from the central star. The rotation forces cause the dust to form an extended ring structure, the dust orbiting in a plane closely linked to the equatorial plane of the star.

The first indirect support for such a process came from T Tauri stars in the 1980s. These are very young stars (perhaps 10^6 years old) of roughly solar mass. It was realised that many (about 1/3) show an "infrared excess" by which is meant the emission at infrared wave lengths is greater than would be expected from their visible emission. This is interpreted as arising from a halo of dust heated by the short wavelength radiation from the star. The dust re-radiates at the longer infrared and radio wavelengths. If spread uniformly, the quantity of dust, as judged from the radiation, is sufficiently large to be opaque to the observer at visible wavelengths so blocking out the star. For the star to be visible, the dust must be restricted to a flattened disc tilted at some angle to our line of sight. The Hubble Space Telescope has been able to resolve at visible wave lengths several dozen discs around T Tauri stars in the Orion Nebula. The protoplanetary discs are called by the contracted name proplyds (PROtoPLanetarY DiscS). The discs are large and the mass they contain is generally more than sufficient to form a Solar System.

The gas cloud collapses when the gravitational energy of the gas (acting to cause collapse) is greater than the kinetic energy of matter opposing it. The mathematical expression for this was given by Jeans and is called the Jeans' criterion: the density appropriate to collapse in a particular case is called the Jeans density associated with the Jeans mass. Compression of the gas due to a nearby supernova, as one example, can trigger condensation by increasing the density. Lower temperatures will minimise the Jeans mass. The usual temperature in a gas cloud is about 10K. The cloud is often in

Table 1.   The changes in the quantities of hydrogen, water ice and rock with distance from the Sun.

Planet	Orbit size (AU)	% Material (by mass)		
		Rock	Water ice	Hydrogen gas
Earth	1	95	5	–
Jupiter	5.2	10	30	80
Saturn	9.6	20	45	35
Uranus/Neptune	19.2/30.1	30	60	10

differential rotation and rotation itself will tend to delay collapse. It will be realised that while some gas clouds will condense to form stars others will not.

The high central compression raises the temperature there and the first thermonuclear energy is released. Radiation now affects the distribution of chemical elements outside. Hydrogen gas is forced away together with the more volatile elements, providing a region of different chemical composition to that outside. Generally, the elements associated with silicates are more dominant in the inner region while hydrogen dominates beyond. As the hydrogen condenses, due to the lower temperature, it forms the major planets. The mass of hydrogen attracted by each condensation is determined by the amount present locally. The lower temperature also causes the water present to condense as ice. Small particles of rocky material will also be attracted. The quantity of hydrogen will decrease with distance from the Sun as is shown in Table 1. The observed distribution of the other chemical elements with distance will depend upon their melting points and the pressures of the initial nebula there. The melting points are known but the appropriate gas pressures are not.

On the basis of this model, the terrestrial planets must have accreted from very small essentially spherical droplets which condensed as the nebula temperature fell. Refractory primitive material is found in chondritic meteorites. These consist of small igneous spheroids called chondrules which were once molten. They must have cooled in space from very high temperatures (1,000–1,500 K) and isolated in space to have preserved their spherical shapes. Eventually they will have made encounters with others and, if the relative speed was sufficiently low, might have been captured, sticking together to form a larger object. It is presumed that some effective sticking took place even is quite large bodies. Once a body reaches the dimension of about 1 km by these encounters which "stuck" it is called a planetesimal.

Computer simulations of these accretion processes suggest a body of the size of the Earth could form after about 10 million years.

The division between the major planets and the terrestrial planets, and specifically between the orbits of Jupiter and Mars, present a special problem. The gravitational attractions of these planets will not allow a planetary body to form and planetesimal pieces are the result. This is the explanation for the asteroid belt. The amount of material there is quite small — collectively no larger than the mass of the Moon.

The details of the theory are very empirical. The initial conditions of the gas are entirely unknown and the accretion times are not subject to quantitative calculation. It is admitted by the proponents of the theory that it is not able to account for the presence of Uranus and Neptune in their present form and barely for the existence of Jupiter and Saturn. Once these bodies are there, on the other hand, the observed obliquities of the planets (the angle between the planet's equatorial plane and the plane of its orbit) can easily be accounted for in terms of the collection of a few large bodies coming together to form each present body. This is characteristic of the theory. It is able to provide qualitative explanations of some observed features but it has not yet been able to provide a true quantitative account from a precise initial model.

(b) A Capture Theory

In the 1930s, Sir James Jeans, of the Jeans mass, suggested that the material that formed the planets could have been torn off the Sun by the gravitational interaction with a passing star. This torn off material contained angular momentum and, so the argument goes, condensed into orbits around the Sun. Close encounters between stars in the main field are rare events and this mechanism would restrict the number of planetary systems that can be expected. It is now known that stars are generally born in groups that may contain a hundred or more members. Some regions are producing many groups such as the nursery in the Orion nebula and in M16. The stars form from Evaporating Gas Globules (EGGs) often from tall pillars of hydrogen gas (called elephants' trunks). This would make encounters between stars common at the early stages of evolution providing the stars with companion systems as they later disperse. The mechanism has one grave difficulty. The streamer of gas drawn off a star must both form an orbit about it and condense. While it may form a series of orbits it will have a high temperature, at least at first, and this will more than overcome the gravitational pull for condensation. Put another way, the streamer will not achieve the critical

Jeans mass. It will not be helped by the high temperature of the gas between the proto-stars at the early stages.

M. M. Woolfson has shown that the mechanism could work, however, if it were turned the other way round. The passing star (for our case the Sun) captures material from the other star during the encounter. Then, the objection to the original Jeans theory do not apply. This capture theory has been developed in some detail and is able to explain many of the characteristics of planetary systems. Viewed from the point of view of multiple stellar births it would suggest that low mass companions are common, as indeed it is being found that they are. It must be admitted that none of the arguments about planetary formation are all embracing and none prepared the scientific community for the unusual companion structures that are being uncovered.

One conclusion from observation is that the planets are not recent additions to the list of cosmic bodies. This is shown, for instance, by the recent Hubble Space Telescope image of a white dwarf (of mass 0.34 solar masses) in M4, 5,600 light-years away. The white dwarf, viewed in ordinary light, accompanies a neutron star (of 1.35 solar masses) found at radio frequencies to be a pulsar, PSRB 1620 – 20. Timing techniques show that there is a third member present which has now been identified as a planetary body with a mass of about 2.5 Jupiter masses. It has a semi-major axis of 23 AU with an orbital period of about 100 years. It appears that the inclination is $55°$. It has been estimated (by NASA) that this trio, including the planet, could be as old as 1.3×10^{10} years, almost as old as the estimated age of the Universe itself. This may be an over estimate but the age must be substantially greater than that of the Sun for the white dwarf star to have formed and stabilised. This shows again that planets and stars are inextricably linked together. It is very likely that globular clusters, which are among the oldest groupings of stars, are the homes of many planets. And planets can be, and can have been, the home of living things.

It appears that planets are an integral part of the inanimate Universe and have been from the earliest times. They may, indeed, be the main interface with the animate Universe. This was suspected over the centuries by thinkers and mystics before modern science had been developed. The pace of discovery increases and many new, unexpected and exciting things will certainly be found in the nearer future. We certainly live in exciting times.

Some Historic Events in the Space Probe Exploration of the Solar System ([1]NASA; [2]Russia; [3]ESA)

Mission	Date of encounter	Spacecraft	Comment
Earth orbit	1957 Sep	Sputnik 1[2]	First unmanned orbit of Earth
Earth orbit	1958 Feb	Explorer 1[1]	Detection of van Allen radiation belts
Moon	1959 Sep	Luna 2[2]	First impact with lunar surface
Moon	1959 Oct	Luna 3[2]	Photographed far side for first time
Earth	1961 Apr 12	Vostok 1[2]	Gagarin in orbit: first man in space
Venus	1962 Aug–Dec	Mariner 2[1]	Flyby of planet
Moon	1964 July	Ranger 7[1]	Close up photos of surface
Mars	1965 July	Mariner 4[1]	Flyby with photos showing craters
Moon	1966 Feb	Luna 9[2]	Lander — photos from surface
Venus	1966 March	Venera 3[2]	Surface impact
Moon	1966 April	Luna 10[2]	Orbit Moon
Moon	1966 June	Surveyor 1[1]	Soft landing
Moon	1966 Aug	Lun. orbiter 1[1]	Photographs from orbit
Moon	1968 Sep	Zond 5[2]	Circumnavigation with life forms
Moon	1968 Dec	Apollo 8[1]	Circumnavigation with human crew
Moon	1969 July	Apollo 11[1]	Manned landing: sample collection
Moon	1970 Sep	Luna 16[2]	Automated return of rock samples
Moon	1970 Nov	Luna 17[2]	Surface rover
Venus	1970 Dec	Venera 7[2]	Soft landing
Mars	1971 May–Nov	Mariner 9[1]	Long-life orbiter
Mars	1971 May–Dec	Mars 3[2]	Soft landing
Jupiter	1972–1973	Pioneer 10[1]	Flyby
Mercury	1973–1974	Mariner 10[1]	Flyby-'74 March, Sep & '75 March

Mission	Date of encounter	Spacecraft	Comment
Venus	1975	Venera 9[2]	Photographs from surface
Mars	1975–1976	Viking 1[1]	Photographs from surface
Venus	1978 May–Dec	Pioneer Venus 1[1]	Long-life orbiter
Jupiter	1977–1979	Voyager 1[1]	Flyby
Saturn	1973–1979	Pioneer 11[1]	Flyby
Saturn	1977–1980	Voyager 1[1]	Flyby
Venus	1984–1985	Vega 1[2]	Atmospheric ballon
Comet	1978–1985	ICE(ISEE 3)[1]	Flyby of comet 21P/ Giacobini-Zinner
Uranus	19771–1986	Voyager 2[1]	Flyby
Neptune	1977–1989	Voyager 2[1]	Flyby
IP/Halley	1984–1986	Vega 1[2]	Photo of nucleus
IP/Halley	1984–1986	Giotto[3]	Close up photo of nucleus
Earth orbit	1986–1998	Mir[2]	First manned orbital space station
Jupiter	1989–1995–2003	Galileo[1]	Orbiter and atmospheric probe
Gaspra	1989–1991	Galileo[1]	Flyby: photo of S-type asteroid
Astrometry	1989–1993	Hipparchus[3]	Hipparchus and Tycho catalogues of stars
Sun	1990–1994	Ulysses[1,3]	Polar measurements
Telescope	1990 Apr 24	Hubble Sp. Tel[1,3]	2.5 m reflector in orbit: H.S.T. Institute
Earth geotail	1992–	Geotail	Study of Earth geotail
Maltilde	1996–1997	NEAR[1]	Flyby: C-type asteroid
Mars	1996–1997	Mars Pathfinder[1]	Soft landing: automatic surf. rover Sojourner
Earth orbit	1998	Internat. Space Sta.	Permanent space station in orbit
Sun	1996–present	SOHO[3,2]	Stationary orbit: Continuous study
Orbit	2003	Spitzer Space telescope[1]	Automatic infrared observatory
Mars	2003–	Spirit Mission[1]	Surface landing
Mars	2004–	Mars Opportunity[1]	Surface landing-opposite side to Spirit
Mars	2004–2005	Mars Explorer[3]	Orbit with instruments
Lunar orbit	2005	Smart 1[3]	Detailed surface studies
Comet Impact	2005 July 4	Deep Impact craft	Impact with Temple 1

Some Useful References

References are a personal thing that must be built up over time but it is useful to have a short list to start from. The speed with which results become available makes it inevitable that the internet references are very important. A few addresses are listed now.

Online Sources

Extra-solar Planets Catalog
http://www.obspm.fr/planets

National Space Sciences Data Center
http://nssdc.gsfc.nasa.gov/planetary

Science at NASA
http://science.nasa.gov

Psychophysiology Lab (NASA-AMES Research Center)
http://human-factors.arc.nasa.gov/ihh/psychophysio

Astronomy Picture of the Day
http://antwrp.gsfc.nasa.gov/apod/

Ciclops Cassini Imaging
http://ciiclops.lpl.arizona.edu/

CELLS Alive
http://www.cellsalive.com

Lists and Plots: Minor Planets
http://cfa-www.harvarch.edu/iau/lists/MPLists.html

Sites for General Information

http://www.google.com

http://uk.dir.yahoo.com/science

Prefixes of Catalogues

M	Messier catalogue of nebulae
NGC	New General Catalogue of nebulae
Arp	Arp Catalogue of Peculiar Galaxies
3C	Third Cambridge catalogue of radio sources
4C	Fourth Cambridge catalogue of radio sources
PKS	Parkes catalogue
PSR	pulsar
QSO	quasar
HD	Henry Draper Memorial Catalogue (original for faint stars but now more general)
HIC	Hipparcos catalogue for stellar distances from parallaxes (measured using space vehicels)
HIP	Hipparcos catalogue of ESA containing 118, 218 stars
Gliesse	(or GJ) catalogue (attempts to include all stars within 20 parsecs (that is 61.72 light years) of the Earth)

Glossary

This is a list of terms of broad interest in planetary science together with general comments and short biographical notes of some people who have added significantly to the subject. It is hoped that the reader will expand the glossary further into a personal document.

A

absolute magnitude *in astronomy* is a standard of brightness of an object, star or planet or other body. For an object within the Solar System it is the brightness an object would show if it were placed 1 AU from both the observer and from the Sun, seen fully illuminated (that is with zero phase angle). For an object outside the Solar System it is the brightness an object would show at a distance of 1 AU from the observer.

accretion is the process by which planetary sized bodies are supposed to form by the continual collection of smaller masses.

adaptive radiation is the name given in evolution theory to the process of the selection of a dominant form of many variants of a species by the mechanism of the survival of the fittest.

advection is the movement of a fluid perpendicular to the direction of local gravity. One example is smoke moving horizontally along a duct.

albedo expresses the reflectivity of a body. The *geometric albedo* is the ratio of the brightness of an object at 0° phase angle to the brightness of a perfectly diffusing disc of the same apparent size and at the same position.

anthropic principle is the statement that the constants of nature underlying our unique Universe have magnitudes which allow animate materials to form. There can be various "strengths" from a weak statement that the formation is possible up to the strong statement that the formation is inevitable. No special place is given to *homo sapiens* and evolutionary arguments accept that this species will ultimately be replaced by others.

aphelion is the point of the elliptical orbit furthest from the star as the focus.

Apollonius of Perga (fl 250 BC) was a Greek mathematician of the Alexandrian school. He studied the conic sections and seems the first person to have introduced the names ellipse, parabola and hyperbola.

Appleton, Sir Edward Victor (1892–1965) was an English physicist who discovered the ionised layer in the Earth's upper atmosphere. It became known as the Appleton layer in consequence. His researches were important in the development of radar.

Archimedes (287–212 BC) was a most distinguished (perhaps *the* most distinguished) Greek mathematician, physicist and inventor who was for many years a resident of Syracuse. He is remembered especially for having solved a difficult problem for the King while in the bath and was so excited that he jumped out of the bath and ran naked through the city crying out in Greek Eureka, Eureka ("I have found it"). The problem was to measure the volume of a complicated body (a royal crown) to find its density. The mass was easy but he realised he could determine the volume by displacing water when he put it in a bath. Simple now but brand new at the time. His many studies led, among other things, to the development of the lever, an understanding of the hydrostatics of floating bodies, and the screw for lifting water. He contributed many inventions for the defence of Syracuse when under siege by the Roman army under Marcellus. These included a giant claw on a lever for lifting hostile ships out of the water and so sinking them. It is believed to have been successful and recent reconstructions of the approach confirm that it could well have worked successfully. Another was a giant mirror to focus sunlight on enemy ships at a distance and so ignite them. Again this seems to have been successful. After the fall of Syracuse in 212 BC, Marcellus sent a soldier to ask Archimedes to join him. Archimedes objected to the soldier walking rudely across his sand slate so destroying his calculations and told the soldier so. The soldier killed him with his sword. There is no record of what happened to the soldier.

arcsec is a measurement of angular distance in the sky. One arc degree is divided into 60 arc minutes and this is divided into 60 arc sec. The factors 60 appear to have originated in the calculations of Sumarian astronomers.

astronomical unit is the radius of a circular orbit around the Sun with an orbital period of 1 Earth year. It is the distance 149,597,870 km and is approximately the mean distance of the Earth from the Sun. (see also light year, parsec and arc sec)

atom is the ultimate unit of a chemical element. It is not a single unit. Rather, it consists of a central nucleus, with a positive electrical charge Z and atomic mass A, enclosed by Z negatively charged orbiting electrons. The atom is electrically neutral. As well as the Z positively charged protons it contains (A–Z) neutrons. It also contains the majority of the mass of the atom. Z is called the atomic number and A is called the atomic mass.

atomic mass is the number of nucleons (protons and neutrons) contained in the central atomic nucleus.

atomic number is the number of negatively charged electrons in orbit about the atomic nucleus. Because the atom is electrically neutral it is also the number of positively charged protons in the nucleus.

aurora is the coloured display produced in a planetary atmosphere by the bombardment of the atmospheric atoms by charged particles from a central star. Aurorae usually appear near the magnetic poles but can appear at lower latitudes if the bombardment is severe.

Avogadro, Amedeo (1776–1856) was an Italian physicist who developed the then entirely hypothetical atomic theory of the structure of matter. He is particularly remembered for his remarkable recognition that equal volumes of gases under identical conditions of pressure and temperature contain the same number of molecules. This is now accepted as a law and is named after him.

axial inclination of the Earth is the angle between the axis of rotation and the plane of the ecliptic. This angle is $27°27'$ and is the cause of the precession of the Earth.

B

bar is a unit of pressure with 1 bar = 0.987 Earth atmosphere, and often taken to be 1.0. In the SI system of units, 1 bar is alternatively 10^5 Pa.

basalt is an igneous rock resulting from the cooling of volcanic lava. It is composed almost entirely of the oxides of silicon, aluminium, iron, calcium, magnesium, sodium and potassium. These are the materials that make up much of the mantle of the Earth and probably of the other terrestrial planets as well.

black hole is a theoretical body which arises as a singularity in general relativity theory. It is to represent the ultimate of dense matter with a gravitational force related to its mass in the usual way. The mass is sufficiently large that the escape velocity is greater than the speed of light. Consequently no matter or radiation can escape from it when once captured. There are a number of bodies believed to be black holes but none has yet been unambiguously identified as such. According to general relativity, the mass M and radius R for a non-rotating black hole must satisfy the inequality $M/R \geq 6.7 \times 10^{26}$ kg/m.

Bode's law — see Titius Bode law.

Boltzmann constant $k = 1.38054 \times 10^{-23}$ J/K.

Boltzmann, Ludwig (1844–1906) was a very influential Austrian physicist and mathematician. He is best known for his pioneering work on statistical physics based on the principles of (the then still hypothetical) atomic structure. By 1871 he had discovered that the average kinetic energy of a molecule is the same in each direction. He very quickly recognised the importance of Maxwell's electromagnetic theory. He was able to provide a derivation of the Stefan T^4 radiation law on the basis of thermodynamics. He will forever be remembered for his statistical interpretation of entropy, S and for his formula $S = k \log W$, which formula is inscribed on his tombstone. His work brought him great controversy with criticisms from scientists who didn't fully understand the probabilistic nature of his arguments. The attacks on him by W. Ostwald were particularly wounding and led him to attempt suicide several times. He was eventually sadly successful in 1906 in a state of depression and ill-health. His theories were proved correct by experiments soon afterwards.

bow shock is a thin layer of electrical current around a cosmic object in the path of the solar wind, or generally stellar wind. The solar wind speed falls across the shock with the result that the plasma becomes deflected and compressed and so heated.

Brahe, Tycho (1546–1601). Distinguished Danish astronomer and astronomer to King Fredrick II of Denmark who financed a substantial observatory called Uraniborg on the island of Hveen near Copenhagen. Brahe had the best astronomical instruments of his day and carried through a massive programme of systematic observations of the heavenly bodies. He found some discrepancies with the measurements of Ptolemy, later suggesting an interpretation in terms of the proper motion of the stars. His observations are still of value today. Of especial mention is the work on the motions of the planets and for this he engaged the services of the young Kepler as a mathematical assistant.

breccia is rock in a finely grained state and fused together during an impact. This refers especially to rock found on the Moon and probably on Mercury.

brown dwarf is a body composed largely of hydrogen but of sufficient mass that thermonuclear interactions occur only between the deuterium atoms. The lower mass is about 2.59×10^{28} kg and the upper mass is about 1.60×10^{-29} kg, which is the mass when H^1 begins to "burn", the body becoming a normal star.

C

caldera is a volcanic crater perhaps the result of the collapse of a shield volcano.

Cassini, Giovanni Domenico (1625–1712) was a Italian/French astronomer. He was born in Perinaldo, Genoa. He began his career at the Observatory at Panzano and discovered Jupiter's Great Red Spot (1655). Also that year he calculated the rotation periods of Jupiter, Mars and Venus. In 1669 he was invited by King Louis XIV of France to be head of the Paris Observatory where he remained for the rest of his life. There he discovered the moons Rhea, Dione, Iapetus and Tethys of Saturn. His son Jacque succeeded him as Head of the Paris Observatory and so did Cassinis in the following two generations, making four generations in all.

Centaurs are small bodies in the outer Solar System. They have been scattered into the Solar System by collisions with Edgeworth–Kuiper belts objects.

centre of mass of a group of particles is the point where the total mass can be assumed to reside in dynamical studies.

chalcophiles are chemical elements that tend to form sulphides.

charge on the electron $e = 1.60210 \times 10^{-19}$ C.

chondrites are a class of stony meteorites most often containing chondrules, which are grains composed of a yellowish or brownish-red magnesium silicate with a fluorine impurity.

chondrites, Type 1 contain no chondrules but have the chondrite composition.

chromosphere is the region of the Sun's atmosphere between the photo-sphere and the corona. The common structure of stars suggests that each star has a chromosphere.

Clairaut, Alexis Claude (1713–1765) was a very eminent French mathe-matician. He was the only member of twenty children to reach adulthood. He was educated at home and showed his aptitude for mathematics very early. At the age of 18 he became the youngest member ever to have been elected to the Paris Academy of Sciences. He joined a small group led by Pierre Louis Maupertius supporting Newton's natural philosophy. He help the Marquise du Châtelet to translate *Principia* into French. In the decade from 1733 he published various findings in the calculus of variations, geodesics of quadrics of rotation and differential equations, proving among other things the exis-tence of integrating factors for solving first order differential equations. In 1736 Clairaut joined an expedition to Lapland led by Maupertuis to measure the degree of longitude, continuing work started by Cassini. The aim was to confirm the Earth to be an oblate spheroid as proposed by Newton (but doubted by Cassini). This led to his famous book *Théorie de la figure de la Terre* in 1743 supporting Newton. (It was said that this work had allowed Newton to flatten both the Earth and Cassini at the same time.) This work also contributed to Maclaurin's work on the theory of the tides. Clairaut next worked on the three body problem and especially on the orbit of the Moon which led, in 1752, to the book *Théorie de la lune*. He also predicted the time of arrival of comet Halley with some accuracy. Following this there was a suggestion in France that it should be called comet Clairaut but this was not followed up. He worked widely in mathematics and his two books *Elements d'algèbre* and *Elements of géometrie* were published respectively in 1749 and in 1765, the year of his death after a short illness.

coesite is a form of quartz which results from subjecting a sample to very high pressure. This material is found in impact craters.

coma is a spherical envelope of gas which surrounds the central nucleus of a comet. It may have a radius as much as 80,000 AU.

comet is a small body in orbit about the Sun composed of ices and silicates. The orbit has an aphelion beyond the orbit of Neptune and a perihelion near the Sun. The volatile materials "boil" off the body when it is near the Sun due to heat radiation. This provides a dust tail and a tail composed of atomic ions. The latter always point away from the Sun due to solar wind pressure.

comet nucleus is the irregularly shaped central core of a comet. It is typically a few kilometres across. It is composed of a mixture of ices, silicates and carboniferous materials.

condrules are small spherical grains found in great numbers in primitive stony meteorites and composed of aluminium or magnesium silicates, together with iron.

conic is a figure formed by cutting a cone. There are three, viz. the ellipse (with the circle as a special case where the semi-major axes are equal), the parabola and the hyperbola. The ellipse is a closed figure but both the parabola and the hyperbola are open, not returning onto themselves.

conservation of angular momentum is the statement that rotational momentum is never lost.

conservation of energy is the statement that energy in its many forms is never lost but merely transformed from one form to another. This conservation is believed to be rigorously true always.

conservation of linear momentum is the statement that the momentum of a set of particles is never lost but only transferred from one particle to another.

constant of Newtonian gravitation $G = 6.67 \times 10^{-11}$ m^3/kgs^2.

continental drift is the continual gradual motion of the continents resulting from plate tectonic processes. Due to this drift the distribution of continents has changed very drastically over geological history.

convection is the process of the transport of material or energy, or the mixing of these entities, by the movement of a fluid. The motion is along the direction of a gravitational field and arises from variations of density in the fluid. A locally low density rises to reach the same value as the environment, while a locally greater density falls.

coronal mass ejection is hot gas expelled from the solar surface by a solar prominence.

Copernicus, Nicholas (1473–1543). A Polish mathematician and cleric who became interested in astronomy while Professor of Mathematics at the Vatican in Rome. He also practised medicine. In Rome he came into contact with Greek thought brought by scholars who had fled the fall of Constantinople in 1453. He became convinced that the Sun and not the

Earth is the centre of the Solar System. His astronomical thoughts and calculations were published in his book *De revoltionibus orbium coelestium* where he developed his heliocentric theory. This was published in 1543 on his death bed. Later understanding of particle mechanics made it clear that the centre of mass of the whole set of Sun and planets is actually the point about which all the bodies revolve. The difference is small: the centre of mass of the Solar System is only a short distance outside the visible surface of the Sun. It does mean, however, that the Sun itself undertakes a small orbital motion about the centre of mass in sympathy with the planets. All stars with companions show the same effect which is important for the detection of exo-planets using stellar motions.

Coriolis force is the apparent force that results from rotation. In the context of a planet it acts on any material travelling from the poles to the equator. It is the cause of cyclonic and anti-cyclonic winds in the Earth's atmosphere.

corona has two different meanings. In solar physics it is the hot highly ionised atmosphere which is very luminous. In the study of Venus it is a circular formation on the surface encircling an area of jumbled terrain.

cosmic rays are very high energy atomic nuclei that abound in space and enter the earth's atmosphere from outside. The composition is the cosmic abundance of the elements which means it is very largely hydrogen nuclei (that is protons).

cosmic abundance of the chemical elements is the distribution observed generally in the cosmos. It is also the solar abundance. It is 98% hydrogen and helium by mass with smaller abundances of oxygen, carbon, nitrogen and neon (in that order): all the remaining elements account for less than 0.5% of the total.

crust is the outer layer of a solid planet.

cryovolcanism is the resurfacing of an ice layer by flowing ice or ice-liquid mixtures.

C-type asteroid is a member of one of the two most common asteroids. They occur most often in the outer part of the main asteroid belt.

current sheet is a two-dimensional surface that separates magnetic fields of opposite polarities. It is found especially within a magnetosphere.

D

d' Alembert, Jean le Rond (1717–1783) was a most gifted French mathematician. He reformulated the vector representation of mechanics developed by Newton in terms of energy. This paved the way for the development of a powerful approach to dynamical problems in the next century. He produced numerous scientific works of which his *Theory of the Winds* and *Precession of the Equinoxes* are most noteworthy.

Darwin, Sir George Howard (1845–1912), the second son of Sir Charles Darwin the author of *The Origin of Species*, was an eminent cosmogonist. In particular he studied the figure of the Earth theoretically. He was elected Plumian Professor of Astronomy and Experimental Philosophy at Cambridge University in 1883.

dendritic describes the tributary system of a river, rather like the branches of a bush.

deuterium is the heavier isotope of hydrogen with a nucleus containing a proton and a neutron. It is sometimes called heavy hydrogen because its atom is virtually twice the mass of ordinary hydrogen. Naturally occurring hydrogen contains 0.015% of deuterium.

differentiation in a planetary context is the process by which planetary bodies form layers of different chemical and minerological composition according to the density. The denser material gradually sinks to the lower regions of the volume.

diopside is a silicate mineral commonly found in meteorites. Its chemical formula is $Ca(Mg, Fe)Si_2O_6$.

doppler effect affects wave motion. When an object is moving away from an observer or is coming towards the observer the frequency of a wave emitted by the object is changed in a characteristic way as far as the observer is concerned. The wavelength is lengthened if the object is moving away or shortened if the object is closing in. This is clear with sound waves, for instance, where the sound of a passing vehicle or train changes as the object approaches the listener and then moves away. It is found also in electromagnetic radiation where, for instance, the light from a source moving away is shifted towards the red (red shift) while one coming closer is moved to the blue (blue shift). The degree of shift depends on the speed of the source so measurements of this type are used to find the speed of a moving object in the line of sight. This technique provides a popular method for finding the motion of an exo-planet. It is used in astronomy to find the motion of galaxies and other objects. If the transverse speed can be measured by observation, finding the doppler speed allows the three dimensional velocity to be determined. The method is now part of spectrometry.

E

earthquake is the explosive release of energy within the main body of the Earth. It is the source of seismic waves which traverse the interior. Such an explosive release of energy has also been detected within the Moon and will also occur in other terrestrial type planetary bodies. The intensity is measured on the logarithmic Richter scale, named after its distinguished inventor.

eccentricity is a numerical value which describes the shape of an ellipse. To calculate it in a particular case, it is the ratio of the distance from the centre of the figure to the focus and the value of the semi-major axis.

ecliptic is the path of the Sun as it moves across the celestial sphere. The path is so named because it is where solar eclipses occur.

ecliptic plane is the plane defined by the Earth's orbit about the Sun.

eclogite is a granular rock composed of garnet and pyroxene.

Edgeworth–Kuiper belt is the name of the region beyond the orbit of Neptune composed of a very large number of icy objects restricted largely to a thin circular region in the ecliptic plane. The sizes of the objects range from those of Pluto (now regarded as a Edgeworth–Kuiper belt object), its satellite Charon and the recently discovered Quaoar down to small objects of comet size. All the objects are composed of a small proportion of silicate materials (Pluto has a proportion of 13.5% by mass and may be typical of the larger members) and water ice. Over 600 such objects had been found by 2002 and more continue to be discovered. This region is believed to be the source of short period comets. Its existence was first proposed by K. E. Edgeworth in 1949 and independently by G. P. Kuiper in 1951. The belt is often referred to as the Kuiper belt in the US.

ellipse is a regular oval figure formed when a cone is cut by a plane making a smaller angle with the base than does the side of the cone. It can also be described as a figure formed when the eccentricity is less than unity.

energy has been described as the universal currency that exists in apparently countless denominations; and physical processes represent a conversion from one denomination to another. Energy is a conserved quantity. The Joule is the SI unit of energy. A convenient unit of energy for domestic purposes is the kilowatt hour (kWh) $= 3.6 \times 10^6$ J. In special relativity theory there is an equivalence between what is perceived as mass, m and what is perceived as energy, E according to $E = mc^2$. The energy equivalence representing the mass of a proton, for example, is $E = 1.7 \times 10^{-10}$ J.

enhanced greenhouse effect is the increased greenhouse effect that results from the presence of particular gases (greenhouse gases) in the atmosphere. Such gases include CO_2 and NH_4 among others. These are given off naturally by plants and animals, and also artificially by the burning of fossil (hydrocarbon) fuels such as coal and oil. (see also **greenhouse effect**)

electron is one of the elementary particles which cannot be reduced to involve a sub-particle. Its mass is 9.10908×10^{-31} kg. Its radius is smaller than anything than can yet be measured and consequently is often treated as a point particle. It carries a unit negative electric charge of 1.602×10^{-19} coulombs. It is accompanied by an anti-particle identical to it except that the electric charge is then positive.

epicentre of an earthquake is the point on the surface of the Earth (or other solid planet) vertically above the focus where the energy is released.

equation of state is a description of the behaviour of a physical system under changes of pressure, temperature, and volume. The simplest system is the dilute gas which approximates closely to the ideal. Here the Boyle and Charles laws apply giving an equation of state in which the pressure, P, and volume, V, are related to the temperature T by the expression $PV = RT$ where R $(= 8.314$ J/molK$)$ is the gas constant. This is the equation for the ideal gas. For greater pressures this simple expression is extended to include a series of terms involving the pressure. (see **virial expansion**)

equinox is one of two locations in the orbit of the Earth where the rotation axis of the Earth points in a direction perpendicular to the Earth-Sun line. These locations lie between the solstices. The Sun is overhead at the equator at each equinox, at the autumn equinox and at the vernal (spring) equinox. At these points lengths of the day and the night are equal. Any planet whose rotation direction is not perpendicular to its orbital plane will show the same effect.

escape velocity is the minimum speed that a body must have if it is to escape from a planet.

Eucrites is a class of meteorites believed originally to have been part of the asteroid Vega. These bodies were released by a collision, presumably with another asteroid.

exo-planets are bodies of planetary mass orbiting stars other than the Sun.

exosphere is the outer region of a planetary atmosphere where the trajectories of atoms and molecules are straight lines essentially unaffected by collisions. These bodies can escape directly from the atmosphere if the speed is above the escape velocity.

F

feldspar is a mineral, one of a number of silicate materials which is rich in aluminium. It is found in rocks on Earth, on other planets and in meteorites

first point of Aries (denoted by γ) is the point in the sky marked by the direction between the Earth and the Sun at the vernal equinox. The point changes slowly with time and is no longer in the constellation Aries.

focus of an earthquake is the point of energy release inside the volume.

focus of an ellipse is one of the two points linked directly by a reflection from any point on the curve. The distance between them by reflection from any point on the curve is the same being twice the semi-major axis. This property is often used to draw the figure.

force is an influence which is determinable and measurable and which causes a physical system to change.

four forces of nature are (i) the electromagnetic force, F_E, describing the interactions between atoms and nuclei; (ii) the weak nuclear force, F_W, which controls radioactivity and the stability of the nucleus; the strong nuclear force, F_S, which describes the binding of the nucleus; and the gravitational force, F_G, which describes the interaction between masses. In terms of the relative strengths $F_S \approx 10^3 \, F_E$; $F_W \approx 10^{-11} \, F_E$; $F_G \approx 10^{-39} \, F_E$. It is seen that the gravitational interaction is relatively weak but acts over the longest distance.

Fourier, Jean Baptiste Joseph (1768–1830) was educated at the École Royale Mititaire of Auxerre where he showed mathematical ability. When he was 21, however, he decided to enter the priesthood although in the event he didn't take religious vows. In 1793 he became a member of the local Revolutionary Council also teaching at the Royal Military College where he had studied. He was imprisoned but released with the political changes that occurred after Robespierre himself was sent to the guillotine. In 1794 Fourier was nominated to study at the École Normale in Paris. There he was taught by Lagrange and by Laplace. In 1797 he succeeded Lagrange as Professor of analysis and mechanics. In 1798 Fourier joined Napoleon's army as a scientific adviser in the invasion of Egypt and was in charge of collating the scientific and literary discoveries made there. He returned to Paris in 1801 to his academic post but was sent instead to be Prefect of the Department of Isère at Grenoble. He was charged with many tasks including working on the Description of Egypt published in 1810. It was here that he published the famous memoir *On the Propagation of Heat in Solid Bodies*. It included the expansion of functions as trigonometric series for which he is famous. The memoir was not well received and was criticised by a number of people including Laplace, Lagrange, Biot and Poisson. The objections were later shown to be false. In 1815 Napoleon awarded Fourier a pension of 6,000 francs beginning on 1 July of that year. Napoleon was defeated on that day and Fourier did not receive any money. A lunar crater has been named Crater Fourier.

G

Gaia hypothesis asserts that the Earth is a self-correcting physical system that provides a corrective response to changes that might occur, either natural or man-made.

Galileo, Galilei (1564–1642) was born in Pisa, Italy, the year of Shakespeare's birth and of Michaelangelo's death. He died the year of Newton's birth. He can be justifiably called the first true scientist in being the first person to consistently draw fundamental conclusions and generalisations on the basis of carefully constructed experiments. His theoretical analyses were expressed mathematically. Among his great experimental achievements was the establishment of the law of inertia, the elucidation of the swinging pendulum and the establishment of the laws of projectiles. He constructed and used a refracting telescope in 1609 after having heard of its construction in the Netherlands. He quickly discovered the four main satellites of Jupiter (since called the Galilean satellites), the rings of Saturn (though it wasn't clear what they were at first) and the phases of Venus. This last observation proved that the Earth is not at the centre of the Solar System and that the Copernican heliocentric system is correct. He observed the Moon and found that its surface is not smooth as an ideal body should be. He was also a man of affairs. He used his telescope to aid marine insurers and he applied his laws of projectiles to the control of the newly developed canons which were then becoming very powerful. It is said that he attempted to sell the Galilean satellites to the Medici family but they declined to purchase them. His astronomical observations were collected in his book *Dialogue on the Two Chief World Systems* which led him into controversy with the Catholic Church and he became technically a prisoner of the Inquisition. While confined to his own house he published secretly in 1638 in the Netherlands the results of his researches into mechanics in the book *Discourses and Mathematical Demonstrations Concerning Two New Sciences Pertaining to Mechanics and Local Motion.* His dispute with the Church was resolved on his terms in the 20th century. His books are still in print.

Galton, Sir Francis (1822–1911) was a cousin of Charles Darwin. Although he was later the founder of eugenics, his early work includes the publication of *Meteorgraphica* which contains the first weather chart. He was among the first to approach biological problems through mathematics. It is of interest that he devised a scheme for the identification of people through their fingerprints.

gamma is an older unit of weak magnetic induction. $1\gamma = 10^{-9}$ T $= 10^{-4}$ gauss.

garnet peridotite is a mineral composed in nearly equal proportions of the oxides of silicon and magnesium (to about 90%). There are many other impurities including the oxides of titanium, aluminium, chromium, iron, manganese, nickel, calcium, sodium, potassium and phosphorus. Different proportions of the impurities provide different colours — the ruby red variety is used as a precious stone.

gauss is the older unit of measurement for a weak magnetic induction. 1 gauss $= 10^{-5}$ T (Tesla).

Gauss, Johann Carl Friedrich (1777–1855) was a German mathematician and astronomer. Of his many mathematical discoveries the method of least squares is of special significance in experimental science. He used this in wide ranging studies of the motions of celestial bodies. He made a number of contributions to magnetic studies and among other things he showed how to separate the Earth's magnetic field at the surface into an internal (intrinsic) component and an external component. This led to the identification of the location of the south magnetic pole. He was at one time Director of the Göttingen Astronomical Observatory and later established a Magnetic Observatory. He became interested in Geodesy and helped to establish a global distribution of observatories to study various properties of the Earth. He was one of the earliest people to accept the validity of non-Euclidean geometry. He was a wise financial investor. The unit of magnetic induction is named after him.

geodesy is the discipline of the determination of the geoid and the study of its properties. Although developed for the study of the Earth it is equally applicable to other bodies.

geoid is the shape (figure) the Earth would take up if its surface were entirely covered by water. The figure is determined by the gravitational field of the body and is a surface of equal gravitational potential (an equipotential surface). Although named after the Earth where all the early work was done the concept is equally applicable to other bodies.

geomagnetic equator is the line across the surface of a planet with a dipole magnetic field mid-way between the north and south poles.

geomagnetic poles are the locations where the intrinsic magnetic field direction is entirely vertical. There is one such point in the northern hemisphere marking the north magnetic pole and a corresponding point in the southern hemisphere marking the south magnetic pole. For the Earth they are each near the correponding geographical pole but this need not be so for other bodies.

geosynchrononous describes an orbit of a secondary body that is the same as the rotational period of the central, primary body. The Moon is in such an orbit about the Earth. It is the natural condition for small satellites about large bodies.

Gilbert, Sir William (1543–1603) is the founder of the study of magnetism and electricity. He was the first to realise that the Earth has an intrinsic magnetism that corresponds to that of a uniformly magnetised sphere. This is equivalent to the field of a bar magnet. The arguments to support his conclusion were included in his treatise *De Magnete* (1600), regarded as the first scientific treatise. The magnetic field of the Earth is now known not to be precisely that of a dipole field but is so to a reasonable approximation. He practices medicine and was appointed physician to Queen Elizabeth I in 1600.

granulation is the spotted appearance of the surface of the Sun. The photosphere is formed by the breaking of convection cells which rise from the deeper interior. Each cell has a characteristic dimension of about 10,000 km.

gravitation is an interaction between masses due simply to the fact that they have mass. In Newtonian mechanics it is a force. In general relativity it is the manifestation of the bending of space-time: mass causes space-time to be distorted and the distortion of space-time is recognised as gravitational mass.

gravitational energy is the ability of gravity to do work. The gravitational energy U of a mass m and radius r is $U = GM^2/r$ where G is the gravitational constant $= 6.67 \times 10^{-11}$ m^3/kgs^2 in the SI unit system.

gravitational force. In the Newtonian system the magnitude of the gravitational force F between two masses m_1 and m_2 at a distance r apart is expressed as $F = Gm_1\,m_2/r^2$. G is the constant of gravitation. The special feature about the law is the mutual relationship between the masses. The inverse square form for the distance simply relates to the expansion of a sphere. Although general relativity provides a more advanced description of gravity, the Newtonian form is always of sufficient accuracy for planetary work.

Great Black Spot is a dark oval covering the northern geographical pole of Jupiter. It was first observed by the Cassini spacecraft (on its way to Saturn) in 2002.

Great Red Spot of Jupiter is a barometric feature on the southern hemisphere of the planet. It is oval in shape with a length of about 10,000 km. It is a long life storm, first observed by Galileo in 1610 and mentioned continuously ever since. It still gives a stable appearance.

greenhouse effect is the raised temperature of the surface of a planet due to the presence of a natural atmosphere which traps infra-red radiation that would otherwise escape to space. It is a simple consequence of the presence of any atmosphere in the absence of living materials and can have beneficial consequences. For the Earth, it raises the surface temperature from below the freezing point of water (from about 254 K) to a little above it (about 290 K), allowing surface water to accumulate and so life to develop. The effect can be enhanced by the presence of particular atmospheric components. (see also **enhanced greenhouse effect**)

H

Hadley cell is the circulation (named after its discoverer) which arises in a planetary atmosphere due to a decrease of heating by the central star with increasing latitude in the atmosphere. Heated gas rises and moves towards the colder poles. There it releases its heat becoming heavier and descends towards the surface whence it returns near the surface to the equator.

haematite is a magnetic mineral being an oxide of iron with chemical formula Fe_2O_3.

Halley, Edmond (1656–1742) was regarded as a worthy companion of Newton in his day. He championed Newton's law of gravitation and predicted the reappearance of the comet that now bears his name — though he wasn't alive to enjoy the event. He made the first magnetic surveys of the Atlantic Ocean in the Royal Naval survey ship Paramour and produced the first magnetic charts. He was appointed Astronomer Royal in 1720, an appointment that he held for twenty two years, until his death.

heliopause is the boundary of the **heliosphere**. It is the location where the pressure of the interstellar medium balances exactly the pressure within the solar wind.

helioseismology is the study of the interior of the Sun by analysing the observed oscillations of the photosphere which originate in the convective zone. The wave pattern is influenced by the internal structure in a way that can be inferred. The method could ultimately be applied to the study of other stars.

heliosphere is the large region, which engulfs the planets of the Solar System and beyond, which is permeated by the plasma and magnetic field of the Sun.

Hipparchus of Nicaea (ca 190–ca 120 BC) was one of the most influential astronomers of antiquity. Although his childhood was spent in Bithynia, most of his life was spent in Rhodes. He wrote many works on astronomy, optics, mathematics, geography and astrology but all have sadly been lost. Ptolemy considered Hipparchus his most illustrious predecessor and made extensive use of Hipparchus' astronomical work in the *Almagest*. Hipparchus made extensive stellar observations between 147 and 127 BC, apparently producing a comprehensive star catalogue. He introduced the classification of stars into magnitudes according to their brightness. He found the distance to the Moon by parallax and discovered the precession of the equinoxes. He explained the different lengths of the seasons by offsetting the Earth from the centre of the planets' orbit. This extended the work of Apollonius of Perga's development of epicycles and eccentrics. Hipparchus used Babylonian astronomical materials linking Greek and Babylonian astronomy. His mathematical approach to astronomy suggests that he probably invented trigonometry.

Hipparchus space mission (1980s) of the European Space Agency used an automatic orbital spacecraft to measure accurately the positions and motions of the nearby stars. It provided unprecedented results which are now available as a library resource for astronomers.

Hirayama family is the group of asteroids which have similar orbital elements suggesting the members had a common origin at some time in the past.

Hooke, Robert (1635–1703) was a most gifted experimental English physicist and inventor. He was also an architect and friend of Sir Christopher Wren. He drew up a grand design for the rebuilding of London after the Great Fire of 1666 but his plans were not adopted. He did design the Monument to the fire in London. He has been greatly underestimated in the past. He was perhaps the greatest experimentalist of his time and fully balanced Newton who was perhaps the greatest mathematician.

Hubble space telescope is the 2.4 m (100 inch) remotely controlled reflecting astronomical telescope in a stationary orbit about the Earth. Being outside the Earth's atmosphere its seeing conditions are especially favourable. It is equipped to cover a wide range of the electromagnetic spectrum. It was constructed jointly by the American NASA and the European ESA and is associated with a controlling Hubble Space Institute.

Huygens, Christian (1629–1695) was an astronomer, physicist and mathematician of the Netherlands. He discovered the true nature of the rings around Saturn. He was a mathematician of great skill. He developed the wave theory of light in opposition to the corpuscular theory developed by Newton. An important invention of his was the pendulum clock with its controller. He was the son of the poet Constantijn Huygens.

hydrocarbons are compounds which include hydrogen and carbon. They are particularly associated with living materials but no entirely so.

hydrostatic equilibrium in a fluid is the condition of balance for a fluid particle between the fluid pressure and gravity.

hyperbola is a figure formed when a cone is cut by a plane making a larger angle with the base than does the side of the cone. It has two symmetrical separated parts called sheets. It can also be described as a figure with an eccentricity greater than unity. In physical applications interest is usually restricted to only one sheet. The curve describes the motion of a passing mass in gravitational interaction with another mass where the gravitational energy is less than the kinetic energy of the motion.

I

inclination of an orbit is the angle between the orbital plane and some reference plane. For a planet in orbit about the Sun this is usually the ecliptic. For a satellite is usually the plane of the equator of the parent planet.

ionosphere is the region of the atmosphere of a planet which is ionised by radiation (and especially the ultra-violet radiation) from the central star.

irregular satellite is one which has a highly eccentric and/or highly inclined orbit. The motion is often retrograde.

isostacy in fluid mechanics is the state in which an object floats in equilibrium in a fluid, the pressure from each side being the same. In geophysics, isostacy is the condition of a continental mass floating on the crust in equilibrium with the buoyancy force and the gravitational force are balanced precisely. This is a special form of the general statement of equilibrium.

J

Jeans mass is the mass of gas that is necessary for it to collapse into a condensed form.

jet stream is a high speed narrow stratosphere stream of air directed from west to east in the northern and southern atmospheres.

Jovian satellites are the major satellites of the planet Jupiter, Io, Europa, Callisto and Ganymede. They were discovered by Galileo soon after he turned his telescope to observe the heavens in 1609.

K

kamacite is an alloy of iron with a small quantity of nickel. It is sometimes found in meteorites.

Kelvin, William Thompson, Lord (1824–1907) was a famous Scottish scientist and inventor who developed electrical equipment and was a leader in the development of submarine telegraphy. His work covered a vast area of activities. He is especially remembered for his introduction of the dynamical theory of heat. He elucidated much of the difficulty associated with the 2nd law of thermodynamics and gave the prescription for defining absolute temperature. His contributions to thermodynamics are commemorated by naming the unit of temperature after him.

Kepler, Johann (1571–1630). Distinguished German astronomer who made accurate measurements of the motions of the planets which led him to formulate his three laws of planetary motion. He began working as a mathematical assistant to Tycho Brahe where he first learned of the planetary motions through the study of the planet Mars. Kepler made accurate tables of the motions of the planest known to him.

kerogen is an insoluble organic material sometimes found in rocks.

Kirkwood gaps. These are gaps in the asteroid belt. They occur where the orbital period of the asteroid according to Kepler's laws of planetary motion is a simple fraction of the rotation period of Jupiter. This shows the strong effect of Jupiter on the dynamics of the asteroid belt.

KREEP is a lunar mineral recognised first by Apollo astronauts. It is rich in potassium (K), the rare earth elements (REE) and phosphorus (P), hence the name.

Kuiper belt is an alternative name of the Edgeworth–Kuiper belt of object beyond the orbit of Neptune.

L

Lagrange Points. These are five special locations in what is called the restricted three body problem. For gravitation, they are points around two orbiting masses where a third, smaller, mass can orbit the larger masses at a fixed distance. Three (unstable) points, called L1, L2 and L3, lie on the line joining the two masses. Two stable points, L4 and L5, lie on the orbit of one mass about the other. Applied to the Earth-Sun system, the point L1 has been chosen as a point to "park" the Solar and Heliospheric Observatory Satellite SOHO between the Earth and the Sun. It was proposed to use the point L2 beyond the Earth away from the Sun to place the Next Generation Space Telescope which was to replace the existing Hubble Space Telescope. These two points are unstable on a time scale of about 23 days so satellites parked there require continual altitude and course corrections.

Lagrange, Joseph Louis, Comte (1736–1813) was a gifted French mathematician and astronomer. He is probably best known now for the gravitational Lagrangian points and for the differential equation that bears his name.

Laplace, Pierre Simon, Marquis de (1749–1827) was a great French mathematician and astronomer. He worked on the motions of the Solar

System and also contributed greatly to the theory of probability. He was called the Newton of France, with some justification.

Laplace plane refers to satellites and rings. The orbit plane is perturbed by several sources including the obliquity of the primary, the Sun and other satellites.

Legendre polynomial is the representation of an arbitrary curve on a spherical surface in terms of characteristic quantities.

libration is a small oscillation, or rocking, which is taken up by a synchronously rotating satellite if the orbit is slightly eccentric. The Moon shows such a motion which is the reason why some 54% of the surface is visible from the Earth rather than the 50% that would be expected without libration.

light year is an astronomical measure of distance. It is the distance travelled by light in a vacuum (that is empty space) in one year. Its numerical value is $ly = 9.42 \times 10^{26}$ m.

lithosphere is the strong upper surface layer of a planetary body. It includes the crust and the top part of the upper mantle.

lithophiles are chemical elements which tend to form silicate materials. Of special importance are silicon, oxygen, aluminium and magnesium.

Lorentz force acts on a charged particle moving in a permeating stationary magnetic field. A moving charged particle behaves like an electric current and this interacts with the magnetic field. The force acts perpendicular to both the magnetic field and the direction of motion of the charge (electric current). This leads to the so-called right hand rule. The interaction is truly three-dimensional, the force and motion not being in the same line as for an uncharged particle.

love wave is a surface shear wave excited by earthquakes or by large surface explosions.

luminosity is a numerical measure of the total quantity of radiation emitted by the body.

M

mafic is a mineral very rich in magnesium and iron.

magnetic equator is the line on a planet surface where a magnetic needle, suspended at its centre of mass, will hang exactly horizontal.

magnetic pole is the place on a planet surface where a magnetic needle suspended at its centre of mass will hang vertically.

magnetopause is the outer boundary of a planetary magnetosphere.

magnetosphere is the region around a planet in which the magnetic energy of the planet dominates that of the solar wind. The magnetosphere is encased in the magnetopause.

magnetotail is the region of the magnetosphere that is pulled downstream by the passage of the solar wind. It may reach a great distance downstream.

magnetosheath is the region between the bow shock of the planet and the magnetopause. The solar wind plasma flows round the magnetosphere in this region.

mantle is the part of the interior of a planetary body between the core and the crust.

mare is an area on the Moon which is a solidified flat lava flow. The same nomenclature is used for the surface of Mars but the region is not necessarily the result of a lava flow.

mascon is a sub-surface concentration of mass that provides a gravity anomaly outside. These were first recognised on the Moon but have also been found in other bodies.

Mercator, Gerhardus (1512–94) was a Flemish geographer who devised his projection to aid navigation. It is the representation on a 2-dimensional plane of details that are actually on a spherical surface. The meridians and parallels of latitude are straight lines at right angles to each other. Representation on a plane distorts details at the higher latitudes but leaves those at smaller latitudes essentially undistorted. For the Earth, as an example, the polar regions are represented as much larger regions than they actually are. Other similar representations have also been developed. For instance, the Peter projection shows countries according to their land area on the Mercator projection.

mesosphere is the region of an atmosphere between the stratosphere and the thermosphere. Its characteristic feature is a broad maximum of temperature.

Messier, Charles (1730–1817). French astronomical observer who was particularly interested in discovering comets. He worked at the Observatoire de Marine, Hotel de Cluny. In 1771, Messier was appointed Astronomer to the (French) Navy. He collected together a catalogue of objects that are not comets but which might be wrongly supposed to be. It contains 110 entries and was published in three stages in 1774, 1780 and 1781. Many of the objects later turned out to be galaxies but some are planetary nebulae and others large gas clouds (such as the Orion nebula). The catalogue is quoted now to identify the nebulae it was designed to help observers avoid.

meteor is a speck of dust from space which enters the Earth's atmosphere at very high speed. Friction raises its temperature so it can be seen. It is usually burned up very quickly although some reach the surface. They are usually at a height of about 60 km when they become visible. Meteors tracks are regularly studied using radar techniques.

meteor shower is an increased flux of meteors over a restricted time. It is caused by the Earth passing through micro-particles distributed along the tail of a comet. The orbit of the comet and that of the Earth have a fixed relationship which makes the meteor display occur at essentially the same time each year and to appear to arrive from the same direction. This gives rise to a number of meteor showers that can be seen, often to the best advantage in the early hours of the morning. They can be studied by radar techniques.

micrometeorite is a meteorite or a piece of a meteorite with a diameter less than 1 mm.

Mohorovičič discontinuity is the discontinuous boundary between the Earth's crust and mantle. It is some 30 km below the continental surface but less than 10 km below oceanic crust. It was named after its Jugoslavian discoverer who first recognised it in the seismic data traces for the earthquake of 1907. It is often called the Moho.

M-type asteriods are those composed principally of minerals with a large proportion of magnesium and iron, called mafic.

N

Newton, Sir Isaac (1642–1727) is generally regarded to have been the greatest man of science ever to have lived. His greatest achievements were probably the discovery of the optical spectrum, the invention of the calculus and (above all?) the law of gravitation. Perhaps the calculus (he called the subject fluctions) is his widest and longest lasting achievement although the modern notation is that of Leibnitz. He re-interpreted Kepler's 2nd law of motion as the conservation of angular momentum in an orbit but showed further that it applies to any central force towards a fixed point and not just to the inverse square law. His work was published in a tract *De Motu* (*Concerning the Motion of Bodies*) published in 1684. This work, together with his several other researches, were collected and published in his great treatise *Principia Mathematica* in 1687 which brings together the different aspects of the science of his day. It incorporates his book *Systems of the World* involving his thoughts on projectiles and the possibility of Earth orbits for small bodies. The publishing of *Principia* was actually instigated and financed by Newton's friend Halley (who eventually got his money back from the sales!). Newton was also interested in alchemy and especially with the transmutation of the chemical elements. Unfortunately, nuclear physics was unknown at that time. Very few areas of science are not indebted to him in some way or another. Apart from his scientific work Newton was Member of Parliament for Cambridge University in 1688 and became Master of the Mint in 1699. He was President of the Royal Society of London from 1703 until his death. He was knighted in 1705. He said himself that if he has seen further than other men it is because he had stood on the shoulders of giants. He did, in fact, live at the time of several men of great eminence. This has been referred to as a golden age of the intellect.

neutrino is an electrically uncharged elementary particle that has an essentially zero mass but carries energy and momentum. It travels with a speed close to that of light. There are some indications that it might, indeed, have a small mass (perhaps a few electron volts) but this is not yet clear. It has three variants together with anti-particles.

neutron is an electrically uncharged atomic particle and a constituent of atoms. Its mass at rest is 1.67482×10^{-27} kg and its radius of the order of 10^{-14} m. It is composed of three quarks (two down and one up) and so is not truly an elementary particle.

neutron star is a collapsed hydrogen star under such compression that the initial protons and electrons form neutrons as the main component of the star. It has a mass in excess of the solar mass but a radius of the order a few tens of kilometres.

nucleon is the collective name of the protons and neutrons constituting the nucleus of an atom. The nucleon is assigned a mass which is the mean of those of the proton and neutron and is usually abbreviated to $m = 1.67 \times 10^{-27}$ kg.

nucleus is the positively charged centre of an atom, composed of protons and neutrons. It contains essentially all the mass of the atom.

O

oblateness is the distortion of a rotating body which is spherical when stationary. The centrifugal force extends the equatorial region and contracts the polar region. The change due to the rotation is the difference between the equatorial and the polar radii. The oblateness is this difference divided by the equatorial radius. This is zero for a truly spherical body but has a positive value if the spherical shape is lost.

occultation is the passage of one body in front of another of smaller apparent angular size. One example is the passage of the Moon in front of a star and another is the passage of an asteroid in front of a star blocking it temporarily from view.

OGLE is derived from the title **O**bservational **G**ravitational **L**ensing **E**xperiment. This is designed to identify companions orbiting solar type stars by the transits of the star.

olivine is a silicate mineral rich in metals. It is common in the Earth and Moon and in meteorites. It is believed to be common also in the other terrestrial planets. Its chemical formula can be written $(Mg, Fe)_2SiO_4$.

Omar Khayyam (1048–1123 AD) was born in Naishápúr, Persia (now Iran). He was an outstanding astronomer, mathematician and philosopher/poet. He was for some years head of the astronomical observatory at Esfaham. He was one of eight learned scholars who successfully reformed the calendar in 1079 to produce the Jalaliera calendar. It surpassed the Julian and approached the accuracy of the Gregorian style. He became interested in the solution of cubic equations (and particularly of the equation $x^3 + 200x = 20x^2 + 200$), but was restricted in his work by not knowing imaginary numbers. He wrote the successful *Treatise on Demonstrations of Problems of Algebra* which included his own work. He summarized his views of the creation of the world and life within it in the well known collection of 101 four line verses called the *Rubaiyat*. It was rendered into English verse by E. Fitzgerald (5th edition (final) 1889).

orbital elements are the five quantities that define the details of an orbit. For an elliptic orbit these are the semi-major axis, the eccentricity, the orbital inclination, the longitude of perihelion and the longitude of the ascending node.

orbital inclination is the angle between the orbital plane of a body and some chosen reference plane. For a planetary orbit the reference plane is the ecliptic plane.

Oort cloud is a broadly spherical region beyond the orbit of Neptune containing a very large number of small icy bodies. It may extend out as far as 100,000 AU from the Sun. It is believed to be the source of long period comets (periods at least 200 years) and of new comets. It is still regarded as hypothetical by some astronomers.

opacity is the ability of a material to block electromagnetic radiations. It can arise by the scattering of photons by electrons (**electron opacity**) or by the absorption of photons.

optical depth is a measure of the ability of a medium to absorb electromagnetic radiation that passes through it. The optical depth is zero for a transparent medium. The medium is said to be optically thin if some radiation can pass through (optical depth < 1) and optically thick if the optical depth > 1.

organic compounds contain carbon and hydrogen. They need not be associated with living materials.

orthopyroxine is a mineral related to olivine with the chemical formula $(Mg, Fe)SiO_3$.

outgassing is the release of gases from the interior of a planet. This is most often associated with volcanic activity but not necessarily so.

P

paleomagnetism is ancient magnetism retained in rocks. It records the magnetic conditions of the environment at the time the rock was laid down (sedimentary rock) or at the time an igneous rock cooled.

palimpset is a closely circular ring on a planet surface that is thought to be the remnants of an earlier crater or rim. It is particularly applied to icy satellites.

Pangea is a hypothetical continent that contained all the land mass before the Triassic period, 225 million years ago. This land mass subsequently broke up, presumably due to plate tectonic forces, to give eventually the distribution of land masses seen today. The disintegration led to larger ocean rims and therefore greater possibilities for marine evolution. The evidence for the movements of the continents comes largely from paleomagnetic studies.

parabola is the figure formed by the intersection of a cone with a plane parallel to its side. If the parabola is rotated about its axis it forms a **paraboloid**. It can be described as a figure with unit eccentricity.

parallax is the angular measure of the apparent displacement of an object brought about by the actual change of the point of observation. The effect is used as a method for deducing the distance of an object using similar triangles.

parsec (pc)s a measure of stellar distances. It is the distance at which a star would have a parallax of 1 sec. of arc. It is 3.09×10^{16} m or 2.53×10^5 AU. It is alternatively 3.27 light years (ly).

perihelion is point of the elliptical path which is closest to the star at the focus being the centre of the gravitation force between star and planet.

phase angle (solar) is the angle of a surface between that of an observing instrument and the Sun.

photometry is the study of the brightness of objects over a range of wavelengths. Modern methods in astronomy involve automatic space vehicles which can allow for observations outside the Earth's atmosphere.

photon is the particle of electromagnetic radiation. It is the carrier of the electric charge.

photosphere is the visible surface of the Sun. It is intensely bright and **should <u>never</u> be viewed directly by an observer either without or with optical aids. The image should always be projected onto a screen.**

plagioclase is a common silicate rock forming mineral involving sodium, calcium, aluminium and silicon. Its chemical formula can be written (Na, Ca)(Al, Si)$_4$O$_8$.

Planck constant of action $h = 6.62 \times 10^{-34}$ Js. It was first defined by his quantum relation between the energy, E, and the frequency, ν, of electromagnetic black body radiation $h = E/\nu$.

planet is a collection of matter under a mechanical equilibrium formed between its self-gravity (the sum of the gravitational attractions between all its component unionised atoms), the electrical attractions between its nuclei and electrons, and the degenerate forces associated with electrons alone. All other forces have only a secondary effect. The result is a set of essentially spherical bodies for which mass M and radius R satisfy the constraint MR^{-3} = constant. There is a minimum mass dictated by the minimum value for gravity, and a maximum mass dictated by the atoms remaining unionised. The radius suffers a maximum value at the maximum mass.

planetary motion, laws of Kepler are a summary of his observations of the motions of Earth, Mars and Jupiter. It became clear later when our knowledge of mechanics had grown that the laws are expressions of three statements. The elliptic motion of a single planet around a more massive star proceeds under the constraints of the conservation of energy and angular momentum ("equal areas in equal times") and the inverse square force of gravity centred on the star and acting between the star and the planet. The motions of two planets around the central star are perturbed to a small extent through gravity. The consequence is that neither planet follows a closed elliptical path exactly although the difference may be very small.

planetesimal is a free floating silicate/ferrous body of no greater than 1 km diameter. It is believed by some workers that the Solar System formed by the accretion of very large numbers of these elementary bodies which formed the initial condensation of a gas cloud.

planetology is the study of the physics and chemistry of the planets.

plasma is a completely ionised gas. It is composed of electrically positive ions and negative electrons in equal numbers to give overall electrical neutrality.

plasma tail is a narrow tail associated with comets swept away from the direction of the Sun by the solar wind. It is presumed the same tails will exist in other exo-systems.

plate tectonics is the study of the gradual movement over the surface of the Earth of continental sized sections of the lithosphere. The energy to drive the movements comes from convection in the mantle.

polarimetry is the study of the polarised light coming from other celestial bodies.

polarisation is the preferential direction of the vibrations of the electromagnetic radiation. Reflection at a surface acts towards the lining up of the vibrations in a direction parallel to the surface. Specific physical processes tend to lead to specific patterns of polarisation and this effect is much used in astronomy.

polaroid is an active transparent medium which allows only electromagnetic waves with a preferential plane of vibration to pass through.

Poynting-Robertson effect is the drag experienced by a small body due to asymmetric scattering of radiation. The radiation is first absorbed and then re-emitted in a different direction.

power is the rate of performing work. The unit of power is the Watt, and 1 W = 1 Joule/second.

Precambrian is the largely unknown era encompassing the earliest history of the Earth. It lasted from the formation of the Earth 4,550 million years ago until about 570 million years ago. The detailed knowledge of the geology of the Earth is seen to involve relatively recent times.

precession is the slow periodic motion of the rotation axis of a body. The rotation axis forms a hypothetical cone in space.

principle wave in seismology is the P-wave or the longitudinal (sound) wave mode propagated within the body due to a discontinuous release of energy (an earthquake on Earth). It has a higher speed of propagation than accompanying the S-wave, or secondary wave.

primordial is that which existed soon after the formation of the Solar System. The System (including the Sun) is believed to have formed as a single unit during the period 5,000–4,550 million years ago.

prograde motion of a planet about a star is an anti-clockwise motion of the planet as viewed looking downwards at the pole.

proper motion of the stars describes the observation that the stars are not stationary as imagined in antiquity but have motions within the galaxy separate from the galactic rotation. The motions are, however, generally sufficiently small to appear negligible at the large distances from the Earth. This allows the constellations to remain stable over long periods of time as viewed from the Earth.

proton is an electrically charged particle with a mass at rest $m_P = 1.67252 \times 10^{-27}$ kg and a positive charge of 1.60210×10^{-19} coulombs (same magnitude of the charge on an electron). It is composed of three quarks (two up and one down) and so is not an elementary particle. It is a constituent of atomic nuclei. Its mass is slightly lower than that of the neutron.

protoplanet is a body in the processes of forming a planetary body.

Ptolemy, Claudius Ptolemæus (AD 85/90–165) of Alexandria. He was an astronomer and geographer of great distinction in antiquity. He is remembered now especially for his astronomical works including the *Almagest* which catalogued the positions of many stars. Tycho Brahe in the 17th century found that some stars had moved from the position recorded in the *Almagest* leading eventually to the recognition of the proper motion of the stars. Ptolemy taught that the Earth is stationary with the Sun and the heavenly bodies rotating around it, the planets being nearer than the stars that are located on the celestial sphere. This interpretation lasted for a thousand years and carries his name. His treatise on geography called *Geographical Outline* is regarded as a great work.

P-wave or primary wave is a seismic wave resulting from an earthquake disturbance. It is a longitudinal wave like a sound wave and passes freely through solid and liquid materials.

pyrolite is a mineral mainly composed of the oxides of silicon and magnesium (to about 83%). Other impurities include the oxides of iron, aluminium, and calcium together with lesser proportions of other elements.

pyroxene is a silicate mineral having a 1 to 1 ratio of a metal oxide to silicon oxide. The metals can be magnesium, calcium or iron.

Pythagoras (c.582–c.507) was a Greek mystic and mathematician. He was born on the island of Samos but moved to Croton in middle age. He died in exile. He believed in a harmonic structure for nature which was an early recognition of the power of harmonic series. He is well known for his work in geometry and especially for his famous theorem for the right angles triangle. His views on "the harmony of the spheres" can be thought of as having some links with the philosophy of modern string theory.

Q

Quaoar (pronounced KWAH-o-wahr) is a trans-Neptunian object of the Edgeworth–Kuiper belt discovered in 2002. It is similar in size (and presumably mass) to Charon, the satellite of Pluto. The name is that of the God of creation of the native American Tongva tribe that initially inhabited Southern California.

quartz is a commonly occurring mineral and is an oxide of silicon with the essential chemical formula SiO_2.

quark is a sub-atomic particle which forms the deeper structure of protons and neutrons and so of matter. There are many forms, none of which are found to occur free.

R

radiation pressure is the small force exerted by photon bombardment of a surface.

radiative transfer of energy is that brought about by electromagnetic radiation without the assistance of matter.

radioactivity is the spontaneous emission of material particles and electromagnetic radiation. The material particles can be electrons, or helium nuclei (α-particles), or positrons. The electromagnetic radiation is generally of high energy such as gamma ray photons. The effect is to transmute an initial atom into a different atom, usually of lower atomic mass. Such materials are said to be radioactive. The radioactivity may occur naturally as in nature, or it might be artificially induced by irradiation by neutrons in a nuclear reactor.

radiogenic heat is the heat produced by radioactive elements in their decay. The heat from potassium, uranium, thorium and other elements are all of importance in various planetary thermal processes, particularly early on in its history.

radionulceide is a radioactive isotope of an element.

rarefaction is a decrease in the density and pressure due to the passage of a compressional (push-pull) wave, such as a sound wave or seismic P-wave. The effect can be severe for waves emitted by a major energy release.

Rayleigh, 3$^{\rm rd}$ Baron (1842–1919) was one of the most eminent British physicists of his time. He was the co-discoverer of argon with Sir William Ramsey. He made numerous contributions to physical theory especially in the area of wave motions and vibrations. He was awarded the Nobel prize for physics in 1904.

Rayleigh scattering of radiation is the scattering of radiation by particles small in comparison with the wavelengths of the radiation present. It is this scattering of sunlight that produces the blue colour of the sky.

Rayleigh wave is a surface wave excited by earthquakes or by large surface explosions. It can be viewed as the analogue of the surface wave in hydrodynamics.

refractory refers to a chemical element that vapourises at high temperatures.

regolith is the highly fragmentary rock that forms the uppermost surface of some planets. It is caused by the continual bombardment by meteors of all sizes over large time periods. Large areas of the surface of the Moon are covered in such materials. It can be rearranged by the bombardment by solar wind particles and is called lunar gardening.

resonance is a state in which energy is passed from one object to another by continual periodic perturbations. In planetary science it refers to the gravitational perturbation of the motion of one orbiting body by another.

retrograde refers to an orbital or to a rotational motion which is of opposite direction to those of other related objects. For instance, the more important of the 57 satellites of Jupiter have orbits which have the same direction as the rotation of the planet but many of the smaller ones (and especially with larger orbital radii) move in oppositely directed orbits. It is generally supposed that these satellites have been captured from outside.

rille is a trench-like valley on the surface of the Moon and other objects. They can be several hundred kilometres long and several kilometres wide.

Roche limit is the critical distance below which an idealised orbiting object with no tensile strength is broken up by the gravitational tidal force of the parent body. It is believed that Saturn's ring system originated in this way. Bodies with sufficient tensile strength are not affected. The effect becomes weaker the smaller the object. Thus, orbiting artificial spacecraft are not in danger of being destroyed by this process.

rock is an assemblage of one or more minerals in solid form.

Rydberg constant has the value $R_\infty = 1.0973731 \times 10^7$ m^{-1}.

S

S-wave in seismology is the secondary transverse wave emitted by an earthquake or any similar explosive release of energy (for instance, an underground nuclear explosion). It propagates only in regions able to sustain shear, that is in solid materials.

S-type asteroid is located in the inner part of the main asteroid belt. It is, therefore, a very common type of asteroid.

scalar is a quantity that has magnitude only but no direction. Temperature is such a quantity. (see also vector)

scale height in an atmosphere is the increase in altitude necessary to decrease the density by the factor e, that is 2.7183.

Sedna is the most distant plenetoid yet found, lying well beyond the orbit of Pluto. Its orbit is extremely elliptic apparently lying between 80 AU and 500 AU from the Sun and with an orbital period of about 10,000 years. It has a probable radius of about 800 km. Surprisingly it shows a distinctly red surface colour rather than the white of water ice that might have been expected. It appears to rotate in about 10 hours. It is not known whether other similar bodies are in orbit in this extreme outer region of the Solar System.

seismic tomography is a technique which allows three-dimensional maps to be made of the interior of a solid planetary body. It requires a considerable body of seismic data and considerable computational power to give even modest detail. The essential feature is to determine local deviations from a standard profile.

semi-major axis is half the longest diameter of an ellipse. In planetary problems it defines a mean value for the size of the orbit. For the Solar System, where the orbits are closely circular, it is essentially the mean radius of the orbit.

shepherd satellite is one that sustains and protects the structure of a planetary ring. It does this through its gravitational influence.

shield volcano is a volcano built up from the repeated non-explosive lava eruptions of a volcano. These tend to be large structures. Examples are the Hawaian Islands on Earth and the Olympus Mons on Mars which is the largest volcano in the Solar System.

short period comet has an orbital period of less than 200 years.

sidereal year is the orbital period of a planet measured with respect to the distant stars. For the Earth, this is 365.2564 days.

siderophile is a chemical element which occurs often in the simple metallic state. Iron is an example.

silicate is a rock/mineral with a crystalline structure composed largely of silicon and oxygen atoms bonded to other elements.

SNC meteorites are three types of body composed of shergottites, nakhlites and chassignites believed to have been ejected from the surface of Mars some millions of years ago. It is thought that this followed impacts on Mars. It is important because it suggests the possibility of simple communication between the planets which could have carried elementary living cells between them.

SOHO is the **SO**lar and **H**eliographic **O**bservatory, a joint ESA/NASA enterprise launched in 1996. It is located in a halo orbit around the L1 Lagrangian point, 1.5 million km sun ward of the Earth. It is therefore able to observe the Sun continuously for the first time. In this way it is possible to gain more knowledge about the interaction between the Sun and the planets. It carries 12 scientific instruments to measure things such as the properties of the corona to the vibrations of the surface which indicate vibrations deep inside. These in turn allow the details of the internal structure to be inferred.

solar activity describes a variety of phenomena that vary with a period of about 11 years. These include sunspots, flares, mass ejections and other features.

solar constant is the total solar radiation received by a unit surface perpendicular to the direction of the radiation in unit time (the radiation flux). For the Earth the mean distance is 1 AU and the solar constant is taken to be 1,367 watts per square metre. Studies from the *Helios* satellite observatory have shown that the solar radiation is subject to small changes.

solar day is the period of rotation of the Earth measured against the position of the Sun. The solar day varies slightly.

solar flare is a sudden release of energy in the region of the Sun's photosphere. One consequence is the acceleration of charged particles and magnetic field into space.

solar nebula is a hypothetical disc-shaped collection of gas and dust which condensed to form the Solar System. While such an origin would provide many of the observed features of the Solar System, there are many features that cannot obviously be accounted for by this beginning.

solar prominence is a cloud of gas from the solar surface held just above the surface by the solar magnetic field. The scale is often huge in terms of the Earth's diameter. A prominence appears dark against the solar surface because its temperature is slightly lower than that of the surface. A typical prominence may last for about 1 month. It might develop into a coronal mass ejection.

solar wind is the high speed outflow of energetic charged particles and entrained magnetic field that is ejected into space from the solar corona. Such ejections are known to be made from other stars and it is believed to be a common cosmic feature of stellar objects.

solstice is the position of the Earth in its orbit round the Sun where the plane of its axial inclination passes through the Sun. There are two such positions giving the summer and winter solstices. Because of the inclination, the zone near one pole will remain continuously in sunlight while the other pole will be constantly in the dark. The parallels limiting these zones, which extend down from each pole by the angle of inclination, are respectively the Arctic and Antarctic circles which are at latitudes $66°33'$ north and south. The hypothetical line joining the positions of the two solstices subtend an angle of $10°$ with the major axis of the orbit.

spectroscopy is the study of the light associated with a body. It may involve emitted light or reflected light.

spectrum is the range of frequencies associated with a particular sample of radiation (colours of the rainbow for visible light) found by spectroscopy. The frequencies may be emitted or absorbed in which case particular frequencies are missing (appear black) in the total spectrum.

speed of light. The speed of light in a vacuum is a constant of nature. Its numerical value is $c = 2.997925 \times 10^8$ m/s. In many calculations it is sufficient to suppose the speed to be 3×10^8 m/s.

spherule is a small spherical particle composed of rock and iron oxide. It is formed in the molten state from a meteorite as it enters the Earth's atmosphere.

Stefan-Boltzmann radiation constant $\sigma = 5.67 \times 10^{-8}$ W/m^2 K^4.

stishovite is a very dense form of silica produced in the very high pressures which occur during a meteorite impact.

stratosphere is the layer in a planetary atmosphere where the temperature remains approximately constant with increasing altitude. For the case of the Earth it is below the ionosphere but above the troposphere.

subduction is the process where one surface lithospheric plate is forced under another. It is the mechanism by which surface rocks are returned to the upper mantle.

sunspot is a relatively small, cool and dark area on the solar photosphere. The area of the photosphere covered by sunspots varies throughout the solar cycle being largest at the maximum of solar activity. Sunspots are also associated with magnetic activity.

super planet is a body beyond the mass where the constituent atoms become ionised but below the mass where the first thermonuclear process begin. This concept is usually restricted to hydrogen which places the mass between the maximum mass of a planet and the minimum mass of a brown dwarf. The case of bodies composed on other materials, such as silicates or ferrous materials, has not generally been considered.

super-rotation is the feature of an atmosphere which rotates faster than the planetary surface supporting it. This is the case for Venus as one example, where the surface rotates in 243.02 days but the top of the atmosphere rotates once every 4 days.

synchronous orbit is one for which the period of the orbit about the gravitational centre is equal to the rotation period of the body about its axis.

synchrotron radiation is the electromagnetic energy emitted when a charged particle is accelerated about a magnetic field line.

synodic period is the time between similar conjunctions of bodies. This might involve orbiting bodies relative to the Sun such as times between consecutive oppositions.

T

tectonism is related to the forces acting on the crust of a terrestrial type planet.

tektite is a small piece of silica glass. It is formed during an impact on Earth.

terrestrial planets are those within the asteroid belt and are Mercury, Venus, Earth and Mars. The term terrestrial is also often used to refer to features of the Earth and the term terrestrial planets can also mean the planets associated with the Earth.

tesla the S.I. unit of magnetic induction.

thermosphere is the upper region of a planetary atmosphere where the temperature increases with altitude due to heating by the star.

tidal heating is the internal heating of a satellite due to the action of tidal forces within it caused by the very strong gravitational field of the primary planet and perhaps other satellites. An excellent example is Io which is heated internally by the tidal action due to Jupiter and other Galilean satellites.

Titius–Bode law is a numerical series giving the approximate orbital distances of the planets of the Solar System. A similar series can give the approximate distances of the four Galilean satellites from the parent planet Jupiter. There have been many attempts to provide a mathematical explanation for the Titius–Bode law but, so far, none have succeeded.

transfer orbit is an elliptical orbit linking the orbits about two planetary bodies. It is of special use in allowing the transfer of a space vehicle between planets to be made in an energy efficient way.

Trojans are asteroids located at the two stable Lagrangian points L4 and L5 of the orbit of a planet about the Sun. The name Trojan comes from the names of the three largest asteroids Agamemnon, Achilles and Hector in Jupiter's orbit. Other Trojans orbit with Mars. Several of Saturn's moons have got Trojan companions. Large concentrations of dust have recently been found at the Trojan points of the Earth-Moon system.

tropical year is the time interval in a planet's orbit between successive vernal equinoxes. The precession of the axis of the body causes this year to differ from the sidereal year.

Tropics of Cancer and Capricorn of the Earth are parallels lying at angles $23°27'$ (being the inclination angle) north (Cancer) and south (Capricorn) of the equator. At the appropriate solstice, the Sun is overhead at all points along these parallels.

tropopause in the Earth's atmosphere is the boundary between the troposphere and the stratosphere.

troposphere is the bottom of the Earth's atmosphere where the everyday weather takes place. In other planets it is the region dominated by convection.

U

uncompressed mean density is the mean density of material at zero pressure which is normally compressed, for instance by being located deep in the interior of a planetary body.

universal constant of gravitation $G = 6.67 \times 10^{-11} \ \mathrm{m^3/kgs^2}$.

V

Van Allen radiation belts are two doughnut shaped plasma belts about the Earth's equator within the magnetosphere. Similar belts will be found around other planets with an atmosphere.

vector is a quantity having magnitude and direction. It is often represented by an arrow where the length indicates the magnitude and the direction of the arrow the direction of the quantity.

vis viva is an energy-like quantity formed by multiplying the mass by the square of the speed and is used as a supplement to the force in early studies in mechanics, especially by Leibnitz and Huygens. The modern kinetic energy is one half the vis viva.

virial expansion is a correction of the ideal gas law to include the effects of increasing pressure. The equation of state has the form $PV = RT \ [1 + BP$

$+ CP^2 + DP^3 + \cdots$]. B is called the second virial coefficient, C the third virial coefficient and so on. The terms account for the interaction between the molecules of the gas in an ordered way. B accounts for the interaction between pairs of molecules, C accounts for the interaction between triplets, and so on. It is rarely sensible to go beyond the quadruplet term even for relatively dense gases. A comparable expansion can be developed in inverse powers of the volume. The word virial is related to word force.

volatiles are chemical elements or molecules with low melting temperatures. Examples are water, ammonia, sodium and potassium.

volcano is a hill or mountain through which molten materials and gas can issue to the atmosphere from the interior of the planet. The largest volcanoes in the Solar System are on Mars or Venus.

volcanism is the term describing the action of volcanoes.

W

wandering stars. The vision of the heavens to the ancient astronomers (without telescopes of course) was some 5,000 stars which appeared fixed in the sky and five stars that moved, together with the Sun and Moon. The fixed stars they set into constellations which remain today much as they were then. The five moving points of light they called wandering stars — we now know them as the planets.

white dwarf is a body which results when the central hydrogen content has been transformed into helium. The equilibrium radius of the body is a balance between the inward force of self-gravity opposed by electron degeneracy force. There is no internal energy source and the body slowly cools to the temperature of its surroundings. It has a mass of the order of that of the Sun but a radius comparable to that of the Earth.

Y

year (standard) is taken to be 365 days except for the leap year when it is 366 days. This differs slightly from the other astronomical years.

Z

zonal jets are high speed winds in an atmosphere which move in a limited range of latitude. They are very evident on Jupiter and to some extent on Saturn.

zodiacal light is the faint glow in the sky seen near dusk and dawn. It is caused by light scattered off interplanetary dust near the plane of the ecliptic.

Problems and Solutions

Problem 1.1. Let T be the time for a planet to make one orbit round the Sun (the orbital period) and a the mean radius of its orbit (semi-major axis). Using the data for the planets of the Solar System, calculate T^2 and a^3 for each planet and derive the ratio $T^2/a^3 \approx$ constant. What conclusions can you draw? (Hint: it is convenient to express the orbital period and semi-major axis of each planet in terms of those for the Earth).

Problem 1.2. Figure I.8 shows that the variation of the apparent diameter of the Sun as viewed from the Earth during one calendar year is about 3.8%. In terms of the elliptic orbit about the Sun (shown in Fig. I.12a), this refers to the different appearance at perihelion and aphelion. With this information, find the ellipticity of the Earth's orbit. Compare your result with the value quoted in the Standard Tables of Constants. Is it possible to gain any knowledge of the semi-major axis of the orbit?

Problem 1.3. A planet of Earth-type mass 6×10^{24} kg orbits a star of solar-type mass 2×10^{30} kg with an orbital radius of 1 AU $= 1.49 \times 10^{11}$ m. What is the strength of the gravitational force between the two bodies? What is the acceleration of the planet caused by the central star? What is the acceleration of the star caused by the orbiting planet? ($G = 6.67 \times 10^{-11}$ m^3/kgs^2).

The acceleration of a body in an orbit of radius r with speed v is v^2/r. What is the acceleration of the Earth-type planet in orbit about the solar-type star? How does this value of the acceleration for the Earth-type planet correspond to the acceleration imparted by the Sun?

Solution 1.1.

The ratio is the same throughout the Solar System, including the distant orbiting bodies. Orbits of Edgeworth-Kuiper objects calculated from Kepler's laws can be assumed reliable.

Planet	$a(E = 1)$	$T(E = 1)$	a^3	T^2	a^3/T^2
Mercury	0.3871	0.24085	0.058	0.058	1.0
Venus	0.7233	0.61521	0.378	0.378	1.0
Earth	1.0	1.0	1.0	1.0	1.0
Mars	1.5237	1.881	3.537	3.538	0.9997
Jupiter	5.2028	11.8623	141.52	140.71	1.006
Saturn	9.5388	29.458	867.92	867.77	1.00(01)
Uranus	19.1914	84.01	7068.38	7057.68	1.001
Neptune	30.0611	164.79	27165.31	27155.74	1.000
Pluto	39.5294	248.54	61767.59	61772.13	0.9999(3)

Solution 1.2.

Referring to Fig. I.12, suppose the Sun is at the focus F_1. It follows that the distance of the Earth from the Sun at perihelion is $F_1 A = a - ae = a(1 - e)$. If the apparent diameter of the Sun then is D_p (remembering that the size of a body appears smaller the further it is away) we can write $a(1 - e) = A/D_p$ where A is some constant. The distance of the Earth from the Sun at aphelion is $F_1 B = a + ae = a(1 + e)$ and the apparent diameter is D_a so that correspondingly $a(1 + e) = A/D_a$. The ratio of the maximum apparent diameter to the minimum apparent diameter of the Sun is, from Fig. I.8, $D_p/D_a = 10.71/10.32 = 1.038$. Dividing the aphelion condition by the perihelion condition and canceling A and a, we can write $(1 + e) = 1.038(1 - e)$. This gives $e = 0.038/2.038 = 0.0186$ for the orbital eccentricity of the Earth. The value quoted in the tables is 0.017, a difference of about 9.4%. The actual orbital distance cannot be found this way because the size of the Sun is not given by the observations. If it were, the calculated orbital eccentricity would allow the solar distance to be found. One of the annoyances of astronomy is that actual distances are difficult to find but relative distances are easier.

Solution 1.3.

$$F = \frac{GMm}{R^2} = \frac{6.67 \times 10^{-11} \times 2 \times 10^{30} \times 6 \times 10^{24}}{(1.49)^2 \times 10^{22}} = 3.605 \times 10^{22} \text{ N}.$$

The acceleration imparted to the Earth is:

$$a_e = \frac{3.605 \times 10^{22}}{6 \times 10^{24}} = 6.01 \times 10^{-3} \text{ ms}^2.$$

The acceleration imparted to the Sun is

$$a_S = \frac{3.605 \times 10^{22}}{2 \times 10^{30}} = 1.303 \times 10^{-6} \text{ ms}^2 \, .$$

For the Earth-type body in orbit, the velocity is

$$v = \frac{2\pi R}{T} = \frac{2\pi 1.49 \times 10^{11}}{3.154 \times 10^7} = 2.968 \times 10^4 \text{ ms}^{-1}$$

$$a = \frac{v^2}{R} = \frac{(2.968)^2 \times 10^8}{1.49 \times 10^{11}} = 5.91 \times 10^{-3} \text{ ms}^{-1} \, .$$

The acceleration imparted to the Earth by the Sun is the same as that of the planet in its orbit and each is about 6×10^{-3} ms^{-1}. The values agree to some 3% in this calculation.

The planet orbits the Sun in order to oppose the acceleration towards the Sun. The energy and angular momentum of the planet in orbit must be maintained and as the planet falls towards the Sun it also moves forward by just the amount to keep orbital characteristics constant. It is this balance between the energies of gravitation and motion that determines the semi-major axis of the orbit for a given orbital speed.

Problem 2.1. The Earth takes 365 days to orbit the Sun at a distance of 1 AU $= 1.49 \times 10^{11}$ m. If the value of the gravitational constant is $G = 6.67 \times 10^{-11}$ m^3/kgs^2 calculate the mass of the Sun. What is the greatest source of error in such a calculation?

Solution 2.1.

From (2.13), $M = 4\pi^2 a^3 / GT^2$ so that $M = 1.985 \times 10^{30}$ kg. The major source of error is the value of G which is still not known to better than about 1 part in 600. To account for this, planetary masses are often expressed in terms of the product GM. For the Sun, $GM = 1.324 \times 10^{20}$ m^{3}s^{-2}.

Problem 3.1. Calculate the transfer data for Earth to Pluto (orbital radius 39.5294 AU) as a Edgeworth-Kuiper object. What conclusions can you draw about the use of transfer orbits in the exploration of the outer Solar System?

Solution 3.1.

The controlling equation for the orbit is

$$\frac{1}{2}mv^2 - \frac{GM_Sm}{r} = -\frac{GM_Sm}{2a}, \tag{3.1}$$

where the symbols have the meaning of Section 3.4.

The orbital distance of Pluto from the Sun is 39.53 AU so that $a = 40.53$ in (3.1). The velocity of the body at the Earth is

$$\frac{1}{2}mv_T^2 - \frac{GM_Sm}{r_E} = -\frac{GM_Sm}{40.53r_E}$$

so that

$$v_T^2 = 1.951\frac{GM_E}{r_E} = 1.951 \cdot v_E^2 = 5.755 \times 10^4 \text{ m/s}.$$

The move to the transfer orbit means the speed of the craft must be increased by

$$(5.775 - 2.95) \times 10^4 = 28.1 \text{ km/s}.$$

The craft moves in the transfer orbit to Pluto according to the expression for the conservation of energy, with v_P being the speed at the Pluto orbit

$$\frac{1}{2}mv_P^2 - \frac{GM_Sm}{39.53} = -\frac{GM_S}{40.53}$$

$$v_P^2 = \frac{GM_S}{r_E}\left(\frac{2}{39.53} - \frac{2}{40.53}\right)$$

$$= 1.24 \times 10^{-3}\frac{GM_S}{r_E}$$

$$= 1.24 \times 10^{-3}v_E^2$$

$$v_P = 0.035v_E = 1.03 \text{ km/s}.$$

The orbital speed of Pluto is 4.74 km/s so the speed of the craft must be increased by the amount $4.74 - 1.03 = 3.44$ km/s.

The ratio of the times for the orbit is, from Kepler's third law,

$$\frac{T_P}{T_E} = \left(\frac{40.53}{2}\right)^{3/2} = 91.16$$

so that $T_P = 91.16$ years. But this is for a full orbit: the journey to Pluto is half this and so is 45.58 years.

The journey times to the outer Solar System are impracticably large. To make them shorter more direct paths would be necessary but this would require the expenditure of unrealistic quantities of energy.

Problem 4.1. Calculate the maximum radius and corresponding mass for a planetary body made of (i) helium ($A = 4$, $Z = 2$), (ii) silicate materials ($A = 30$, $Z = 15$) and (iii) ferrous materials ($A = 56$, $Z = 28$).

Problem 4.2. Let H be the highest equilibrium elevation of a mountain on a terrestrial-type planetary surface while g is the acceleration due to gravity at the surface. Show that H and g satisfy the relation $Hg = $ constant. How do the predictions of this relationship compare with the measured data for the terrestrial planets? What special information can be obtained about the dynamic structure of the near surface from a quantitative application of your arguments.

Solution 4.1.

(i) For helium,

$$R(\text{max}) = 1.47 \times \frac{2^{2/3}}{4} R_J = 0.56 R_J = 3.98 \times 10^4 \text{ km}$$

$$M(R\text{max}) = 2.1 \times 10^{27} \frac{2^3}{4^2} = 1.05 \times 10^{27} \text{ kg}.$$

(ii) For silicates,

$$R(\text{max}) = 1.47 \times \frac{15^{2/3}}{30} = 1.47 \frac{6}{30} R_J$$

$$= 1.47 \times 0.2 R_J = 2.10 \times 10^4 \text{ km}$$

$$M(R\text{max}) = 2.1 \times 10^{27} \cdot \frac{15^3}{30^2} = 7.87 \times 10^{27} \text{ kg}.$$

(iii) For ferrous

$$R(\text{max}) = 1.47 \times \frac{28^{2/3}}{56} R_J = \frac{9.22}{56} \times 1.47 R_J = 1.73 \times 10^4 \text{ km}$$

$$M(R\text{max}) = 2.1 \times 10^{27} \frac{28^3}{56^2} = 1.47 \times 10^{28} \text{ kg}.$$

Solution 4.2.

The mountain is maintained by the strength of the material below. The maximum height occurs when the ground underneath cannot support even the addition of one single additional molecule at the top of the mountain. The effect of adding such an atom at the top will be to cause two molecules in the base to be torn apart. In terms of the energy, the potential energy of the atom at the top, ε_H, will be sufficient to supply the energy needed to over come the interaction energy, ε_i, between the molecular pair at the base. Expressed another way, we require $\varepsilon_H = \varepsilon_i$.

The atom added to the top will have mass $m = Am_P$, where $m_P = 1.67 \times 10^{-27}$ kg is the mass of the proton and A is the number of protons and nucleons in the nucleus of the atom. We can write $\varepsilon_H = mgH = 1.67 \times 10^{-27} \, AgH$. We suppose ε_i is known so that

$$1.67 \times 10^{-27} \, AgH = \varepsilon_i$$

so that

$$gH = \frac{\varepsilon_i}{1.67 \times 10^{-27} \, A}.$$

ε_i and A are constant for a given planet which means that $gH = $ constant.

For the Earth, $A \approx 30$, $1.67 \times 10^{-27} \, A = 5 \times 10^{-26}$ kg and $\varepsilon_i \approx 3 \times 10^{-20}$ J. This gives $gH = 6 \times 10^5$ m^{2}s^{-2}. Since $g = 9.8 \approx 10$ ms^{-2}, $H \approx 6 \times 10^4$ m $= 6$ km. This is similar to the height of Mt Everest. For Mars, g is smaller so H can be expected to be greater. For Venus g is only slightly smaller than for Earth allowing the equilibrium height to be a little larger.

The heights calculated this way are for an equilibrium balance but sub-surface internal motions can allow for greater heights. Detailed calculations, together with a quantitative knowledge of A, could allow the sub-surface dynamics to be investigated in a particular case.

Problem 7.1. Get settled in the garden on a dark night and count the number of shooting stars (meteorites) that you can seen in an hour. Repeat this at different times of the night and at different times of the year.

Solution 7.1.

You should see one or two an hour generally with the largest number appearing before dawn. At certain times of the year the counts will be very much greater for a few nights. This will be a meteor stream which is composed of minute fragments from the tail of a comet. At their height these streams can provide several hundred shooting stars per hour.

Problem 11.1. 72% of the surface area of the Earth is covered by oceans. Calculate the area of the Earth's oceans. Accepting that the mean depth of the oceans is 3800 m, calculate an approximate value for the mass of the oceans. How does this compare with the masses of the smaller bodies of the Solar System?

Problem 11.2. Measured data show that the heat flow from the surface of the Earth is 5×10^{-2} W/m². The mean radius of the Earth is 6.4×10^3 km and the mass 5.98×10^{24} kg. The mass of a representative interior particle can be approximated by $m = 40\, m_P$ and $m_P = 1.67 \times 10^{-27}$ km while the mean interior temperature is 3000 K. The energy per interior particle, ε, is given by the classical formula $\varepsilon = 3kT$ (the Dulong-Petit law), where k is the Boltzmann constant. Show that the present heat flow can be maintained for about 1.4×10^{10} years.

Solution 11.1.

The area of the Earth's oceans is, then, 3.66×10^{14} m². Taking the mean density of the oceans as 1.0×10^3 kg/m³, the total mass becomes 3.66×10^{17} kg. This is of the same order of magnitude as the mass of the larger comets.

Solution 11.2.

The surface area of the Earth is 5.10×10^{14} m² so the total heat flow through the surface is $E_S = 5.10 \times 10^{14} \times 5 \times 10^{-2}$ W $= 2.55 \times 10^{12}$ W. On the other hand, the mass of a particle is $m = 40 \times 1.67 \times 10^{-27}$ kg $= 6.68 \times 10^{-26}$. This gives the total number of particles within the Earth as $N = 5.98 \times 10^{24}/6.68 \times 10^{-26} = 8.95 \times 10^{49}$. The energy per particle is $\varepsilon = 3 \times 1.39 \times 10^{-23} \times 3 \times 10^3 = 1.25 \times 10^{-19}$ J so the total internal energy is $E_V = 1.25 \times 10^{-19} \times 8.95 \times 10^{49} = 1.12 \times 10^{31}$ J. An estimate of the time which the heat flow can continue is $\tau \approx E_S/E_V = 1.12 \times 10^{31}/2.57 \times 10^{13} = 4.36 \times 10^{17}$ s. There are 3.17×10^7 seconds in a year so $\tau \approx 4.36 \times 10^{17}/3.17 \times 10^7 = 1.4 \times 10^{10}$ years. This is something like 3 times the present age of the Earth.

Problem 12.1. The equatorial radius of Mars is 3.40×10^6 km and the equatorial radius of the Earth is 6.4×10^6 km. There are no oceans on the surface of Mars but approximately 72% of the surface of the Earth is covered by oceans. Presuming that the surface figures of Earth and Mars are perfect spheres, show that the total surface area of Mars is similar to the continental surface area of the Earth, that is that part not covered by the oceans.

Solution 12.1.

The surface area of Mars is found to be 1.45×10^{14} m^2. The surface area of the Earth is 5.15×10^{14} m^2. The continental area is 28% of the whole giving it the value 1.49×10^{14}. This is very close to the value for Mars.

Problem 13.1. The highest point on the surface of Venus is 12 km above the accepted radius 6. of the planet while the lowest point is 2 km below the mean radius. What is the percentage roughness of the surface of the planet? How does this value compare with the roughness for the Earth where the maximum height is 8 km (Mount Everest) and the maximum depth of 12 km (the depth of the Mariana Trench in the Pacific Ocean)? Are these planets smoother than an apple?

Solution 13.1.

The percentage roughness for Venus is $(14/6075) \times 100 \approx 0.2\%$. The Earth gives a similar value. An apple of diameter 8 cm (say) would need to have creases of value 0.016 cm or 0.16 mm. These would be much smaller than the observed marks. The surfaces of Venus and Earth are smoother than that of an apple.

Problem 14.1. Jupiter reflects 45% of the incident solar radiation while the Earth reflects 39% of the radiation incident on that. The Earth's effective temperature is 246 K. Jupiter is 5.2 times further from the Sun than the Earth so the temperature of Jupiter at the surface is less. What, on this basis, is the temperature to be expected for the visible surface of Jupiter. The actual temperature is measured to be 130 K. Comment on the comparison between these two temperatures, one calculated and the other measured. (It is necessary for this calculation to remember that the radiation flux is proportional to the fourth power of the temperature).

Why is the cooling of Jupiter likely to be slower in the longer term than the simple conduction of heat might suggest?

Solution 14.1.

The equilibrium temperature of the surface of a planet results from the balance of the energy between the radiation received from the Sun and that radiated by the surface. The radiation from the Sun falls off as the square of the distance away. The comparison is with the Earth, where the distance $R = 1$ AU and the measured radiation per unit area at this distance is $F = 1376$ kW.

The radiation from the total surface is $4\pi a^2 \sigma T_p{}^4$, where a is the radius of the planet and T_p is the surface temperature. The radiation received from the Sun is $\pi a^2(1 - A)F/R^2$, where A is the fraction of the radiation reflected by the planet surface (the\albedo). Energy is not lost so these two expressions must be equal, which means $4\pi a^2 \sigma T_p{}^4 = \pi a^2(1 - A)F/R^2$. For the Earth, $R = 1$ and $A_E = 0.30$: for Jupiter, $R_J = 5.2$ and $A_J = 0.58$. Given that $T_P(E) = 256$ K it follows that $T_E(J) = 98$ K. The measured temperature at the visible "surface" is 130 K, which is higher. This means that Jupiter has an internal heat source.

It was seen in the text that heat arises from the separation of liquid helium from the hydrogen/helium mixture comprising the inner structure. The solubility of helium in hydrogen falls as the temperature becomes lower meaning the separation of helium proceeds faster as the planet cools. This provides a longer term internal heat source which supplements the ordinary heat flow to the surface.

Problem 22. The Sun is a spherical body with a radius of 7×10^8 m. It has a measured surface temperature of 5.8×10^3 K and an estimated central temperature of 1.5×10^7 K. What is the mean gradient of temperature across the interior? How can the thermal state of the interior be described? If it is realised that the main heat generation occurs in the central 10% of the volume how is this description changed?

Solution 22.

The mean temperature gradient is given by $\frac{1.5\times10^7}{7\times10^8} = 2.1\times10^{-2}$ km^{-1}, remembering that the surface temperature is negligible in comparison with the central temperature. This is a very small gradient and conditions in the vicinity of each point inside are indistinguishable from those of thermodynamic equilibrium to a high degree of approximation. The situation where there is a gradient of temperature over all (the Sun is radiating energy which is not an equilibrium activity) but where every point within has its own local equilibrium state is an example of local thermodynamic equilibrium. The various thermodynamic formulae apply everywhere locally even though conditions differ from point to point.

Taking account of the case where the central region contains the main energy source, the temperature can be expected to be approximately uniform there. The gradient outside, however, will be only slightly larger than before ($\frac{1.5\times10^7}{6.3\times10^8} = 2.3 \times 10^{-2}$ km^{-1}). This will lead to an approximate model of the interior where the central region is radiating energy and so is far from thermal equilibrium but where the surroundings, which carry the energy away, are in an effective local thermodynamic equilibrium. This perhaps strange arrangement arises because the size of the body is so great on a laboratory scale.

Index